Philosophical Issues of Human Cyborgization and the Necessity of Prolegomena on Cyborg Ethics

Ivana Greguric
University of Zagreb, Croatia

A volume in the Advances in Human and Social Aspects of Technology (AHSAT) Book Series

IGI Global
PUBLISHER of TIMELY KNOWLEDGE

Published in the United States of America by
IGI Global
Information Science Reference (an imprint of IGI Global)
701 E. Chocolate Avenue
Hershey PA, USA 17033
Tel: 717-533-8845
Fax: 717-533-8661
E-mail: cust@igi-global.com
Web site: http://www.igi-global.com

Library of Congress Cataloging-in-Publication Data

Names: Greguric, Ivana, author.
Title: Philosophical issues of human cyborgization and the necessity of prolegomena on cyborg ethics / by Ivana Greguric.
Description: Hershey PA : Information Science Reference, [2022] | Includes bibliographical references and index. | Summary: "This authored book discusses the philosophy of technology, technoethics and other areas of research scholarship and practice that revolve around the challenging intersection of humans and new emerging technologies and offers guidance for moral actions and protocols for preserving the vitality of life within a rapidly changing technologically society"-- Provided by publisher.
Identifiers: LCCN 2021023100 (print) | LCCN 2021023101 (ebook) | ISBN 9781799892311 (hardcover) | ISBN 9781799892328 (paperback) | ISBN 9781799892335 (ebook)
Subjects: LCSH: Cyborgs--Moral and ethical aspects. | Robots--Moral and ethical aspects. | Robotics--Moral and ethical aspects. | Cybernetics--Moral and ethical aspects. | Transhumanism.
Classification: LCC TJ211.28 .G74 2022 (print) | LCC TJ211.28 (ebook) | DDC 174/.9629892--dc23
LC record available at https://lccn.loc.gov/2021023100
LC ebook record available at https://lccn.loc.gov/2021023101

This book is published in the IGI Global book series Advances in Human and Social Aspects of Technology (AHSAT) (ISSN: 2328-1316; eISSN: 2328-1324)

British Cataloguing in Publication Data
A Cataloguing in Publication record for this book is available from the British Library.

Advances in Human and Social Aspects of Technology (AHSAT) Book Series

Mehdi Khosrow-Pour, D.B.A.
Information Resources Management Association, USA

ISSN:2328-1316
EISSN:2328-1324

Mission

In recent years, the societal impact of technology has been noted as we become increasingly more connected and are presented with more digital tools and devices. With the popularity of digital devices such as cell phones and tablets, it is crucial to consider the implications of our digital dependence and the presence of technology in our everyday lives.

The **Advances in Human and Social Aspects of Technology (AHSAT) Book Series** seeks to explore the ways in which society and human beings have been affected by technology and how the technological revolution has changed the way we conduct our lives as well as our behavior. The AHSAT book series aims to publish the most cutting-edge research on human behavior and interaction with technology and the ways in which the digital age is changing society.

Coverage

- Digital Identity
- Computer-Mediated Communication
- Gender and Technology
- Information ethics
- Technology Adoption
- Activism and ICTs
- Public Access to ICTs
- Cultural Influence of ICTs
- End-User Computing
- Cyber Behavior

IGI Global is currently accepting manuscripts for publication within this series. To submit a proposal for a volume in this series, please contact our Acquisition Editors at acquisitions@igi-global.com or visit: https://www.igi-global.com/publish/.

The Advances in Human and Social Aspects of Technology (AHSAT) Book Series (ISSN 2328-1316) is published by IGI Global, 701 E. Chocolate Avenue, Hershey, PA 17033-1240, USA, www.igi-global.com. This series is composed of titles available for purchase individually; each title is edited to be contextually exclusive from any other title within the series. For pricing and ordering information please visit https://www.igi-global.com/book-series/advances-human-social-aspects-technology/37145. Postmaster: Send all address changes to above address.

Titles in this Series

For a list of additional titles in this series, please visit: www.igi-global.com/book-series

Technological Influences on Creativity and User Experience
Joshua Fairchild (Creighton University, USA)
Information Science Reference • © 2022 • 305pp • H/C (ISBN: 9781799843542) • US $195.00

Human Factors Issues and the Impact of Technology on Society
Heather Lum (Embry-Riddle Aeronautical University, USA)
Information Science Reference • © 2021 • 333pp • H/C (ISBN: 9781799864530) • US $195.00

Ubiquitous Technologies for Human Development and Knowledge Management
Hakikur Rahman (Institute of Computer Management and Science, Bangladesh)
Information Science Reference • © 2021 • 396pp • H/C (ISBN: 9781799878445) • US $195.00

Technological Breakthroughs and Future Business Opportunities in Education, Health, and Outer Space
Angus Hooke (Australian Institute of Higher Education, Australia)
Business Science Reference • © 2021 • 353pp • H/C (ISBN: 9781799867722) • US $225.00

Machine Law, Ethics, and Morality in the Age of Artificial Intelligence
Steven John Thompson (University of California, Davis, USA & University of Maryland Global Campus, USA)
Engineering Science Reference • © 2021 • 266pp • H/C (ISBN: 9781799848943) • US $295.00

Examining the Socio-Technical Impact of Smart Cities
Fenio Annansingh (York College, City University of New York, USA)
Information Science Reference • © 2021 • 231pp • H/C (ISBN: 9781799853268) • US $190.00

Human-Computer Interaction and Technology Integration in Modern Society
Hakikur Rahman (Ansted University Sustainability Research Institute, Malaysia)
Engineering Science Reference • © 2021 • 347pp • H/C (ISBN: 9781799858492) • US $195.00

Information Technology Applications for Crisis Response and Management
Jon W. Beard (Iowa State University, USA)
Information Science Reference • © 2021 • 294pp • H/C (ISBN: 9781799872108) • US $195.00

IGI Global
PUBLISHER of TIMELY KNOWLEDGE

701 East Chocolate Avenue, Hershey, PA 17033, USA
Tel: 717-533-8845 x100 • Fax: 717-533-8661
E-Mail: cust@igi-global.com • www.igi-global.com

"Nurturing serious thought and dialogue among philosophers, scientists, and many publics about the ever greater 'cyborgization' of living beings, including human beings, is urgent. Ethical thinking is always hard, but it is more necessary than ever."

Donna Haraway
Distinguished Professor Emerita
History of Consciousness Dept.
University of California at Santa Cruz

"It is a good idea to develop a robust ethical analysis of the possibilities and limits of human modification, whether biological or electronic."

Stuart Russell
Professor of Computer Science
Scientific advisory board The Future of Life Institute
Director of the Center for Intelligent Systems

"The coming age of artificial intelligence, machine learning and cyborgs will raise a new set of ethical, social and legal issues. It is important that we start to discuss these issues now, before it is too late."

Peter Singer
Professor of *Bioethics* at Princeton University
Laureate professor at the Centre for Applied *Philosophy* and Public Ethics at the University of Melbourne

"Increasingly, we live in an age of cyborgization. We're designing smarter and smarter algorithms and machines. We're introducing them into our environments, our ways of life, and even our bodies. Foreseeably, we will give machine intelligence more control—of our vehicles and factories, of our social planning and our daily lives. As the process accelerates, human intelligence will merge with machine intelligence. All this sets us challenges that are ethical as much as logistical. What strange paths will our species take as we cyborgize? We must develop new ethical concepts and social practices to govern our choices."

Russell Blackford
Editor-in-chief, Journal of Evolution and Technology
Australian philosopher and critic
Fellow of the Institute for Ethics and Emerging Technologies

Table of Contents

Foreword: To Cybernetic Beings

There can be no doubt that the age of the cyborg is upon us. But what does this mean and what effect will it have on all of our lives? In particular how do robots fit in to this and what will it mean to human life expectancy and human capabilities? Attempts to even consider answering such questions can be felt to be extremely problematic. Perhaps the subject is too embedded with sci-fi predictions, maybe the technology involved is, for a number of reasons, difficult to understand, even the way it is presented in the media often has a tongue in cheek approach to it.

What this book by Ivana Greguric attempts to do is to tackle the subject head on in a way that is readily understandable to everyone who, at least some of the time, uses their ability to think. I believe that she succeeds in accomplishing her task to the extent that this book could easily become a standard reference for all things cybernetic, particularly as they relate to the future. In the past I have been critical of some articles in the field written on this, or a related topic, in that the author lacks the necessary technical knowledge and as a result the conclusions drawn are often so far removed from reality as to be laughable, although they may be what some people want to hear. I do not have such a criticism of this book. Underpinning much of the western philosophy of today is the concept of our existence and what form this takes. Descartes, Kant and others attempted to answer questions such as Who am I and how do I know that I exist? Early in this book Greguric tackles this philosophy head on in terms of what it all means as far as cybernetic beings are concerned. In particular with a brain that is possible part human part machine network, does this mean that 'I' becomes 'We'? Potentially, as the author points out, it could be the end of philosophy as we know it.

Concepts such as transhumanism and cyborgs are examined in terms of the different forms possible. For those unsure of what a cyborg is, this in depth analysis readily parades the variety of possibilities that exist. Each one is critically examined in turn in terms of what it actually is. In doing so, integrated bionics, animal cyborgs and even insects are discussed. At the same time the work of movers and shakers in the field is investigated, principally that of Stelarc, Eyeborgs, Steve Mann and, I am delighted to say, my own work.

Overall, as you will see, the book is extremely readable and you do not need a degree in philosophy or technology to follow what's going on. In bringing this about, Greguric successfully weaves into the text popular terminology that has become widely used. Examples of this would be Hacking humans, Biohackers and Neurohacking. Also, topics of considerable interest, such as how the research being carried out affects life expectancy, are treated as a natural progression.

On reading the book I found some topics that may well come as new, but strongly related, input for many. I would include in this organ exchange and organ replacement, particularly when considering using technology as a form of replacement. But also matter and energy exchange comes into this category, as does, what I find most interesting of all, information exchange organs. The culmination of this is however a more general concept of human enhancement and what this can lead to.

What I also like about the book though is that Greguric also considers some of the social effects of cyborgisation. These include the effect on sports, both in terms of future developments but also effects that we are starting to see now. Arts also are considered here, particularly in terms of performance arts. In considering such topics there is a gentle introduction to the sexy side of robots and robotics to investigate what's really possible and what might be possible in the future. Of course as far as the future is concerned we cannot go far without looking at what all this will mean as the onset of cyborgs and robotics directly affects humans. How will humans co-habit with cyborgs and robots? Will the singularity definitely happen? What are artificial brains? Is immortality in a sense possible? As a novel addition to this section of the book Greguric includes transcripts from a number of one-to-one interviews she carried out with experts in the field. These succeed in putting much of the book's contents into context.

For me the book in its entirety is an excitingly modern, comprehensive coverage of the field of cyborgs and human enhancement. It gives a plethora of opinions as to where things are headed and what this might mean. It employs philosophy, technology and futurology in a very friendly and readable way to present, what could be a difficult, and to some people a very different, topic in a palatable form. In that way the book can be read through rather in the form of a storyline or alternatively it can be dipped into for specific pieces of information. Enjoy!

Kevin Warwick
University of Reading, UK

Foreword: Cyborg Ethics We Must

It is a pleasure to write the Preface for Ivana Greguric's book *Philosophical Issues of Human Cyborgization and the Necessity of Prolegomena on Cyborg Ethics*. I first met the author at the 2013 Computer Ethics: Philosophical Enquiry (CEPE) on *Ambiguous Technologies: Philosophical Issues, Practical Solutions, Human Nature* held in Lisbon. At that time I was struck by the timeliness and importance of her work on cyborg ethics. Over the last five years, her philosophical inquiry into this new area has deepened and evolved. The may stones uncovered in this current volume attest to the dedication and thoroughness of a young scholar whose work has truly blossomed, making a contribution to Philosophy of Technology, Technoethics and other areas of research scholarship and practice that revolve around the challenging intersection of humans and new emerging technologies.

How do we go about questioning the essence of our emerging cyborgised reality of the age of *scientific humanism?* This is a both a challenging and popular topic at a time where controversial technological changes are outpacing our ability to navigate their possible impact on human life and society. This helps situate the discussion on the transtechnical essence of technology and changing assumptions humanism and transhumanism which the author contends, "show that the cyborgisation of man through actions for improving and reshaping his body and mind has the goal of establishing the man as the appropriate means of *scientific work* and the life boundaries of the cyborgised community of *scientific humanism* as *naturalism* set forth by it." The author builds a strong case for the necessity of cyoborgoethics to help guide moral actions and protocols for preserving the vitality of life within a rapidly changing technologically society.

The book begins with a retracing of Hegel's system of philosophy and Marx's criticism of the absolute notion of Hegel's system of philosophy. Of particular importance is the emphasis placed on the nature of historical reality and the essence of alienation that arises in humans during periods of societal change. Chapter 1 and Chapter 2 in this volume draw attention to the important role of philosophy in guiding society to rethink the ontological and anthropological importance of human and the creation of new forms of life with a unique non-biological ontological basis to deal with. This sets the stage the discussion (Chapter 3) on the dialectic transformation of machine into machinery with an eye towards transhumanist perspectives. This early ground work helps situate the contemporary treatment of cyborg ethics within a historical and philosophical framework. This is important in guiding researchers in exploring the essence of *scientific work with an eye towards the* cyborgisation movement which is reshaping human life and society.

Following the historical and philosophical grounding of the cyborg, Chapter 4 begins drawing attention to what it means to be living in a cyborg era and what variations and applications are out there. This chapter covers key considerations including: Conceptualizations of cyborgs, replacement procedures, normalization procedures, improvement procedures, reshaping procedure, and types of cyborgs (animo cyborgs, cyborg insects, other animal cyborgs, homo cyborgs). It also deals with targeted areas of human-cyborg enhancement including: Bionic arms, bionic fingers, bionic legs, bionic ear, bionic eyes, bionic hearts, bionic skin, bionic tongues, bionic brains, and bionic clothes. The chapter coverage extends to discussing the importance of computer interfaces and key applications that brought undergraound cyborg research into the mainstream, namely, Kevin Warwick's cybernetic projects, cyborg popularization in the arts, Steve Mann's portable computers, augmented reality goggles, and robo-cyborg.

What volume on the topic of cyborg ethics would be complete without address concerns about hacking human beings. The book discusses the various areas of human enhancement application where such concerns arise including physical enhancements (self-identification via chip, payment via chip, hearing enhancement via magnetic headsets, infra-red spectre vision enhancement, and chips for monitoring bodily functions), cognitive enhancement (deep brain stimulation, memory chips, neuron manipulation, and neuron dust), and mood enhancements (mood improvement, neuroprosthetics) and life extension technologies (cyborg digital mind, cyborg digital body, and cryonic body preservation). This rapidly evolving landscape of ethical hacking possibilities maps out one of the most promising areas of Technoethics on the horizon.

The later chapters (Chapter 6-9) deepen the discussion and extend the case for cyberethics. The importance of virtual reality in human life, avators, and communicating with digital characters demonstrates the pervasiveness of technology's penetration into our lives, not only physically, cognitively, and emotionally, but also environmentally. The 'cyborgisation' of humans in virtual space in Chapter 6 extends the landscape of the discussion on cyborgethics. Following a popular trend in current Philosophy of Technology and Technoethics (see *Techne Journal* for more), the body is revisioned as a cybernetic organism in Chapter 7 as the text delves into cyborg anthropology and the new types of information, matter, and energy exchanges taking place that need to be appreciated. This harkens back to the pioneering cybernetic work of Norbert Weiner and extends and extends into current and future reconfigurations of man-machine relations that are shaping human life and society from human enhancement to driverless cars. The discussion of cyborgization in the arts, increasing integration of robots in society, and future possibilities covered in the remaining chapters entrenches the case for cyborg ethics by demonstrating the popularization of cyborgs in areas of life and society. This helps show that cyborg normalization may be just over the horizon. The expert interviews from internationally acclaimed leaders in the field, Richard Walker, Kevin Warwick, and Amal Graafstra, provide additional insight into how cyborgization came from and where it is, or can be going. Overall, the book makes a strong case for the necessity of cyborgethics.

Rocci Luppicini
University of Ottawa, Canada

Preface

This is a monograph about the cybernetic sciences – metaphysics whose main focus are the new ontological and anthropological facts, and cyborgoethics, aimed at the values and protocols for moral action of man in the age of *scientific humanism* as *naturalism*. Our goal is to show the manifestations of scientific and technological alienation of man and rethink the metaphysical articulation of the human reality and the ability to think.

Cyborgs are already among us and we have all, in one way or another, become cybernetic beings through our immersion in information and communications technology. It remains to be seen what effect science and technology will have on our lives as they are controlled by the cybernetic science, planned and designed through this act of controlling, and ultimately, as the essence and existence of everything ontologized/functionalized through *scientific work*. The pinnacle of this persisting functionalisation of everything that exists is cyborgization of man through improvement and reshaping of his mental and physical capabilities and functions, and establishment of life on a new scientific and technological basis in the cyborgized community of *scientific humanism* as *naturalism*.

Today, even the non-thinking eye can clearly see that, alongside procedures for restoring or replacing lost limbs or human bodily functions to establish the body's natural balance, it is all too common to see procedures through which a healthy human body and mind is reshaped, enhanced or established as a cybernetic system. By offering the magic vision of removing all the shortcomings of the disposable and incomplete natural human bodies and enhancing the mind, and by offering the new basis of life in a scientifically controlled history, *scientific work* enhances and reshapes man by making him a partially artificial being – a cyborg which functions in the scientific community as the means and purpose of *scientific work* in an effective and programmed manner. In the future, man's physical and mental capabilities will be partially or completely replaced by functions artificially created and managed by particular software. For the biological body of the human being, this represents a one-way street.

There is no doubt that man has gone through changes over the course of history, that he created his human world with the aid of science and technology, so the progress in and of itself is not an issue if the purpose of the historical evolution is internal to man himself. However, the procedures of enhancing and reshaping involve a science-directed evolution of man as the drive of *scientific work* as the creator of *scientific history* in which man is the means and not the purpose of development. In this case, it is not man who evolves as a natural and mental being trying to live a meaningful life and use his personal will to decide with his fellow men on the principles of ethical relationship towards the world and other human beings, but it is his artificially created bodily implant that evolves, the implant which can be controlled from outside the human body and mind and without his consent, in a similarly created artificial historical, or non-historical reality of *scientific humanism* as *naturalism*.

The cybernetically organized, programmed and directed cyborg beings will not have the free will of human beings, nor will they have any need or possibility to think about the purpose of their existence as *scientific work* as the nihilistic being of the scientific history will manifest itself most directly openly through their essence and existence. The analysis of the scientific character of work and its drive as the way of creating life and organizing the community, it is shown that every product of *scientific work*, in our analysis either a cyborg or a robot, is scientifically predetermined in all the stages of their creation and survival. The scientific characterisation of work does not create complete products which would cause its driving force to halt, as completion cannot be the condition of its survival, its *will to power*.

As man and everything human are within the scope of the absolute establishment of the *scientific work* and since the possibility of abolishing his mental capabilities, autonomy of will and overcoming the biological bodies and replacing them with technical ones is looming, we find it necessary to start the dialogue on the issues and principles of cyborgoethics, advocate for legislature of moral actions which represent a key part of this book. Without a thoughtful and intersubjective dialogue on the essence of *scientific humanism* as *naturalism* and its internal possibility to create something new and different from everything before, man and possibility of ethical life and life-appropriate wisdom will cease to exist.

The enhanced man, the man-machine does not carry in itself the primal vitality of life, has no meaningful relation to nature and non-human beings, has no gender determination, self-awareness, personality, no emotion and no free will. A partially or completely artificial being is not an authorized *self-being*[1] as its life power does not stem from the interplay of the *empire of freedom* and the *empire of necessity*[2] but from the new basis set forth by the *scientific work*.

We are facing a time of major challenges for the survival of man and his nature as the speed of our self-purposeful scientific and technological progress and its effect on our lives surpass our capabilities of qualitative control. We must think and decide how human beings can coexist with enhanced cyborgs and robots and determine the line which should not be crossed. We firmly believe that in the depths of cybernetic metaphysics there still lies a possibility of transition towards a meaningful post-scientific life, thought and action. The cyborgoethic principles which we advocate here necessarily involve help in the form of an intersubjective dialogue of all social subjects (individuals, scientists, physicians, politicians, economical lobbyists, etc.) in order to preserve the vitality of life for all living beings and determine a protocol for the behaviour of scientific and technological complexes and new cybernetic beings. We have to be aware that the cybernetic complex has its own purpose of progress, independent of the interest lobbies, and that the potential emergence of superintelligence will establish its own strategic goals in which man and nature will no longer be the means of progress but mere raw materials.

This book is primarily aimed at the students of social sciences and humanities, students of information and communications sciences, and all other individuals who want to expand their knowledge with possible new philosophical insights about the future of man in the society of *scientific humanism* as *naturalism*. The book does not strive to offer ready-made solutions, but seeks to use the facts from the current reality to scrutinize the philosophical, ontological and anthropological understanding of man and his cyborgized reality and indicate potential dangers for the future of man in the age of *scientific humanism*. In this respect, the book illustrates that the key ingredient of the future of mankind will be a philosophical dialogue about the possibility of thought and an intersubjective dialogue of all parties about the ethical issues involved, so that we as human beings could withstand the transhumanist and posthumanist intentions of *scientific work*.

REFERENCES

Burger, H. (2008). *Ljudska moralnost: Studije o antropologiji i etici*. Naklada Breza.

Marx, K. (1867). *Capital: A Critique of Political Economy* (S. Moore & E. Aveling, Trans.). Progress Publishers.

ENDNOTES

1 *Self-being* is a concept from H. Plessner which denotes the internal world of man present within him regardless of his awareness of it. It is the relation of self towards self, which represents the starting point of human being (Burger, 2008, p. 55).

2 *The empire of freedom* and *the empire of necessity* are concepts taken from Marx's *Capital*. Marx utilizes the two-part dialectic to create the possibility of stepping out of the dialectic circulus of the capitalistic society – from the *empire of necessity* into the *empire of freedom* which can only flourish from the empire of necessity as its basis. However, the liberated labour without a thought-out essence perpetuates the *empire of freedom* into *scientific work* which establishes the empire of *scientific humanism as naturalism* (Marx, 1867).

Acknowledgment

I would like to express my gratitude to all my professors and colleagues who supported me in the process of writing this book, specially to Emeritus Igor Čatić who had an initial influence on the direction of my scientific development. I am particularly thankful to prof. Kevin Warwick, Emeritus Professor, Deputy Vice-Chancellor of Research at Coventry Universities (UK), prof. Rocci Luppicini, Associate Professor, University of Ottawa (CA); Stelarc, Australian performance artist; Amal Graafstra, founder of the company *Dangerous things* (WA); Richard Walker, spokesperson of the Human Brain Project (CH); Ian Pearson, futurologist at the *Futurizon* (UK) and Jaime Mayor Oreja for their cooperation during my scholarly work.

Special gratitude is also owed to prof. emerit. Donna Harraway, PhD (University of California, USA); Russel Blackford, editor-in-chief at the *Journal of Evolution and Technology* (UK) at the time; prof. Stuart Russell, PhD, scientific advisor at the *Future of Life Institute* and Peter Singer, Professor of *Bioethics* at Princeton University. I am also indebted to the Brocher foundation for their grant which allowed me to gain insight in some aspects of research necessary for writing this book.

Many thanks to the leading media company 24 hours (24 sata) who funded the translation of the book and to the translator Frane Malenica.

Finally, I would like to thank my family – my father Franjo, my mother Danica and my daughter Laura for their unconditional love and immense faith in my success. I dedicate this book to them!

Chapter 1
Philosophical Foundations for Establishing Scientific Humanism as Naturalism

ABSTRACT

In this chapter, the author presents the Marx-Hegel dialogue, a retracing of Hegel's system of philosophy and Marx's criticism of the absolute notion of Hegel's system of philosophy. Of particular importance is the emphasis placed on the nature of historical reality and the essence of alienation which arises in humans during periods of societal change. In this chapter, attention is drawn to the important role of philosophy in guiding the society to rethink the ontological and anthropological importance of human beings and the creation of new forms of life with a unique non-biological ontological basis.

BACKGROUND

The discussion on the philosophical aspects of cyborgization of man and social relations in the era of *scientific humanism* as *naturalism* stems from the New Age understanding of thought and being of entities and Hegel's synthesis of both this relation and the overall philosophical tradition into systematic knowledge of the *absolute idea* as the substance-subject of world events and thought. The philosophers operating in the post-Hegel era, following Hegel's philosophy as the *absolute science of absolute spirit* (the absolute idea) which encompasses everything in its essence and existence, were faced with the question of whether it is possible to think in philosophical terms after Hegel. In order to answer this question, this book will seek to establish a dialogue of post-Hegel scholars trying to make a radical departure from the tradition of metaphysically-based thought, preparing for its end and the inception of a possible transmetaphysical thought. Marx's expectations that the *abolition* of form and *realization* of Hegel's philosophy in a non-philosophical reality of the world will elevate this reality onto the level of a self-aware subject of world history who enables freedom, *humanism* and *naturalism* of man, perpetuating into *scientific history*[1], in which being means to be established by the *scientific work* as the subject,

DOI: 10.4018/978-1-7998-9231-1.ch001

which is its own subject-object, whose form is the substance-subject and man, an item and the means of its self-confirmation.

The philosophical dialogue about the final frontiers of philosophy as metaphysics leads us to two questions:

1. When will the supply of human beings and natural resources run out?; and
2. When will we stop to passively accept the ubiquity of the world of science and technology and its rationalizing mind and awaken from the deep slumber of metaphysics to reach an epoch of new, other and different kind of thinking and living?

In the depths of philosophy as metaphysics there lingers a "solution" of the current problems of civilization of science and technology, of *scientific humanism as naturalism*. As we shall see in our research, the powers of the scientific and technological mind can be opposed by a way of thinking which illuminates the very essence of cyborgizing realization of life, man, nature, and society, as well as the ideologized anthropocentric and transhumanist ethics.

2.1 HEGEL'S SYSTEM OF PHILOSOPHY

We begin the discussion about Hegel's philosophy with his thoughts from the foreword to the second edition of *Encyclopaedia of the Philosophical Sciences*:

Free and genuine thought is inwardly concrete; hence it is Idea, and in all its universality it is the Idea or the Absolute. The science of it is essentially a system, since what is concretely true is so only in its inward self-unfolding and in taking and holding itself together in unity, i.e., as totality. Only through the distinguishing, and determination of its distinctions, can what is concretely true be the necessity of these distinctions and the freedom of the whole. (Hegel, 1991, pp. 38-39)

From his earliest works, Hegel has focused on the issues of methodology, truth, freedom, and the philosophy of history, through *The Phenomenology of Spirit* and *Science of Logic,* to his more mature approaches to philosophy as systematic knowledge in the *Encyclopaedia of the Philosophical Sciences.* In this monograph, we are interested in those aspects of Hegel's philosophy which approach the ontological structure of history as the immanent structure of historical action, through which the absolute spirit as the substance subject of everything reaches self-consciousness, the absolute truth and freedom in the scientific history. We shall use the *Encyclopaedia of the Philosophical Sciences* to map out Hegel's system of philosophy, as it contains all key ingredients of Hegel's entire system.

In §14 of the *Encyclopaedia*, Hegel claims:

The thought, which is genuine and self–supporting, must be intrinsically concrete; it must be an Idea; and when it is viewed in the whole of its universality, it is the Idea, or the Absolute. The science of this Idea must form a system. For the truth is concrete; that is, while it gives a bond and principle of unity, it also possesses an internal source of development. Truth, then, is only possible as a universe or totality of thought; and the freedom of the whole, as well as the necessity of the several sub–divisions, which it implies, are only possible when these are discriminated and defined. (Hegel, 1830, pp. 10-11)

What is apparent in this segment is Hegel's intention to turn philosophy into a systematic scientific discipline, as this is the only way to capture the whole, that is, to reach the absolute scientific truth. In the same section, Hegel also makes the following comment: "Unless it is a system, a philosophy is not a scientific production. Unsystematic philosophising can only be expected to give expression to personal peculiarities of mind, and has no principle for the regulation of its contents" (Hegel, 1830, p. 11).

To create the system of philosophical sciences which consists of the *Science of Logic, Philosophy of Nature, Philosophy of Spirit*, Hegel uses the speculative[2] dialectic method which is, according to him, the soul of creation and exposition of the system of science as a true whole. It is impossible to build a system of philosophy without the speculative dialectic method as the absolute method of mediation, it is the mover and the creator of the system and all the categories within the system, in terms of both their individuality and generality. In this tripartite circle of circles, it builds everything that is in its whole and which thus represents the truth. The truth is, in other words, the substance subject, the movement of the substance "of positing itself, or is the mediation of its self-othering with itself" (Hegel, 1977, p. 10). The speculativity of the dialectic method shows us that Hegel sees in everything material something logical or ontological, that which can be understood only in the philosophy of as the absolute science. Hegel's entire system of philosophy of science rests on this method: "The idea in its pure form (as the Subject of the system) is given in the *Science of Logic*, which represents the affirmation; in its otherness (*Anderssein*) in the *Philosophy of Nature*, the idea is the negated materially, in space-time; in the Spirit, the idea finally returns to itself through respect for the otherness and in this return it makes a circle of circles" (Sutlić, 1994, p. 8). The spirit is the realized form of the substance becoming Subject. It has its structure in the ontology, which also represents theology for Hegel and falls under the purview of history. Applied regional ontologies are the nature and spirit and, thus, applied theologies. Nature stands as a testament to the existence of God, the spirit stands as a testament to the existence of God, and God is defined as an ontological structure, a circle from the most abstract category of being to the most concrete category of the idea. This circle is the divine life, and the divine life is reflected in the nature and the spirit through a medium that is alien to god as an ontological structure. This represents the boundary of Hegel's philosophy. Hence, Hegel makes a somewhat unexpected claim at the beginning of the *Philosophy of Nature* that the idea falls into nature as its otherness. That what is other in the otherness, what Hegel cannot overcome from the position of the absolute idea is the banal fact, a mere occurrence, almost identical to apparition. But even this attempt to reconcile essence and concept, immanence and transcendence, ultimately ends up as the reconciliation of immanence and transcendence in transcendence. Even though Hegel identifies the real concept and the absolute concept with reality and elevates it to the status of the substance subject, it is only given in the philosophy in the element of concept which denotes the entire destiny of the concept from abstraction to the absolute. This attempt to reconcile essence and concept, immanence and transcendence, ultimately remains in the domain of transcendence, in the element of thought and this is where Hegel reaches the limits of his philosophy. In this circle of circles, the closed metaphysics reaches its endpoint and it needs to turn into itself, find its place in the reality, life, the material side of the historical time. The material, everchanging action, this fundamental transformation must have its own internal order, lest it should not fall apart. For Nietzsche, this is the eternal recurrence of the same, and for Marx it is the endless *practice of work* as the scientific history.

Hegel portrays the development of the system as a whole by stating:

Each of the parts of philosophy is a philosophical whole, a circle that closes upon itself; but in each of them the philosophical Idea is in a particular determinacy or element. Every single circle also breaks through the restriction of its element as well, precisely because it is inwardly [the] totality, and it grounds a further sphere. The whole presents itself therefore as a circle of circles, each of which is a necessary moment so that the system of its peculiar elements constitutes the whole Idea-which equally appears in each single one of them. (Hegel, 1991, p. 39, §15)

This clearly shows that the system of philosophical sciences and the method for developing and portraying it are intrinsically connected – one cannot be possible without the other.

Before describing the first circle of Hegel's system of *Science of Logic*, it is necessary to rethink the terms science and logic. Hegel conceives logic as "the science of the pure Idea, that is, of the Idea in the abstract element of thinking" (Hegel, 1991, §19) in which

Thinking immediately involves freedom, because it is the activity of the universal, a self-relating that is therefore abstract, a being-with-itself that is undetermined in respect of subjectivity, and which in respect of its content is, at the same time, only in the matter [itself, note: IG] and in its determinations (Hegel, 1991, p. 55, §23).

The very science belongs to the logical, and for Hegel this means "the ontological structure of everything that is, the science as the momentum of *logos* as the ontological structure of everything that is (Sutlić, 1994, p. 7). The science is also the consciousness of this action that knows itself as such. Philosophy as the science of the Absolute (mediated by man) in the totality of everything that is in that what is, is the true medium of truth and freedom. Since it is the science of the Absolute, which is both the subject and the object of its action and has its own consciousness, it is also the absolute science of the absolute in which the beginning is actually the end of movement (advance), and so is the result the beginning which is "the underlying ground of all that follows without vanishing from it" (Hegel, 2010, p. 49).

2.1.1 The science of logic

Hegel examines the *Science of Logic*, the first major circle of circles of the *Encyclopaedia of Philosophical Sciences*, through the science of being, which leads us through the science of essence to the science of concept. These are the intermediary and constitutive moments of the *Science of Logic* which is in constant systematic interaction with the circle of the *Philosophy of Nature and the Philosophy of Spirit* on their road towards building and developing the absolute idea in world history. All three sciences and their categories are portrayed through and developed in the tripartite dialectic sequence: thesis – antithesis – synthesis, which completes the circle of *Logic*. The about-self and for-self idea portrayed and developed in this manner transitions into nature as its otherness. This transition, however, is not something that comes after logos in the temporal succession, but simultaneous in terms of transcendental priority, logos, nature and spirit for all empirical matter (Sutlić, 1984). Marx (1857) will note, quite insightfully, that this method of moving upwards from abstract to concrete is a way for thought to incorporate the concrete and reproduce it as spiritually concrete, but is certainly not a process of creating the concrete itself. At the very end of *Science of Logic*, Hegel claims:

Considered according to this unity that it has with itself, the Idea that is for itself is intuiting and the intuiting Idea is Nature. But as intuiting, the Idea is posited in the one-sided determination of immediacy or negation, through external reflection. The absolute freedom of the Idea, however, is that it does not merely pass over into life, nor that it lets life shine within itself as finite cognition, but that, in the absolute truth of itself, it resolves to release out of itself into freedom the moment of its particularity or of the initial determining and otherness, [i.e.,] the immediate Idea as its reflexion, or itself as Nature. (Hegel, 1991, § 244)

Hegel perceives the *Science of Logic* as a system of interaction of the whole and its parts in constitution of sciences as a singularity created through thought. Without the dialectic, it would not be possible to build such a system. It moves and constitutes this system and the categories within, with their collectiveness and individuality. In the process of cyclical movement of the dialectic sequence, all affected entities emerge, persist, and cease to exist as parts of a whole, which is a symbol for the endless movement and existence, i.e. the truth set in the sphere of the absolute spirit. Hegel himself compares this movement forward to being

[…] led back to the absolute knowledge which is its innermost truth. This truth, the ground, is then also that from which the original first proceeds, the same first which at the beginning came on the scene as something immediate. – It is most of all in this way that absolute spirit (which is revealed as the concrete and supreme truth of all being) comes to be known […]. (Hegel, 2010, p. 49)

The *Science of Logic* is essentially the philosophy of logos and its structure is shown as the onto-theology. As ontology, it collects everything relevant for philosophy as a system of science. Its categories stand outside of their material determination but give determination and sense to every material substance within the philosophical system of science, at the end of *Logic*. The absolute idea as a true term is conceived as an iconic subject of creation transformed into nature as its reflection. By returning to life, it merges with its own self in its own reality (the otherness as a reflection of logical and ontological structure). In terms of Marx's criticism of Hegel's *Logic*, we can note that the transition of *Logic* into *Philosophy of Nature* is a transition from abstraction into observation caused by desire for fulfilment with content.

2.1.2 The Philosophy of Nature

According to Hegel's interpretation

Nature is implicitly a living whole; more closely considered, the movement through its series of stages consists of the Idea positing itself as what it is implicitly, i.e. the Idea passes into itself by proceeding out of its immediacy and externality, which is death. It does this primarily in order to take on living being, but also in order to transcend this determinateness, in which it is merely life, and to bring itself forth into the existence of spirit, which constitutes the truth and ultimate purpose of nature, and the true actuality of the Idea. (Hegel, 1970, p. 216).

The idea as nature is reflected in mechanics, physics and organic physics – the three are perceived as the elements of the tripartite dialectic circle in the *Philosophy of Nature*. The circle of the *Philosophy of Nature* ends within the organic physics with the "death of the individual of its own accord" (Hegel, 1970, p. 209), which marks the synthetic transition from the *Philosophy of Nature* towards the spirit and the *Philosophy of Spirit*. For Hegel, the death of a man as an individual represents a confirmation of generality, which enables the transition into the spirit:

With this, nature has passed over into its truth, into the subjectivity of the Notion, whose objectivity is itself the sublated immediacy of singularity, i.e. concrete universality. Consequently, this Notion is posited as having the reality which corresponds to it, i.e. the Notion, as its determinate being. This is spirit. (Hegel, 1970, p. 211)

This transition of idea into spirit and *Philosophy of Nature* into *Philosophy of Spirit* follows the same pattern of simultaneity in terms of "transcendental priority, logos, nature and spirit for all empirical matter" (Sutlić, 1994, p. 15).

It is important to note that Hegel finds the spirit to be the Idea, that what constitutes the reality of the reality. The *Philosophy of Nature* gives onto logos the historical space and time so that the onto-theological structure is filled with content. In other words, the *Philosophy of Nature* reflects all nothingness of nature and immaterialness of the metaphysically perceived object, which can get its ontological meaning only through the spirit and the *Philosophy of Spirit*.

Given the fact that we live our lives in the age of *scientific humanism* as *naturalism,* it should be noted here that the objectivization of nature by the modern natural sciences has its foundations in Hegel's Science of Logic, i.e. the rule of concept in the form of operative logic. This view is proposed by Georg Picht, who finds Hegel and his logic to be the empire of pure thought based on natural sciences as all things which take place in nature are subordinate to logic, insofar as they follow the mathematical sciences. Thus, logic is the one science whose truth implies the entity of empire of thought and the empire of nature (Picht, 1989, p. 82). In that respect, Hegel's system of identity of nature, concept, and history represents the foundation of the all new-age natural sciences which devaluate and destroy nature. Picht believes that the logic of natural sciences, and therefore Hegel's logic as well, represents the main source of the modern ecological crisis as the climax of the crisis in the man-nature relationship: "The fundamental assumption of the new-age empirical history – equating the nature and concept, shows its true colours today – the new-age logic is a theory of power, the identification of nature and concept is the violent act which allows the new-age European man to subdue a specific part of nature" (Picht, 1989, p 83). Ultimately, the operative logic of natural sciences engulfs man himself; not only his organic entity but also the social aspects of his existence.

2.1.3 The Philosophy of Spirit

With regards to the concept of Spirit, Hegel claims the following:

From our point of view mind has for its presupposition Nature, of which it is the truth, and for that reason its absolute prius. In this its truth Nature is vanished, and mind has resulted as the 'Idea' entered on possession of itself. Here the subject and object of the Idea are one – either is the intelligent unity, the notion. This identity is absolute negativity –for whereas in Nature the intelligent unity has its objectivity perfect but externalized, this self–externalization has been nullified and the unity in that way been made one and the same with itself. Thus, at the same time it is this identity only so far as it is a return out of nature (Hegel, 2001, p. 2).

Hegel describes the development of the spirit through three dialectic circles: subjective, objective and the absolute spirit. At the end of the dialectic circle of circles, the spirit as the being transitions from its substantive form through mediation to subjectivity – the absolute of world history, which reaches the absolute knowledge in the philosophy. In Hegel's interpretation, the absolute marks everything absolute which has returned to itself in its form but whose form is one with its content.

The subjective Spirit is "In the form of self-relation: within it, it has the ideal totality of the Idea – i.e. it has before it all that its notion contains: its being is to be self-contained and free. This is Mind Subjective." (Hegel, 2012, §385). The subjective Spirit is the focus of three sciences: anthropology, phenomenology of spirit and psychology. "While the subjective spirit in anthropology and phenomenology dealt with the moments of soul and consciousness – it is only here [in psychology, note by IG] that we are interested in the spirit that defines itself within itself; it is not a reflection of self nor a reflection within self but the subject, the medium of everything that is." (Sutlić, 1994, pp. 16-17).

The objective Spirit is "in the form of reality: realised, i.e. in a world produced and to be produced by it: in this world freedom presents itself under the shape of necessity" (Hegel, 2012, §385). The objective Spirit transitions into the objective reality to form it shape it to its liking and to fill itself with content. The objective spirit involves formal or abstract law, moral and customs. The abstract law encompasses Hegel's determination of man as a human being. It possesses the subjective, free will, but the real freedom of man in the civil society is based on the external sphere – private property. Hegel claims that possession does not represent mere multiplication, generality and permanence of ownership, but that is the only guarantee for man to be a concrete individual entity in a civil society (Hegel, 1952). The second moment of the objective Spirit is the morality that has "its existence inside it" (Hegel, 2012, §487). The third moment is the "substantial will, made actual in the subject and conformable to its concept and rendered a totality of necessity, it is the ethics of actual life in family, civil society, and state." (Hegel, 2012, §487).

Ethics as substance is the civil society as a community of opposed individual interests. Ethics is the synthesis of the subjective spirit (individuals in the developments of its spirituality), morality, customs and the civil society which reach its fulfilment in the state.

The state is the actuality of the ethical Idea. It is ethical mind qua the substantial will manifest and revealed to itself, knowing and thinking itself, accomplishing what it knows and in so far as it knows it. The state exists immediately in custom, immediately in individual self-consciousness, knowledge, and activity, while self-consciousness in virtue of its sentiment towards the state, finds in the state, as its essence and the end-product of its activity, its substantive freedom". (Hegel, 2012, §257)

Only the people aware of themselves as the embodiment and manifestation of the notion of customs are people in the true sense. Participating in the state is the maximum of customs in time and space, not only in the ontological structure but also in the temporal structure of history. The ontological structure of state in time is shown as the world history as there can be no state without the freedom to shape history.

With regards to achieving freedom as a recognized necessity, Hegel argues:

The state is absolutely rational inasmuch as it is the actuality of the substantial will which it possesses in the particular self-consciousness once that consciousness has been raised to consciousness of its universality. This substantial unity is an absolute unmoved end in itself, in which freedom comes into its supreme right. On the other hand, this final end has supreme right against the individual, whose supreme duty is to be a member of the state. (Hegel, 1952, §258)

Those who use their freedom towards others and not towards themselves enjoy freedom as a recognized necessity whose real place of realization is the state as a framework of public actions and the balance of interests. The state as the mental about-self and for-self in the structure of the objective spirit is the medium of freedom which enables the highest form of art, religion and philosophy as manifestations of the absolute spirit. The state is not complete if it has no knowledge of itself, as this is the element of its substantive freedom. Because of this, philosophy, art and religion provide the sense of purpose of freedom for the state and the individual, if they participate in these areas of the absolute spirit and have awareness of its place in history. If an individual has no awareness of his place in history, he does not exist in the historical sense, because even though he is, there is no historical reality of the world without scientific reflection about it.

On this circular path from the absolute knowledge, its immediacy or freedom, the spirit must mediate itself through movement of its activities. It is its own path illuminate by consciousness. This is the only way for it to make itself what it really is – the absolute knowledge of the absolute. "The subjective and the objective spirit are to be looked on as the road on which this aspect of reality or existence [of absolute spirit, note: IG] rises to maturity" (Hegel, 2012, §553). The absolute spirit is the "identity returning and ever returned into itself: if it is the one and universal substance it is so as a spirit, discerning itself into a self and a consciousness, for which it is as substance" (Hegel, 2012, § 554). It is reflected in the history, revealed in religion and philosophy. The absolute spirit in the arts is the image in arts, senses in religion, and free thought in philosophy, though which is *about-self* and *for-self*. Philosophy as a reflection in its inner ontological temporality closes the circle of circles in the concept of the absolute idea which perceives itself. The absolute idea has thus made the transition of self-development from logos, through nature to absolute spirit. Through its inner ontological structure, the idea has formed in the reality of world history, and through philosophy, it has reached the self-consciousness of its own freedom. "In philosophy, what *is* is not just in the substantive form of what is, but is also in the subjective form of what is about self and for self" (Sutlić, 1994, p. 104). The being is organized as the absolute mediation of substance as subject which mediates itself in the otherness and returns to itself, maintaining the higher unity of the opposition.

We can say that Hegel's philosophy as systematic science about the whole, as the absolute science of the absolute, truth and freedom constitutes from itself a pure being, i.e. the being without an entity. A philosopher is a mediator for the absolute spirit reaching the absolute knowledge and awareness of its freedom. In this respect, one might conclude that man participates in the freedom of the absolute as the medium of its revelation in the world history. However, Hegel's entire system shows that man is a mere moment and object of self-action of the absolute on its road to itself. Man is reduced to pure being, devoid of its essence, humanity, and reduced to a worker participating in the creation of the pure being. The continuity of Hegel's metaphysics and its realization in the reality of the world is reflected in establishment of man – worker – scientific worker – cyborg into the centre of the activity of scientific work as the absolute of scientific history.

2.1.4 The Absolute Spirit Of World History

The state is a "self-aware substance in which the spirit had developed to the level of organics, i.e. to the middle moment of subjective logic and the last moment of the Philosophy of Nature. The synthesis of the Philosophy of Spirit, that is, of the objective spirit is given in the state" (Sutlić, 1994, p. 20). The state regulates its internal structure through constitution and as an individual among other states its defines it be external state law. When we talk about world history in which all nation take place "these particular minds [minds of individual nations, note: IG]" (Hegel, 2012, §536) are just moments in the development of the spirit in its reality – world history. The world spirit here acts as the substance subject of historical activity in which the state is only the final stage of history [logical, not temporal, note: IG], an actualization of the concept of ethics. On the one hand, it is the reality of the ethical idea in the real time of world history, and on the other hand, its ontological structure represents the synthesis of all moments of the objective spirit.

The limited and finite national spirits are intermediary steps and moments in an unbound development of the true world spirit, which it performs in the world history as its court of judgement of individual and limited national spirits. The court represents the dialectic unity of universal, particular and individual:

This movement is the path of liberation for the spiritual substance, the deed by which the absolute final aim of the world is realised in it, and the merely implicit mind achieves consciousness and self-consciousness. It is thus the revelation and actuality of its essential and completed essence, whereby it becomes to the outward eye a universal spirit—a world-mind. (Hegel, 2012, §549)

We can, thus, conclude that when it [world spirit, note: IG]:

[...] is realized, i.e. when the time as the external, outer is united with logos, or better, with onto-logic of the objective spirit, this world spirit becomes the world court (Weltgericht) which decides the faith of nations by determining their place and function in realization of one of the moments of the ontological structure which is revealed in its entirety in the philosophy of law, from the abstract natural law, through morality, to customs. (Sutlić, 1994, p. 24)

Hegel in particular notes that "world history is not the verdict of its power, i.e. the abstract and non-rational inevitability of a blind destiny. On the contrary, since mind is implicitly and actually reason, and reason is explicit to itself in mind as knowledge, world history is the necessary development, out of the concept of freedom of the mind alone, of the moments of reason and so of the self-consciousness and freedom of mind. This development is the interpretation and actualisation of the universal mind" (Hegel, 1952, §342). Since mind rules over history, it has a strictly defined logical structure, i.e. its purpose, sense and continuity of progress. The world history represents the true essential and substantial freedom in progress, or in other words, the ontological structure of the state and the objective spirit in space and time. This way, the history of the world spirit becomes

[...] its own act. Mind is only what it does, and its act is to make itself the object of its own consciousness. In history its act is to gain consciousness of itself as mind, to apprehend itself in its interpretation of itself to itself. This apprehension is its being and its principle, and the completion of apprehension at one stage is at the same time the rejection of that stage and its transition" (Hegel, 1952, § 343).

Falling into historical time and space, the spirit grasps itself in itself as the unity of universal, particular and individual, of reality and concept, of necessity and freedom. By falling through space and time, it fill itself with content, but the space and time do not define the world spirit. Both time and space are primarily alien to the spirit. In other words, "as the idea falls from the element of logos into nature and spirit to reflect in them and return filled with meaning, so does the objective spirit, the world spirit, needs history to reflect in it and comprehend what is to thus become free" (Sutlić, 1994, pp. 31-32). The structure of history, as described by Hegel in the *Philosophy of Right*, materializes through the historical activity. The internal structure of history, this abstract right of the state as the last form of ethics, manifests itself in the predominantly foreign medium of time. This represents the ontologization of the real temporal process. Put differently, all possible events, all possible human passions which think they are the subjects of history can amount to no more than being entities in the service of the spirit. The spirit becomes real only through national spirits and states as the agents of its realization. One might say that the world spirit is not just established, it represents the movement of its action towards self-action. The spirit is complete in movement, on the path to itself it is the path that makes itself what it actually is – the subject substance of world history. When the idea and the reality of all moments of the objective spirit merge into one, the world spirit reaches the end of its path, it reaches its freedom and the knowledge of this freedom. If the

[...] world history is the progression of spirit in the self-awareness of its freedom, then the philosophy is the freedom itself. In philosophy, what is is not only in the substantive form of what is about itself but in the subjective form of everything what is about self and for self. The absolute spirit in the element of philosophy befits the world corresponds with world spirit in the medium of the objective spirit, i.e. in the temporal medium of historicity. (Sutlić, 1994, p. 104)

Hegel's conception of world history, thus, ends when the external temporality and spatiality of history coincide with the internal completion of the ontological structure of history regardless of the fact that this history is still taking place in time as it is by its highest determination past. In other words,

When philosophy reaches the maturity it received with Kant and post-Kantian developments in Fichte, Schelling and Hegel himself, the crown of world history, the world spirit as the historical form of the absolute idea leaves behind the historical medium in which it appeared and now exists in its pure form as the absolute spirit of the absolute concept, the absolute idea. In other words, the absolute is recognizable in arts, religion, and philosophy in different forms and at different stages the way it is, not just in the medium of history. (Sutlić, 1994, p. 103).

For Hegel, this is the history of the absolute spirit standing above and outside history, it is its first and its last, and man is a mere participant and interpreter of its arrival to the absolute knowledge of itself in philosophy. In Marx's interpretation of history, we find the productive powers and relations in the production which take the place of the absolute spirit.

2.2 MARX'S CRITICISM OF THE ABSOLUTE NOTION OF HEGEL'S SYSTEM OF PHILOSOPHY

2.2.1 Philosophy and Historical Reality

After Hegel's *Encyclopaedia of Philosophical Sciences*, which incorporated the historical time into thought and showed the absolute spirit as the substance subject of the ontological structure of history, Marx was faced with a crucial question – is it there any space left for original philosophical thought in the post-Hegel world since every post-Hegel thought must respect his entire philosophy? Marx believed that philosophy as a science which encompassed the totality of the historical reality in the conceptual element could be overcome and that new ideas can be put forth only if we abandon the field of philosophy in its entirety. The original thought must be started on new foundations, which can serve to survey the shortcomings of Hegel's philosophy, which arise from philosophy as such. These new foundations are neatly laid out in the following statement: "The philosophers have only interpreted the world, in various ways; the point, however, is to change it" (Engels, 1969, p. 65).

Marx finds these new foundations of thought in the historical reality of the world, with the condition of *abolishing* and *realizing* philosophy itself: "you cannot transcend philosophy without actualizing it." (Marx, 1977, p.136). This is an inevitable path towards the act of raising the reality onto the level of philosophy as "Consciousness [philosophy, note: IG] can never be anything else than conscious existence, and the existence of men is their actual life-process (Marx and Engels, 1968, p. 11). The actual life-processes consist of "the real individuals, their activity and the material conditions under which they live." (Marx and Engels, 1968, p. 7).

The world processes and the historical reality of man as the new philosophical criteria are in Marx's analysis regarded as an estranged activity and an estranged reality. Opposed to it, above it, and against it stands Hegel's total philosophy which has raised historical reality to the status of the absolute concept and has given it its meaning only in the element of philosophical thought. By creating everything from thought, Hegel's speculative philosophy sees life only through space and time in which the absolute idea as the imaginary substance subject of creation will be filled with content, leaving reality without its real ontological structure. As Marx very correctly points out, Hegel uses reality only as a proof of position of the speculative ontological structure.

The position or self-affirmation and self-confirmation contained in the negation of the negation is taken to be a position which is not yet sure of itself, which is therefore burdened with its opposite, which is doubtful of itself and therefore in need of proof, and which. therefore, is not a position establishing itself by its existence-not a position that justifies itself; hence it is directly and immediately confronted by the self-grounded position of sense-certainty. (Marx and Engels, 1988, pp. 144-145)

For Marx, this difference between thought and reality symbolizes the alienation of both thought (philosophy) and reality, which is why thoughtful action in the world requires abolishment and realization of Hegel's philosophy for the sake of fulfilling the reality with meaning: "It is not enough that thought strives to actualize itself; actuality must itself strive toward thought." (Marx, 1977, p. 138). Marx is interested in reconciliation of thought and being in reality, in thought which is based carried by the being as man's activity as a species, which is why he proclaims

[man] is a species-being, [...] and has to confirm and manifest himself as such both in his being and in his knowing. [...] And as everything natural has to come into being, man too has his act of origin – history – which, however, is for him a known history [...] (Marx and Engels, 1988, pp. 155-156)

The synthesis of reality of thought and being is "Where speculation ends, where real life starts, there consequently begins real, positive science, the expounding of the practical activity, of the practical process of development of men. Empty phrases about consciousness end, and real knowledge has to take their place" (Marx and Engels, 1998, p. 43). However, as Hegel's total philosophy does not correspond to the reality of the world, the reality needs to be critically approached and fundamentally changed. Philosophy for the sake of reality, not reality for the sake of philosophy – that is the task that Marx has set for himself. In this sense "the world becoming philosophical implies philosophy becoming worldly [...] its realization is also its downfall" (Sutlić, 1974, p. 27).

In the critical analysis of the historical reality, Marx abandons the philosophical terminology but retains Hegel's dialectic method as a way of conducting research, illustrating and developing empirical concepts. The key ideas of this transition towards the empirical sciences had already been laid out in *The German Ideology* in the section dedicated to the critique of speculative philosophy in general. "The premises from which we begin are not arbitrary ones, not dogmas, but real premises from which abstraction can only be made in the imagination. They are the real individuals, their activity and the material conditions under which they live, both those which they find already existing and those produced by their activity. These premises can thus be verified in a purely empirical way" (Marx and Engels, 1998, pp. 36-37) and described using the terminology of empirical sciences.

In the same work, Marx emphasizes that speculative philosophy as the ideology of the ruling class represents the ruling ideas which, in order to conceal their actual being, "must bring an order into this rule of ideas, prove a mystical connection among the successive ruling ideas [...] To remove the mystical appearance of this "self-determining concept" it is changed into a person – 'Self-Consciousness' – or, to appear thoroughly materialistic, [...] into the 'thinkers', the 'philosophers' (Marx and Engels, 1998, p. 70). This is the way Hegel himself "has only found the abstract, logical, speculative expression for the movement of history; and this historical process is not yet the real history of man-of man as a given subject, but only man's act of genesis-the story of man's origin" (Marx and Engels, 1988, p. 145). Hegel himself admits at the end *of Philosophy of History* that he "has considered the progress of the concept only" and has represented in history the "true theodicy" (Marx and Engels, 1998, p. 70). In contrast to Hegel and the ruling ideologies, Marx starts from the premise of reality.

The first premise of all human history is, of course, the existence of living human individuals. Thus, the first fact to be established is the physical organisation of these individuals and their consequent relation to the rest of nature. [...] All historical writing must set out from these natural bases and their modification in the course of history through the action of men. (Marx and Engels, 1998, p. 37)

Thus, history and world history are not

[...] a mere abstract act on the part of the "self-consciousness", the world spirit, or of any other metaphysical spectre, but a quite material, empirically verifiable act, an act the proof of which every individual furnishes as he comes and goes, eats, drinks and clothes himself. (Marx and Engels, 1998, p. 59)

This does not mean that Marx rejects the results of philosophy. Using the dialectical analysis of the reality, he imbues philosophy with content which he shows in accordance with the results of the empirical analysis of the real world. In this material of reality given in the element of thought, instead of Hegel's absolute spirit as substance subject of the world history, Marx notices the productive powers of labour on the one hand, and the proletariat on the other end as the real subject which is by its position in life and determination invoked and interested in thoroughly changing the material world and its ideological content as it represents the maximum of the alienation of man. The proletariat and philosophy are in a position of mutual dependence: "Just as philosophy finds its material weapons in the proletariat, so the proletariat finds its spiritual weapons in philosophy" (Marx, 1977, p. 142). Thus, the proletariat which "can thus only exist world-historically" (Marx and Engels, 1998, p. 57). However, the role of the proletariat for untangling the contradictions of philosophy and reality is not our focus here.

Taking into consideration the essence of Marx's thought and his critique of Hegel's philosophy of the absolute concept as the foundation and the absolute subject of the whole philosophical system and the substance subject of world history, we are constructing our own philosophical position based on Marx's determination of work, productive powers of work and science as the paradigmatic form of productive powers of labour. It is on these terms that Marx bases his transition of philosophy of metaphysics into a systematic meaning of the world and gives it the status of world history of humanism and naturalism as the subject of world history. In order to take this path, to "be radical is to grasp matters at the root. But for man the root is man himself" (Marx, 1977, p. 137). This brings us to the question of all questions, the question of the essence of man. What is man?

2.2.2 The Essence and the Alienation of man

We have so far shown that Marx's understanding of the reality of man and his essence is no longer just a philosophical matter as it also transitions into the practical and starts from the empirical dialectic analysis of reality and its portrayal, as "man is no abstract being squatting outside the world. Man is the world of man, the state, society" (Marx, 1977, p. 131). Here in man's world, in which he is the creator of everything on the one hand, and just a man on the other hand, one must rethink the differences between essence, existence, and alienation of man as an entity in historical time. It is clear for Marx that man is at the same time a natural and a historical entity, which differs from other animate and inanimate entities not in terms of its qualities and attributes, but in terms of his relationship towards itself, the world and his fellow man. As an essential entity, man is able to create history and within his historical time change his natural environment – to create his human world and be aware of this creation. What gives man the power he has and what enables him to interact with his fellow human beings, and animate and inanimate beings alike, is labour as his free and conscious activity, or self-activity.

Marx emphasizes that labour is man's "*life-activity, productive life itself* [...] is the life of the species" (Marx and Engels, 1988, p. 76) since his own life is an object as his activity is "free activity", i.e. self-activity". It is precisely this life activity that makes him different from other living beings:

Where there exists a relationship, it exists for me; the animal does not "relate" itself to anything, it does not "relate" itself at all. For the animal its relation to others does not exist as a relation. Consciousness is, therefore, from the very beginning asocial product, and remains so as long as men exist at all. (Marx and Engels, 1998, p. 49-50)

What we can see here is Marx's intent to survey the essence of man in his relation towards self, towards other men, towards nature and being, i.e. to understand him as a self-aware and self-relating entity – man is a social being. Expanding the horizon of understanding labour as the being of man, Marx argues:

In creating an objective world by his practical activity, in working-up inorganic nature, man proves himself a conscious species being, i.e., as a being that treats the species as its own essential being, or that treats itself as a species being. [...] man produces universally. [...] even when he is free from physical need and only truly produces in freedom therefrom. (Marx and Engels, 1988, p. 76-77)

However, in the movement of labour as man's activity even "nature appears as his work and his reality" (Marx and Engels, 1988, p. 77). Through work, man acts like a force of nature and throughout his labour he consumes what nature consumes within him (natural energy). Man can never control nature as the basis of life as it is the condition of every control. To be natural is to be human, and for Marx this means to work.

If we look at Marx's philosophy in total, one might argue that man in his interpretation is the social entity of labour – it is a working entity as his every activity (labour) is a conscious self-activity, which reaches its freedom at the end of history. In Hegelian terms, labour represents self-movement, progress in the transition of substance into the absolute subjectivity. Starting from the unthought and neutrally defined concept of labour as man's activity – self-activity, Marx fails to "understand man itself in his essence from the perspective of essence of the technologically defined labour and science" (Sutlić, 1974, p. 278). Because of this, Marx fails to understand that *humanism equals naturalism* in terms of work, i.e. *scientific work*. The liberated scientific labour, puts forth scientific *humanism as naturalism* as space-time of its-self-activity – the production for the sake of production itself.

In the *Phenomenology of Spirit*, Hegel defines self-consciousness as "the Self's pure knowledge of itself" (Hegel, 1977, p. 482). In the abstract and self-conscious subject – the Spirit, Hegel sees the essence of man and the essence of everything that is. For him, spirit represents the true essence of man and the true form of the spirit is the thinking spirit, the logical speculative spirit. Man appears only as self-consciousness, and different forms of alienation noted by Hegel recur only as forms of estranged consciousness and self-consciousness. Marx acknowledges that Hegel's *Phenomenology of Spirit* understands the self-realization of man as the process of creation through labour that he has understood the true, material, actual man as the result of his labour. But Hegel's perception of labour is abstract and represents a speculative concept. To paraphrase Marx, we might argue that Hegel's philosophy regards as true alienation of man the very act of alienation of man's entity, the self-consciousness. On the other side of speculation, Marx places true work as the activity of man through which his essence and self-consciousness is expressed. We might say "It is not consciousness that determines life, but life that determines consciousness" (Marx and Engels, 1998, p. 42).

In the civil society as the space of dialectical empirical analysis, Marx shows the real existential (ideological, political, economic) and ontological alienation of man and his reduction to the role of a worker in the world of estranged labour, the estranged essence of man. In this civil society, man as a worker labours to change the whole world (natural and historical), turning into a complete production line of capital. "Capital is thus the governing power over labour and its products" (Marx and Engels, 1988, p. 36). In the period of history of the civil society, the capital as the otherness of man's activity (labour) is the absolute subject-substance of the new age activity which has to destroy, both dialectically and practically, in the name of man. The products of man's labour, his materialized living labour turn him into a dependent creature, lost in the time space of the capitalist society, which has transformed all historical time into labour time. In the reality dimensions of estranged labour, the worker does not fulfil his human essence:

As a result, therefore, man (the worker) only feels himself freely active in his animal functions – eating, drinking, procreating, or at most in his dwelling and in dressing-up, etc.; and in his human functions he no longer feels himself to be anything but an animal. What is animal becomes human and what is human becomes animal. (Marx and Engels, 1988, p. 74)

Thus, estranged labour estranges from man "(1) nature, and (2) himself, his own active functions, his life-activity, estranged labour estranges the species from man. It turns for him the life of the species into a means of individual life" (Marx and Engels, 1988, p. 76). The concept of estranged labour (Marx also uses the term engendered labour and life), was borrowed from Hegel's speculative philosophy. However, he illustrates it in the actual empirical analysis of the national economy as a result of privatization of factors of the productive process and the division of labour which includes the division of conditions for labour into means of production, the object of labour, the capital and the workers' labour. With an ever-increasing division of labour grows the fragmentation of privatized factors of the productive processes, so that the "labour itself can only exist on the premise of this fragmentation [...] the productive forces appear as a world for themselves, quite independent of and divorced from the individuals, alongside the individuals: the reason for this is that the individuals, whose forces they are, exist split up and in opposition to one another [...]" (Marx and Engels, 1998, p. 95). Marx concludes that

[...] on the one hand, we have a totality of productive forces, which have, as it were, taken on a material form and are for the individuals no longer the forces of the individuals but of private property [...] On the other hand, standing over against these productive forces, we have the majority of the individuals from whom these forces have been wrested away, and who, robbed thus of all real life-content, have become abstract individuals, but who are, however, only by this fact put into a position to enter into relation with one another as individuals forces in the intercourse and association of these individuals" (Marx and Engels, 1998, pp. 95-96).

Their individual work is their only connection with the estranged productive powers as general labour. The work activity has been reduced to the means of sustaining the bare minimum of life, and man and his labour to merchandise. All of man's natural and historical powers have become the powers of capital as the substance subject of civil means of organizing life. Consequently, Marx advocates the abolition of market-based mediation of productive factors, private property and the division of labour as capitalist productive relations, as this is a prerequisite for uninterrupted development of the productive powers of work that are only becoming what they were in their essence – the means of man's self-activity and his return to his free anthropological essence: "With the appropriation of the total productive forces through united individuals, private property comes to an end." (Marx and Engels, 1998, p. 97). The abolition of private property also represents "the complete emancipation of all human senses and qualities [...] but it is this emancipation precisely because these senses and attributes have become, subjectively and objectively, human". Marx and Engels (1998) regard this historical existence in which thought, being and essence of man have reached a point of reconciliation as the real truth, the truth of being, the real historically achieved humanism and naturalism of man in which total development of the individual is made possible.

The comparison Marx's and Hegel's determinations of the ontological structure of history of the civil has provided us with the following insights. By using Sutlić's (1972) interpretation in *Bit i suvremenost* (Essence and modernity) we can argue that Hegel's speculative system of philosophy raised a specific mediation of production of life in the civil society to the status of being of man in general and made the historical structure and time of this imaginary being the absolute in terms of Nietzsche's eternal now and *eternal return of the same*. As a Young-Hegelian, Marx is aware that Hegel did not resolve the issue of being as an apparition. The apparition has no being by and of itself but in something else, which makes

Hegel's entire philosophy of the absolute concept as the substance subject of the whole human history speculative and an essential falsity.

By using empirical terms, Marx argues that the specific civil means of production of life, which is also a form of community, rests on private ownership (capital) and estranged labour as a historically transient method of maintaining life and cannot be declared a universal foundation of historical life of man. His dialectic analysis of reality

[...] shows that history does not end by being resolved into "self-consciousness as spirit of the spirit", but that each stage contains a material result: a sum of productive forces, a historically created relation to nature and of individuals to one another [...] This sum of productive forces, capital funds and social forms of intercourse [...] is the real basis of what the philosophers have conceived as "substance" and "essence of man", and [...] revolt against as "self-consciousness" and the "unique". (Marx and Engels, 1998, pp. 61-62)

Through this determination, Marx destroys the totality of circle of the nihilistic self-establishment of the absolute spirit as the deceptive being of history and replaces is with the openness of free labour (undeterred development of productive powers) and man's conscious and free self-activity in historical time. From today's perspective, addressing Marx's viewpoint as the decisive perspective at the tipping point between periods is of vital importance to rethink the key terms of his thought (work/labour as the essence of man, productive powers of work, scientific work, nature, *humanism = naturalism*, history), which he tacitly assumes without thinking them through or thinking them out beforehand. Thus, the reconciliation of the idea and the reality of the world of disalienated labour, which Marx's entire critical attempts strived for, places before all thinkers a question of historical liberation of work and his scientific character in the period of *scientific history.*

2.3 THE MARX – HEGEL DIALOGUE

Rethinking the philosophy of Hegel and Marx in their mutual dialogue requires that we, before anything else, understand what was said and written in their works. Sutlić would say that to think and to speak through the dialogue of their works means to treat what was reached through dialogue, without being depleted but leaving futile many topics for philosophical orientation in the essence of historical time.

Within the Marx-Hegel dialogue, our philosophical dilemma is resolved – whether to adhere to philosophy which justifies the existence of world of estranged labour and man or whether to turn to the "idea of the future in crisis, which correlates the philosophical triumphs of philosophy with realization of the new world" (Sutlić, 1972, p. 111). In the framework of philosophical discussion, ontology and metaphysics are sometimes regarded as synonyms. The critique of metaphysics and its domination of path towards its abolishment have long dominated the philosophical discourse.

This starts with Immanuel Kant, although he uses the term metaphysics in the affirmative sense of the word. However, he wants to establish metaphysics as science, and thus crucially needs to question whether it is possible at all. In the Critique of Pure Reason, he unequivocally denies the possibility of knowing that what is beyond the experience ("thing-in-itself" Ding an sich) by replacing metaphysics with gnosology, which shapes Kant's criticism. If one wants to understand the foundations of Kant's abolishment of as-

sumptions which classic metaphysics is based upon, one must take into consideration the concept of his breakthrough and that is the objectivity of the object (Gegenständlichkeit des Gegenstandes). Through mediation of this term, Kant questions the possibility of an object, not just the possibility of knowing the object, but the possibility of the object to be as such, to be ascribed being (in the classical logical syllogism this ascription of being is done with the copula be), even though he does not derive any logical consequences of this issue, staying, formally, at the gnosological level of scrutinizing the objectivity of the object. (Veljak, 2013, pp. 2-3)

The "common denominator of all variants of classical metaphysics is the determination of being as the eternal, unchangeable, self-equivalent, *is* which allows all entities to *be*, determining that they be exactly and exclusively what each of them according to their essence (and inherent capabilities, or more precisely, the extents of this inherent capabilities, specific to the species that the entity belongs to) can become (Veljak, 2005, p. 772). The classic Kantian metaphysics reaches its limits. Fichte, Schelling and Hegel

[...] took over Kant's transcendental overturn, which is reflected in replacement of metaphysics [...] Hegel's Science of Logic (in which the identity of identity and non-identity are crucial terms) represent, by normal standards, the pinnacle of shown (or perhaps outlined) critique of classical metaphysics, which stems from Kant's transcendental shift. But, according to some interpretations, this work by Hegel represents the last major system of the western metaphysics, the synthesis of realistic and constructivistic ontology (Veljak, 2013, p. 4).

By moving our focus towards the possibility of in the world of today and by comparing the key philosophical concepts, we begin our dialogue of the two great thinkers whose philosophies have marked the end of an era in philosophy.

2.3.1 Philosophy and Historical Reality

In Hegel's system of philosophy, only that what is mental can be real and what is mental is real. The fact that something exists is no guarantee of its real existence. Only that what was engulfed by the movement of the spirit and filled with logos has the status of real. The entire movement of the spirit takes place in the tripartite sequence thesis–antithesis–synthesis. The subjective spirit represents the thesis, the objective spirit represents the antithesis, and the absolute spirit represents the synthesis of the *Philosophy of Spirit*. It is important to explain Hegel's reasons, real of philosophical, for his construction of system of philosophical sciences which will, according to him, encompass in thought all time (historical time) of human history and nature. Hegel's thought is faced with the reality of the civil world based on the important contradictions of sustaining life, on separation of essence form existence, and philosophy as thought of essence of this reality is proven incapable of reconciliating the philosophical idea of sense with the estranged and purposeless reality in which the being is perceived as nothing. Sutlić argues that the impotence of every philosophy, including Hegel's, stems from its internal capabilities (or incapabilities) as metaphysics. In order to overcome this impotence of philosophy and give it new power and importance, Hegel builds a system of philosophical sciences in which the speculative dialectical method and the great synthesis of circle of circles – *Science of Logic, Philosophy of Nature and Philosophy of Spirit* – will create a philosophical bridge over the emptiness of reality and imbue it with ontological sense in the element of scientific thought – philosophy as the absolute knowledge of the absolute, which has overcome the subject-object relation, which had burdened previous philosophical realizations.

We return to the philosophical origin of impotence of metaphysics. Aristotle's first philosophy or ontology scrutinizes entity as entity, it wonders about his essence. It is

[...] the study of being of entities or the study of the essence of entities, the study of entity as entity (Lat. ens qua ents) and it is also the study of the highest entity, the most powerful all-encompassing entity, about what is how it is in its activity and all its immutability. That what is the highest, the supreme Aristotle calls the divine. (Sutlić, 1994, p. 94)

The Greeks see the being as the presence of what is present. The entity is the present and the presence, that what is present, here and now, that what is what is as is now (Sutlić, 1994, p. 95). With the new age begins the process that lasts "from Descartes to Kant and to which Kant gives the definitive formula – the basic discipline is no longer ontology, but transcendental philosophy (Sutlić, 1994, p. 94). On the other hand, transcendental denotes that what enables every experience, that what is beyond experience. The question of objectivity of the object or subjectivity of the subject was in Kant's philosophy shown to be the question of transcendental consciousness, i.e. the subject "even if everything that is is nothing else but material for this subject" (Sutlić, 1994, p. 95). Hegel's answer to the question of what an entity is as entity is as follows:

[...] entity as entity means spirit, and the spirit is the process of substance returning to itself in the form of a subject; the substance becomes about-self and for-self in the subject. The absolute knowledge in the Phenomenology of Spirit is the substance that has become subject, i.e. is aware of itself and is therefore free. (Sutlić, 1994, p. 96)

Based on the experience of Kant, Fichte and Schelling, Hegel "realized that to be means to be set by the subject, and the subject is nothing more than a subject vis-à-vis object, its own subject object, its own material whose form is the substance subject" (Sutlić, 1994, p. 97). This way, Sutlić argues, Hegel has reduced the possibility that something is to that what it is, but of the highest rank – the highest entity of the substance subject, the absolute. Entity as entity is the spirit which establishes itself on its way to reaching substance within itself in the form of subject in the time of world history. Transience and nothingness of everything that is, the truthfulness and temporality of the totality of history and the very possibility of entity as entity remains unthought.

Taking Sutlić's analysis of Hegel's philosophical system into consideration from the position of rephilosophication of Marx's thought, we might argue that the concrete for Hegel is something what is only when all three dialectic circles are within it – the general, the special and the individual. The speculative method ascribes creativity to the idea as the imaginary subject and the reality remains the mindless object of its development. Since the absolute idea in its development allows everything that is in its essence and existence, this means that to be means to be established by the subject which is its own subject object. In the historical space and time, the absolute idea as an entity is the world spirit which has established itself and has realized itself in this path the *absolute science of the absolute*. The self-establishment of the absolute as the subject from its substantial form to its substance subject form represents its arrival to freedom and self-awareness – the knowledge of itself. Once the absolute idea as the world spirit reaches itself, once the ontological and external temporality overlap, the world history becomes past by its highest determination. Now the absolute spirit is free to move in the content of arts, religion and philosophy established by its own activity.

The absolute-established philosophy as the *absolute science of the absolute,* which is its own subject-object, is the medium of freedom as everything that is in it is not just in substantial, but also in subjective form. Since philosophy is also the science of the totality of world activity, it is also the science of truth. Overcoming the nothingness of the reality through speculative dialectic synthesis of history ends with the internal ontology of development of the absolute in itself, and the real historical reality is still left with the eternal presence, duration through recurrence of the same. This way, Hegel's logic in the element of thought takes on itself all life and movement. The historical reality and man himself are left with the form of transience, no historicality and no sense. Since the world spirit has produced and realized itself as consciousness in the historical time and space, we can say that Hegel deals with the self-conscious reality (in the element of thought) and not separation of consciousness and reality, as Marx argues.

In the critique of Hegel's speculative philosophy, Marx quite lucidly points out:

But because Hegel has conceived the negation of the negation from the point of view of the positive relation inherent in it as the true and only positive, and from the point of view of the negative relation inherent in it as the only true act and self-realizing act of all being, he has only found the abstract, logical, speculative expression for the movement of history; and this historical process is not yet the real history of man-of man as a given subject, but only man's act of genesis-the story of man's origin. (Marx and Engels, 1988, p. 145)

This way, Hegel has included the *nothing* into the optimal ontological functioning of the world and has left it as a real, albeit a covert measure of the world and the essence of entity. As an essential entity, man becomes the moment in the movement of the absolute, without the possibility of relating to the world on its own, closed in itself and the emptiness of being. Man still governs over *nothing*. Thus, philosophy as metaphysics based on the estranged non-essential reality of the civil society as its self-awareness reaches its own emptiness and existence without sense. From the context of Marx's critique of speculative philosophy and philosophy in general, one could say that philosophy cannot scrutinize the basis of alienation as the alienation is its own basis. As an expression and measure of alienation, as an ideology of civil society, it is its own obstacle towards fulfilling its own task [freedom, note: IG]. This is why Marx can say that philosophy and reality should be critically faced and fundamentally changed, one for the sake of the other.

Instead of philosophy in crisis of orientation in the essence of the alienated time, devoid of essential life content, Marx argues for re-thinking of the reality which stems towards reaching the sense of life, historical time and freedom of man. He starts from the real man as he is in history: "the first premise of all human existence and, therefore, of all history, the premise, namely, that men must be in a position to live in order to be able to "make history". [...] The first historical act is thus the production of the means to satisfy these needs, the production of material life itself" (Marx and Engels, 1998, p. 47). The real power and the actual subject of history is "where speculation ends -- in real life -- there real, positive science begins: the representation of the practical activity, of the practical process of development of men" (Marx and Engels, 1998, p. 53). Marx concludes that "The production of life, both of one's own in labour and of fresh life in procreation, now appears as a double relationship: on the one hand as a natural, on the other as a social relation [...]" (Marx and Engels, 1998, pp. 48-49). This real man in Marx's time lives in an estranged civil society which is in its essence a negation of everything human and natural, which is why Marx is not interested in uncovering the sense of what is, but in making that what is speak with sense immediately by and of itself. In contrast to Hegel who saw in the state the "actuality of the substantial will which it possesses in the particular self-consciousness" (Hegel, 1952, p. 230), the realization of the absolute spirit as the being of civil community, Marx regards the state as an instrument of the capital, which is the otherness of labour, the real being of the actual community of the civil society upon which rests the production of life and a form of a completely estranged community.

Marx overcomes Hegel's philosophy as a reflection of the estranged reality with the *only science – the historical science*, which is the actual and empirically testable equivalent of Hegel's *absolute science of the absolute*. Instead of Hegel's absolute idea of the world spirit, the substance subject of all activity, Marx advocates the concept of labour – the actual labour, which will in evolve in his later analyses into the concept of *scientific work* as the paradigmatic form of productive powers of work. In other words, the work of the concept (Hegel) as the cause – energy, development of world spirit which produces itself, is its own object and its own end purpose is replaced by the concept of work, empirically testable and self-explanatory subject substance of world history, which is also a concept which produces itself, is its own object and end purpose.

Unlike Hegel, Marx sees the reconcilliation of essence and existence in disalienation of reality and not in the element of concept. The truth is a whole, but in the element of reality in which work realizes itself as the actual substance subject. Even though he advocates the language of positivistic sciences as a more appropriate one for the dialectical analysis of reality, Marx still uses the language [explains and explores the concepts of reality, note: IG] of Hegel's metaphysics. This way, philosophy in its abolished form is contained in its own critic of political economy, which sees the economic sphere of society as the real absolute of history out of which all historical spheres grow and whose character they have. Raised to the status of philosophy (ontologic), the economic sphere in Marx's view has an important role, not purely philosophical nor economical. However, the key concept, the concept of labour, left unthought, persists somewhere between philosophy and political economy for Marx, between essence and existence, without its true essence coming into question.

For Hegel, the substance subject of world history is the absolute spirit which recognizes itself in the actual institutions and is aware of itself this way. In contrast to Hegel, who "fell into the error, therefore, of considering the real as the as the result of self-coordinating, self-absorbed, and spontaneously operating thought, while the method of advancing from the abstract to the concrete is but a way of thinking by which the concrete is grasped and reproduced in our mind as a concrete" (Marx, 1904, pp. 293-294), Marx emphasizes the concrete substance subject of history, not denying that the concrete totality of history as a meaningful totality is a product of thoughtful contemplation, but "a product of thought, of comprehension; not however in the sense of a product of a self-emanating conception which works outside of and stands above observation and imagination but of a mental consummation of observation and imagination (Marx, 1904, p. 294). Consciousness as a conscious being is only "the theoretical shape of that of which the living shape is the real community" (Marx and Engels, 1988, p. 105). The reality of productive powers and relations in the production is "the real basis of what the philosophers have conceived as "substance" and "essence of man"" (Marx and Engels, 1998, p. 62) and as estranged awareness, compared with the estranged entity of man with the real life as the superior power – the absolute knowledge of the absolute as subject of world history. Only through the liberated labour as the subject of world history (with the action of the proletariat) can the historical reality reach the unity of essence and existence and realize itself in this unity.

Hegel's authority is philosophy, Marx's authority is the *one and only science, the science of history.* Since Marx understands the term *science of history* as Hegel's philosophy realized as metaphysics [science of history, note: IG], the one and only science of history marks the true introduction of metaphysics (saved content) into the immanence of historical reality. This is a victory for metaphysics, despite Marx's efforts and through Hegel's mediation of speculative dialectal possibilities. Sutlić argues that only the realization of philosophy, its completion and establishment represents the absolute victory for philosophy and metaphysics in the form of its abolishment. Hegel's ontology of the world spirit is thus replaced with the scientific reality – the *scientific history*, and non-philosophical reality of with the *technology of labour.*

This is the "scientific reality, and scientific reality is the true subject of all activity. Thus, science takes the place of world spirit as the paradigmatic form of productive power of labour" (Sutlić, 1994, p. 79). To understand the relationship between Hegel and Marx "means to understand this whole process of abolishment, preservation, raising to a higher level – abolishment of philosophy by form, preservation of its content and raising this content to a higher level of reality; abolishing the existing reality, preserving the reality of reality, and raising this reality to a higher level, the level which philosophy carries within itself" (Sutlić, 1994, p. 157).

2.4. ESSAYS ON PHILOSOPHICAL ORGANIZATION OF MARX'S THOUGHT AND THE POSSIBILITY OF THOUGHT

In his monograph *Ogledi uz filozofijsko ustrojstvo Marxove misli* (engl. *Essays on philosophical organization of Marx's thought*), Sutlić has set his goal to "understand Marx's accomplishment in its entirety [...], to rethink the philosophical structure of his thought in the transition of the philosophical system into a systematic meaning of modern activity of the world" (Sutlić, 1994, Foreword) and lay the foundations for the possibility of historical thought. Throughout the years, this study has been my inspiration and a guide through the relationships between Hegel and Marx for our understanding of establishment of *scientific work* in the age of scientific *humanism as naturalism*, which is why I feel the need to put forward the basic outlines of the ontological background from Sutlić's *Praksa rada kao znanstvena povijest* (engl. *Practice of work as scientific history*) as it represents a part of the structure of our understanding the causes for enhancing and reshaping of man in the age of *scientific humanism as naturalism.* In this part of the analysis, we will use Sutlić's interpretation of the main ontological claims.

In the critique of Hegel's philosophy as the absolute science, which tried to overcome the *nothing* of the historical world with a dialectical synthesis within the scope of the thought, Marx's starting point is the reality itself in which he reveals the real subject, which will by its ontological status be capable of revolutionizing the reality and thought at the same time, one in the name of the other so that reality is thought out and the thought is made real. Marx's critique of "everything that persists" should be understood in the sense to

[...] uncover and liberate all primeval, primal, original activity – events that precede the separation of the 'idea' and 'appearance', 'real' and 'persisting', 'essence' and bare 'survival' etc., and all allow to be what they are [...] and to abolish them as such with a revolutionary shift of conditions of their differences. (Sutlić, 1974, p. 8)

This is why the critique from the "'opposition' of the primeval" (Sutlić, 1974, p. 9) is not Marx's "manifestation of a 'position' of any kind, but the primal action of the crisis which caused the separation of the 'idea' and the 'persisting' [...] which is why the primeval is shown as birth and growth itself" (Sutlić, 1974, p. 9). The critique is "looking for the third, which is not situated after but prior to this difference" (Sutlić, 1974, p. 9). This third, which is prior to this difference, is the concept of work/labour. According to Sutlić, what Marx does to the concept of work is "[...] by methodically securing it – we insinuate (German *Unterstellen*) work in the form in which it belongs exclusively to man" (Sutlić, 1974, p. 10). This work is older than

[...] every social form [...] it is a precondition for understanding every particular, concrete concept of work – such as the 'general nature' of the work itself is to produce, 'form', 'values of use', condition of the existence of man, the 'eternal natural necessity', 'mediation of exchange of matter between man and nature'. (Sutlić, 1974, pp. 10-11).

Marx uses the very concept of nature quite ambiguously. The nature formally and materially determines work as such, "in short, nature is genuinely the same as human activity and vice versa – human activity is genuinely the same as nature" (Sutlić, 1974, p. 11). In that respect, Sutlić concludes:

[...] the essence of man – the process of work, does not only include nature as an object but is natural in and of itself; it is like nature, we might say, a paradigmatic case of nature. In the working process, man is natural because, firstly, he relates to other natural powers as 'a single natural power', and secondly, because he works as the nature itself. (Sutlić, 1974, p. 11)

In other words, "nature in the sense of totality of all things (all entities) and man (in the sense of one particular entity in this whole) are the nature which produces and works" (Sutlić, 1974, p. 12). The condition for man's existence and essence is that he works as nature. This line of reasoning clearly shows that "just like 'nature' is ambiguous between entity and being (*rens* and *esse*), so is work ambiguous between the work of man and the work of mankind....However, work is man's *essentia*, that what (*quid*) makes him different from other entities, a 'condition of existence' of man " (Sutlić, 1974, p. 13). But it is at the same time the being of man and all natural entities, "from man's perspective and from nature onwards, the incursion of nature (as being) into the essence of man and the incursion of essence of man into nature (as being)." (Sutlić, 1974, p. 14). Sutlić claims that

[...] mutual incursion (in immanent transcendence of nature into human nature and human nature into nature itself) contains two elements – first, that everything that is is the object of work, the 'value of use', and all things are 'objects of work found by nature', that their 'affiliation' = their 'objectivity, and secondly, that man is also a natural being, precisely when he 'actualizes' and 'realizes' his naturality as the object of this objectivity. Through man's labour, nature is made conscious, it takes an appropriate form – historically transformed nature is an appropriate form of nature. The objectivity of the object, as well as subjectivity of the subject from the perspective of man versus nature, are one and the same – Nature as Work. (Sutlić, 1974, p. 15)

In the environment of capitalistic production caused by the division of labour and privatization of the productive factors of a single working process (which consists of a purposeful activity, object of labour, means of labour, and products of labour), the alienation of all aspects of work as postulated by Marx occurs and it includes the essence of man and the being of all entities in historical world. This conception of labour and the revolutionary task of abolishing the capitalist productive processes which limit its unhindered development of productive powers which functions as its own purpose leaves Marx within the domain of metaphysics as he did not try to find

new means of production, a new relationship between man and everything that is [...] a new basis, 'its own basis' (Grundlage) upon which the productive powers will mark a new are not just with their quantitative growth but with their qualitative determination – a new relation not implicitly contained in what Marx regards as simple 'elements', simple moments of work (Sutlić, 1974, p. 17).

This approach leaves Marx within the domain of realizing the contents of Hegel's metaphysics, but with a crucial difference that the place of the absolute spirit of human history "is taken by *scientific work*, working science [...] whose core is technology [...] technological management – computation, upon which all modern thinking and doing is based" (Sutlić, 1974, pp. 146-147). Thus, instead of a new element, Hegel's onto-theology is replaced by the *techno-logic* of labour, which is also anthropology, cosmology, and possibly theology. The *techno-logic* does not imply that the starting point is the logos – thought, but the rational, calculated thought which aims to triumph over everything. The *techno-logic* is a collective essence of preparing everything for an operative procedure. Marx saw technology as a completely neutral way of maintaining life, and all his attempts were aimed at liberating the productive powers of work as a condition for achieving freedom in real history. Sutlić rightfully calls him a thinker of the liberated technology, which has proved to be in significant contrast with the concept of liberty that Marx strived for.

Communism as realized *humanism* and *naturalism,* the *empire of freedom* and the *empire of necessity* with all its humanistic expectations remain in the cul-de-sac of metaphysics as *practice of work of scientific history*. In the *practice of work as a scientific history*, the absolute work is the self-established Work which is also the object of its own activity, which makes it free. Man is the mediator and the moment between particular capability of the substance subject and the final actualization in which it turns into substance subject about-self and for-self, i.e. freedom. The contradictory position of man in the absolute work of *scientific work* makes him, on the one hand, the subject of the whole process of historical movement in the world of labour (work as a man's self-activity), and on the other hand, man is the moment and object of scientific work in his ontological self-development in which every object grasped (even man himself) is rationally established in what it is. This position of co-activity through the absolute gives man the possibility of absolute and unlimited power in the eternal presence, but on the other hand, man as a moment and object of self-movement of the absolute is also powerless as his human essence is not within him but it stems from the being as an apparition. The power of man is the impotence of his being which grounds the world in an empty word into *nothing*. The metaphysics of *practice of work* is shown as the completion of an old world in which the *practice of work* as the essence of *scientific history* enables stability in change as it reconciles the contradictions in reality in the same manner that Hegel did in the absolute spirit of world history. The world of *practices of work* collects in itself both the essence of essence and the essence in its bare from of existence from which the difference has evaporated. In this practice which wants to make all change superfluous, *humanism* (and the human) is reduced to the moment of activity of practice in its self-realization as freedom. "[Labour] established this way in its unity and simplicity [...] must become the questions of all questions [in order to, note: IG] to be experience and [...] within communism liberated and something more fundamental and simpler than man – worker in the world of work" (Sutlić, 1974, p. 18).

By comprehending Marx's viewpoint on history, the possible interpretative principles of the trans-philosophical though, Sutlić argues that the history of today must be thought of in a trans-philosophical manner if we want to say something relevant about it.

In the uniform world of labour in which the oblivion of being is extended to the complete nihilism (being = nothing; entities, including the human beings, have lost all traces of their self and have become actualizations and personifications of organization and its institutions) [...] and still a barely visible way of thinking different than metaphysics and its nihilistic end takes place (Sutlić, 1994, p. 161).

The shift in the essence of *practice of work* as the essence of *scientific work* assumes gaining insight into the historicality and the epochality of practice, and thus the historical genesis of man. The historical event requires a shift, growth of significantly different productive powers of work, whose essence will enable a different historical relation towards being, essence of man, entities and nature. Philosophy must seek philosophical pathways, ways of surpassing philosophy and assumptions of reality upon which it halted, abandon the yearning for wisdom and situate itself (not within *one and only science* as is the case with Marx, nor within the *absolute science of the absolute* as is the case with Hegel, note IG) in wisdom itself. Sutlić regards the transphilosophical thought to be historical thought as in the time of being as nothing, which has completely occupied the historical time, in the depths of old seeks the foundations of new, travels with no roadmap, often through the backroads of metaphysics. Historical thought is not always visible, it is easy to oversee in interpretative actions. It stems from the *one and only science*, the discontinuity of historical eras, which Marx saw and ignored in the name of Hegelian understanding of continuity in the development of freedom. In the name of historical relation of man, being and essence of entities, it guards over the possibilities of historical events. This thought must be situated in the metaphysically unfounded, which is older than the first assumption of all human history, and it must speak the language of the living not of metaphysical persistence. Historical thought, Sutlić claims, must abandon the rules of formal logic, the language of the metaphysical speech and its linguistic structures. Marx and Hegel "are the crucial condition for recognizing and initiating historical thought, but with a critique of the metaphysical character of their thought" (Sutlić, 1994, p. 172). Life makes all questions worth asking, so we are invested in the shift in the hidden reality of reality itself. For this, we need more thought more life-oriented than philosophy – we need the historical thought. Through his wanderings and sketches of historical thought, Sutlić took the path of surpassing philosophy as metaphysics in the openness of historical time. His thought remains an illuminating spark that his life made in the pathways of philosophy.

REFERENCES

Hegel, G. W. F. (1830). *Encyclopaedia of the Philosophical Sciences* (W. Wallace, Trans.). https://www.marxists.org/reference/archive/hegel/works/ol/encycind.htm

Hegel, G. W. F. (1952). Hegel's Philosophy of Right (T. M. Knox, Trans.). Clarendon Press.

Hegel, G. W. F. (1970). *Hegel's Philosophy of Nature* (M. J. Petry, Trans.). George Allen and Unwin Ltd.

Hegel, G. W. F. (1977). *Phenomenology of Spirit* (A. V. Miller, Trans.). Oxford Univ. Press. https://www.marxists.org/reference/archive/hegel/works/ol/encycind.htm

Hegel, G. W. F. (1991). *The encyclopaedia logic, with the Zusätze: Part I of the Encyclopaedia of philosophical sciences with the Zusätze* (T. F. Geraets, W. A. Suchting, & H. S. Harris, Trans.). Hackett.

Hegel, G. W. F. (2001). *Philosophy of Mind* (W. Wallace, Trans.). Blackmask Online. https://www.marxists.org/reference/archive/hegel/works/ol/encycind.htm

Hegel, G. W. F. (2010). *The science of logic* (G. Di Giovanni, Trans.). Cambridge University Press. http://public.eblib.com/choice/publicfullrecord.aspx?p=554736

Hegel, G. W. F. (2012). *Hegel's Philosophy of Mind* (W. Wallace, Trans.). Clarendon Press. https://www.marxists.org/reference/archive/hegel/works/ol/encycind.htm

Marx, K. (1857). *Grundrisse. In Foundations of the Critique of Political Economy* (M. Nicolaus, Trans.). Marxists Internet Archive.

Marx, K. (1904). *A Contribution to the Critique of Political Economy*. Charles H. Kerr & Company.

Marx, K. (1969). Chapter. In F. Engels (Ed.), Ludwig Feuerbach and the End of Classical German Philosophy (pp. 59-61). Progress Publishers.

Marx, K. (1977). *Critique of Hegel's 'Philosophy of right'* (J. J. O'Malley, Trans.). University Press.

Marx, K., & Engels, F. (1968). *A Critique of The German Ideology*. Progress Publishers.

Marx, K., & Engels, F. (1988). *Economic and philosophic manuscripts of 1844* (M. Milligan, Trans.). Prometheus Books.

Marx, K., & Engels, F. (1998). *The German ideology: Including Theses on Feuerbach and introduction to The critique of political economy*. Prometheus Books.

Picht, G. (1989). *Der Begriff der Natur und seine Geschichte*. Klett-Cotta/J.G.Cotta'sche Buchhandlung Nachfol.

Sutlić, V. (1972). *Bit i suvremenost*. Izdavačko preduzeće Veselin Masleša.

Sutlić, V. (1974). *Praksa rada kao znanstvena povijest: Ogledi uz filozofijsko ustrojstvo Marxove misli*. Kulturni radnik.

Sutlić, V. (1994). *Uvod u povijesno mišljenje*. Demetra.

Veljak, L. (2005). Ontologizacija rodne diferencijacije kao apstraktna antiteza klasičnoj metafizici. Filozofska istraživanja, 4(99), 772.

Veljak, L. (2013). Praktična i praktička filozofija. *Filozofska istraživanja*, *4*(132), 603–610.

ENDNOTES

1 *Scientific history* is the unity to equivalence of theory, practice and *poiesis*, contained and merged within the scientific news and a command, a telling word as information, intention, decision and history. Scientific history is analogous to Nietsczhe's *eternal recurrence of the same*, just as the *practice of work* is analogous to Nietzsche's *will to power* (Sutlić 1974: 141-161).

2 For Hegel, the term *speculative* implies that one is reflected in the other, the reflected in the reflecting.

Chapter 2
Transtechnical Essence of Technology

ABSTRACT

This chapter introduces the reader to the stage of the discussion on the dialectic transformation of machine into machinery with an eye towards the transhumanist perspectives. With the help of Heidegger, the discussion is focused on the transhumanist technology and the ideologues of transhumanist improvement of human evolution. This early groundwork helps situate the contemporary treatment of cyborgoethics within a historical and philosophical framework. This is important in guiding researchers in exploring the essence of scientific work with an eye towards the cyborgization movement that is reshaping human life and society.

3.1. MAN AND THE MACHINE

Throughout his relationship with nature, man has developed various means of labour to survive in the harsh natural environment and ensure the stability of life and historical development. One such means of labour are the machines – mechanical contraptions which transform one form of energy into another or allow the work process to be performed automatically or semi-automatically (Anić, 2003, p. 1246). Where the basic means of labour differ from the machines is the position of the living worker in the work process. The basic means of labour are directly manipulated by the workers who use their physical powers and skills to process the raw materials and turn them into the end product. In the machine labour, the living labour is "subsumed under the total process of the machinery itself, as itself only a link of the system, whose unity exists not in the living workers, but rather in the living (active) machinery, which confronts his individual, insignificant doings as a mighty organism" (Marx, 1857, p. 615). This quote illustrates the historical lesson that transformation of means of manufacture labour into industrial machinery also transforms the living labour of the worker into a living addition and means of the machine itself. The machine represents a mechanical reconstruction of unity of the productive process that was destroyed by the manufacture division of labour. As materialized living labour, the machine allows the living labour to vacate the working process in order to be replaced by the machine in the processing of

DOI: 10.4018/978-1-7998-9231-1.ch002

objects of labour. From a the functionalist perspective on the means of labour, there is no fundamental difference between a stone axe and a computer-oriented cybernetic machine, as both serve the same function as work aids. The machine still functions as an aid in the contemporary sophisticated and digital versions, but its social essence has a different dimension altogether, to which we will return in §3.2.1 *Dialectic transformation of machine into machinery.*

There is a good reason why we should also mention the role of those means of labour which serve as instruments in scientific and technological research. Science requires technological means which increase in refinement and physical power with each passing day as a tool which science itself creates through, or requires from, technology. Science and technology can only live together or die together. It is worth emphasizing that sophisticated means of labour used in the scientific research experiments have the capability of probing into the depths of the matter at hand and allow us to reach previously unimaginable levels of knowledge. Science as the general mind, realized through the scientific and technological means of labour, adapts to the known pathways of technological means as the new scientific discoveries depend on them. The newly uncovered micro-, nano-, and quantum worlds can no longer be compared to the visible world around us, and our understanding of the world lags far behind the recent discoveries. The unknown and the unforeseen carries within it the spirit of modern technology and science, which is already reflected in the post-humanist philosophy.

The *abolition* and *realization* of philosophy as metaphysics in the reality of the world has left a philosophical void in the whole understanding of the world and its truth. The "*Ende* is akin to completion (*Vollendung*), the transformation into the final possibilities... the final transition (*Ausfalltung*) of philosophy towards independent sciences which interact with each other represents the legitimate finalization of philosophy" (Heidegger qtd. in Sutlić, 1974, pp. 63-64). Natural sciences are thus given priority over philosophy, but due to their enclosedness into specialized disciplines, they are unable to secure the desired development of society, the prerequisites of which are integrated knowledge and sciences. Cybernetics becomes the new basis of science as the general theory of systems management and communication. It offers a synthetic model of the scientific quest for truth of the world, and opens up the pathway for the *scientific work* as the *techno-logic* towards the creation of the only and total truth in the *only science – the science of history.*

Cybernetics has enabled the scientific and technical construction of a machine whose behaviour mirrors that of the living beings. This does not entail that the machine is a living being in the sense of organic life, but if we envisage the cybernetic organism as an entity which uses information to communicate with its environment to reduce the entropy of the nature and the Universe, the cybernetic machine fulfils this aim. The father of cybernetics Norbert Wiener believes that operating a cybernetic machine means transferring certain messages to the machine in the form of an order to work: "The cybernetic machine acts intelligently, provided two important conditions are met: it is presented with a problem and it is provided with a need to sensibly solve it with precisely defined goal" (Ectors, 1971, p. 153). The cybernetics thus gives the machine the ability to think by setting a goal in front of it: "The commands through which we exercise our control over our environment are a kind of information which we impart to it (Wiener, 1950, p. 17). The cybernetically shaped information does not consist of words, but pieces of code and represents a coded information measured in bits and bytes. The very "process of receiving and of using information is the process of our adjusting to the contingencies of the outer environment, and of our living effectively within that environment. [...] To live effectively is to live with adequate information. Thus, communication and control belong to the essence of man's inner life, even as they belong to his life in society" (Wiener, 1950, p. 17-18).

From the perspective of cybernetic features, the nature and the universe represent an environment subject to constant entropy (disorganization). Every living organism utilizes its natural constitution to slow its downfall towards the thermodynamic balance by decreasing entropy. The struggle of organisms to achieve a higher level of order which the organisms strive to achieve through metabolic processes is actually a desire for reducing entropy which keeps the organism a step away from death. In this respect, a similar anti-entropy position is taken by the cybernetic machines, which aim to establish such an organization of production and life of man in the historical reality of *scientific humanism as naturalism* in which entropy is nullified. They are designed to be controlled by programmes, i.e. its "*actual* performance rather than its *expected* performance [...] known as *feedback*, and involves sensory members which are actuated by motor members and perform the function of *tell-tales* or *monitors* that is, of elements which indicate a performance" (Wiener, 1950, p. 24). The monitors control the tendency of the machine for entropy and this is why Wiener believes that living entities and the machines are "are precisely parallel in their analogous attempts to control entropy through feedback" (Wiener, 1950, p. 26).

Man receives the information from the real world through his senses (eyes, ears, smell) and relates to his environment based on these information. In cybernetic machines, this role is performed by various sensors, motors, and photocells, which not only inform them "what the existing circumstances are, but enable them to record the performance or nonperformance of their own tasks. This last function, as we have seen, is called feedback, the property of being able to adjust future conduct by past performance." (Wiener, 1950, p. 33). For all types of behaviour, there have to exist "central decision organs which determine what the machine is to do next on the basis of information fed back to it, which it stores by means analogous to the memory of a living organism" (Wiener, 1950, p. 33). For cybernetics, there is a parallel between the nervous system of a man and the memory of an automatic machine as both the organic system and the mechanic memory operate on the decision-making principle "on the basis of decisions they have made in the past" (Wiener, 1950, p. 33). The memory program which enables the machine's decision-making processes also allows it to create (much like a man defying his mortality) its own area of organization in the world which gravitates towards entropy. Based on this, Wiener concludes that entropy and progress co-exist in this world side by side.

Progress can be observed from two perspectives. From the first point of view, the significance of belief in progress stands as a testament to the omnipotence of scientific and technological discoveries and the procedures of *scientific work* which will create the possibilities for complete control of development of the scientific and historical reality in the age of *scientific humanism as naturalism.* Secondly, progress has an ethical character as the right of man to grasp the eternal life through scientific and technological means and through acts for improving and reshaping his own body and mind – in other words, to create the social relations which will be a reflection of heaven on earth. This enables the cybernetics as ontology and anthropology to determine both the ethical matters of life and the theological matters of ethics of death. As a systematic science which aims to manage both the living organisms and the machines, cybernetics sees progress as a means of reducing entropy achieved by *scientific work* as the substance subject of *scientific reality* in the age of *scientific humanism as naturalism.* Man as the mover of the absolute must adapt to this, as the adaption will allow him to fight entropy more easily as an organic being in synthesis with mechanical parts or a mechanical organism in the age of *scientific humanism as naturalism.* The cybernetic machine is presented as the condition of survival and the condition for the life of man himself, but it is actually the first step in the creation of non-human beings and the new principles of life in the cybernetic community. Like Marx, Wiener did not foresee the ontological and anthropological significance of scientific work that cybernetics would bring about, which puts the realized metaphysics of *scientific work, techno-logic* and *ergo-logic* on the same level plane. For Marx, the machine represents the means of labour in terms of accumulated knowledge of the society (conscious labour) which replaces the unconscious labour of nature. This conscious extension of man pushes the border between nature and culture in favour of the cultivated, humanized nature, human nature, *humanism as naturalism*. In this correlation, Marx is left with a hidden change of the ontology or metaphysics of human nature, which has a new scientific and technical basis (as is the case with Wiener). The cyborgized reality of the *scientific humanism as naturalism* shows that the enhanced biological origin of the human body in collaboration with the machine and the cyborgized reality is drifting towards a new sphere of life of non-human beings and a new concept of subjectivity of perception. Not only does the enhancement of human beings lead to dehumanization of "nature" of man but the whole "nature" within him gets dehumanized a well. The existence of biological man loses all meaning and in this loss, it gains its full scientific and technical metaphysical truth.

Berdyaev provides a rather illustrative quote about the relationship between man and the machine:

Says man to machine: 'I need you in order to ease my life and increase my power,' and the machine retorts, 'I don't need you at all. I shall do everything without you; you may perish so far as I am concerned.' (Berdyaev, 1934, p. 40)

3.2 ASSUMPTIONS OF REALITY OF HUMANISM AND TRANSHUMANISM

3.2.1 Dialectic Transformation of Machine into Machinery

Hegel's dialectic as the all-encompassing science of the immanent self-development of spirit, reality and man is at the same time logic and ontology. It consists of the triad thesis – antithesis – synthesis, the elements of which constantly overcome and abolish one another and in which the progress of concept, life and reality takes place. Every lower step (thesis) is abolished (antithesis), preserved and overcome by a higher-order step (synthesis) which generates thesis as the beginning of a new circle. This very same dialectic is used by Marx to develop and portray the concepts of reality of the civic society, which implies creating the image of the perceived dialectic transformation of machine into machinery. Marx depicts the machine as a mechanical reconstruction of realized living labour (mental and physical) whose creation enables the division of labour. Thus, the "worker's activity, reduced to a mere abstraction of activity, is determined and regulated on all sides by the movement of the machinery, and not the opposite" (Marx, 1857, p. 614). The machine as a means of labour undergoes a metamorphosis in capitalism, the last of which in Marx's time was the automated system of machinery, which essentially represents the dialectic transformation of means of labour to self-purposeful machinery.

While the machine as a means of labour represents a mechanical reconstruction of the work process, machinery in Marx's view represents the mechanical reconstruction of unity of the production process into labour, land and machines, which were unified by the labour market and capital by the time the machinery comes to life. For Marx, the dialectic analysis of the reality of civil reproduction of life shows that "Capital posits the production of wealth itself and hence the universal development of the productive forces, the constant overthrow of its prevailing presuppositions, as the presupposition of its reproduction." (Marx, 1857, p. 466). It shows that contrary to the irrationality of the market, machinery establishes an immediate, rational, functional and organic unity of productive factors: "Machinery appears, then, as the most adequate form of fixed capital, and fixed capital, in so far as capital's relations with itself are concerned, appears as the most adequate form of capital as such" (Marx, 1857, p. 616).

Marx shows that the overthrow of presuppositions of production through scientific and technical development of means of labour as a form of fixed capital also represents an internal contradiction of the capital itself as the social productive relationship and establishes a tendency towards new scientific and technological means of production. In this respect, Marx makes a vague and perhaps not sufficiently thought out, and even contradictory claim that:

In the machine, and even more in machinery as an automatic system, the use value, i.e. the material quality of the means of labour, is transformed into an existence adequate to fixed capital and to capital as such; and the form in which it was adopted into the production process of capital, the direct means of labour, is superseded by a form posited by capital itself and corresponding to it. In no way does the machine appear as the individual worker's means of labour. Its distinguishing characteristic is not in the least, as with the means of labour, to transmit the worker's activity to the object [...]. (Marx, 1857, p. 614)

In this respect, Marx highlights the crucial difference between the machine, as realized living labour, and the machinery as a realized form of social production of relations. A conclusion is drawn from this that the machinery is the production in and of itself, a manner of production: "Machinery uses labour, nature and the machines, i.e. means of production as means of its own purpose, and the ultimate purpose and the true essence of machinery is an insatiable lust (*apetitus*) for eternal self-reproduction." (Rodin, 1974, p. 55). The speculative dialectic transformation of means of production into machinery was described quite concisely in Rodin's study *Marxova misao zajednice* (engl. *Marx's community thought*). We will use his work to reinterpret several crucial ideas that will serve as an introduction into our discussion on assumptions of reality of Marx's understanding of *humanism as naturalism* through question on the essence of technology.

The parallels and differences between the machine and the machinery are based on the assumptions of scientific and technological development of means of production and further development of capital based on it. In Marx's view, this leads to the overthrow of assumptions upon which reproduction is based. The machine is transformed "into machinery in the process of mechanization of the capital under the assumption of abolishment of civil form of private ownership over factors of the productive process. This decay of the private ownership of productive factor takes place parallel to the decay of the market-mediated correlation between productive factors" (Rodin, 1974, p. 55). This fact also stands as a testament to the deterioration of market as an intermediate form of integrating productive factors into the productive processes. The immediate and planned merger of productive factors through technification of the capital does not represent factual abolishment of ownership in general:

Quite the contrary, after the abolishment of the classical form of private civil ownership over fundamental factors of production and, consequently, over production itself, new forms of privatizations arise, no longer outside the presuppositions of production but as a function within the productive process as the only authentic source of the overall historical existence of man. (Rodin, 1974, p. 87)

The existence of an individual and the technical community no longer depends on the individual will of the workers and capitalists, it becomes an integral part of the scientific and technological process of labour, *scientific work* as the substance subject which predefines for itself all means of its self-reproduction. The physical work becomes the means of the machinery, as a result of which:

The production process has ceased to be a labour process in the sense of a process dominated by labour as its governing unity. Labour appears, rather, merely as a conscious organ, scattered among the individual living workers at numerous points of the mechanical system; subsumed under the total process of the machinery itself, as itself only a link of the system, whose unity exists not in the living workers, but rather in the living (active) machinery, which confronts his individual, insignificant doings as a mighty organism. In machinery, objectified labour confronts living labour within the labour process itself as the power which rules it; a power which, as the appropriation of living labour, is the form of capital. The transformation of the means of labour into machinery, and of living labour into a mere living accessory of this machinery, as the means of its action. (Marx, 1857, p. 615).

The realization of this tendency represents the dialectic transformation of means of production into machinery as the production process itself, as the productive power out of which a new type of scientific technical and historical community will arise, which Marx saw as the accomplished *humanism = naturalism*, but is in fact *scientific humanism as naturalism.*

The machinery as means of production and the most appropriate form of capital has come to existence within the capitalist ways of production of life:

The full development of capital, therefore, takes place – or capital has posited the mode of production corresponding to it – only when the means of labour has not only taken the economic form of fixed capital, but has also been suspended in its immediate form, and when fixed capital appears as a machine within the production process, opposite labour; and the entire production process appears as not subsumed under the direct skilfulness of the worker, but rather as the technological application of science. [It is,] hence, the tendency of capital to give production a scientific character; direct labour [is] reduced to a mere moment of this process. (Marx, 1857, p. 618)

One can note that man-worker becomes an instrument and a mere accessory to means of labour which are dialectically transformed into the production process, a self-purposeful productive power from which a new scientific and technological community shall rise, a community no longer based on market mediation but on machinery.

The final purpose and the final aim of capital is its scientific and technical self-regulation which is "both discipline, as regards the human being in the process of becoming; and, at the same time, practice [Ausübung], experimental science, materially creative and objectifying science, as regards the human being who has become, in whose head exists the accumulated knowledge of society." (Marx, 1857, p. 631). Through fulfilment as the completion of the dialectical unity of means of labour and objects of labour in the self-purposeful activity, productive power, *scientific work*, work of science, it also represents the unity towards the identity of *theoretical, practical* and *poietic* activity. For Marx, this means the highest purpose of the liberated labour, production as its own purpose. Today, we interpret Marx's words about the *empire of freedom* in which such changes had taken place that the new emerging man contains the accumulated knowledge of the society is interpreted contrary to his expectations. In the *empire of freedom, scientific work* is liberated and its actions establish both partially and completely artificial beings, cyborgs, robots, avatars and illusions which contain the accumulated knowledge of the society, while man is empty and devoid of his alienated essence and existence. By cyborgizing the historical and natural reality, *scientific work* is draining the life away as the first and the last reality of the world, replacing it with artificial features.

The dialectic transformation of means of production into machinery, a self-purposeful activity, the liberated labour – *scientific work* – transitions from the substantive into the subjective form and it completes the internal dialectic of the civil society as it reaches its ontological end, though it will persist much longer:

Society has been replaced by a realized system of connections (communication) among the isolated and functionally fixed monads which are distributed throughout the productive process due to their technical features and needs. (Rodin, 1974, p. 63)

3.2.2 On the Essence of Technology

The Greek sources show that the word *téchne* is present in the Greek culture, especially in Plato's works, as a word denoting skill or craft. *Téchne* is the capability, ability or realization and the ability or achieving this capability through a particular action or production, governed by the type of knowledge which is often perceived as *téchne* itself – craft. On the other hand, this word is also related to man and his shortcomings and can be understood as reasonable, astute use of powers of nature against nature itself, that is, against its forces that might endanger man. It is well known that the fate of mankind has had an unbreakable bond with the power and craft of invented tools ever since the Prometheus's gift of fire. It is possible that the ancient Greek understanding of skill represents the starting point for the possibility of establishing a more adaptable, resilient and technically controlled man-machine which will use the *scientific work* as means of its self-development.

Aristotle uses the term *τέχνη* for the productive activity of man, i.e. for the artistic *poiesis*, architecture, medicine, but also for manual labour. For him, *téchne* is not the knowledge of the goal but of means by which the subject wants to achieve a particular aim. What is established in *téchne* does not stem from itself but from the one who produces (the poietic activity). Thomas Aquinas revives the term *téchne* in the 12th century when he claimed that man had to develop different technologies to survive in the hostile natural environment. The modern usage of this term is related to the industrial revolution and the New Ages during which the ancient Greek term *téchne* is reduced to *technica*, which denotes a methodical way of mastering and governing nature. Gehlen believed that the true origin of technology lies in man's natural shortcomings, in the inability of his bodily organs and functions to adapt to the hostile natural environment, which forced him as the use his capabilities to build his world of technology. Friedrich Dessauer explains that technology:

[...] refers to objects and activities that do not belong to nature as they never existed in it, but have occurred in historical time only by virtue of human, i.e. spiritual functions. The objects and activities which we denote technology share one feature with nature – they can be observed. (Dessauer, 1944, p. 7)[1]

In recent times, "the word *technology* is used as a part of the name or the key trait of the current period [...]Technology denotes a system or an image of a world built through technology" (Selak, 2013, p. 41). Croatian scholars, namely Šarić and Čatić (1998) and Selak (2013), make a distinction between different uses of the word *technology* in everyday speech and professional and scientific discourse. This distinction is particularly prominent in Croatian in which two different words are used to denote similar concepts – *tehnika* and *tehnologija*. *Technology* when used as *tehnika*, according to Selak (2013), denotes art, agility, skills and crafts, which encompasses the meaning of the ancient Greek word *téchne*. The word *technology*, used in this sense, also denotes a group of models or scientifically or experientially formulated solutions for applying and conducting scientifically defined or practice-driven insights by meeting the needs and the demands of human life, in commercial use and trade, in science, medicine, healthcare, artistic endeavours, at home, or in organizing the personal life etc. (Selak, 2013, p. 42).[2] Technology (as *tehnika*) is also a field of science which analyses the application of scientific accomplishments in physic, chemistry, biology etc. In other words, it represents technical sciences. The word *technology* (when used as *tehnologija*) stems from J. G. Beckmann's 1777 definition of technology as an all-encompassing science of interrelation between technology, economy and society (Selak, 2013, p. 42).

Up until the New Age, philosophy has regarded the term *technology* (used as *tehnika*) as the art of seeking and relating towards the nature and the world. New Age philosophy abandons the mental maturation of the essence of this concept and sees its purpose in the practical and the pragmatic sense. Acting as knowledge about beings, it wants to provide a rational insight into the structure of the world, into the rules of society and nature to increase the power of man in the world. Philosophy has undertaken the task that was previously demonstrated by technology with some success as a theoretical, rational and exact science which has allowed intervention into the historical reality as a whole with its insights and technical means. In order for philosophy to be able to act within its own historical time, Hegel decided to put philosophy on a path closer to science, shedding off its name denoting 'love for knowledge' and becoming true knowledge itself. Thus, philosophy as the science of the absolute becomes the highest method of reaching truth.

In the dialectic transformation of means of production into productive power carried by the application of science and technology in the capitalist production, Marx brought the civil society to the end of its contradictory development and overcame in *practice of labour* as the self-purposeful activity the negative ontological structure of estranged labour and capital as its real otherness, introducing the world to real history of science and technology upon which the new way of managing life is based – the *scientific history*. It is because of this that the question of essence of technology has a central place in the metaphysical, transmetaphysical, transhumanist and postmodernist thought of today's world.

In the previous chapter, we have shown the historical dialectical connection between man, the machine, the productive powers and the dialectic transformation of means of production into the productive power of the capital that represents the basis for the new post-capitalist form of the science and technology community with man in the centre of it. We could also see that man as the medium of technology is placed into its essence and it uses him as means for its end. How is it possible that technology as a means for achieving particular goals, not only in the teleological movement of thought in Hegel's *Science of Logic* but in the actual life, breaks away from man's control and transitions from being a means to an end into a self-purposeful entity using man as means of its ends? According to Rodin, the best answer comes from the structure of teleological movement of thought. It shows that if a man stops being an external practical purpose that uses technology (poietic activity) as means for its end, he loses his control over technology. This means that "if man loses control over means of purposes that he created, there is a chance these means are operated by a different purpose or a different will" (Rodin, 1974, p. 68).

What is this will that can establish control over technology? Marx's hopes were aimed at the proletariat as the general will of the subject of world activity which would use the technology – the productive powers – as means for its goals of establishing the relations of unity of essence and existence in the reality of true world history. Historical rejection of Marx's expectations about the collective subject and the general will that will use the technology as means for its goals stems from his incomplete acceptance of the term work – productive powers of labour – *scientific work* as the "the necessary condition for effecting exchange of matter between man and Nature; it is the everlasting Nature-imposed condition of human existence, and therefore is independent of every social phase of that existence, or rather, is common to every such phase" (Marx, 1867, p. 130). In this historical reality, however, this simple work process, *scientific work* as the being of the *empire of freedom* and the *empire of necessity* establishes the proletariat as the means and the constitutional moment of structure of its self-movement. The proletariat is not moved by the will which would establish control over scientific work as technology but is a means of its own *will to will.*

The New Age technology is thus developed within the capital society "into an existence adequate to fixed capital and to capital as such" (Marx, 1857, p. 614). It is no longer the *poiesis* but the unity of *theoria, praxis* and *poiesis*, which is characterized, in Heidegger's words, by the extortion of nature to deliver the energy to be stored. This extortion as revelation functions as management, regulation and security. In this circle, everything is established, everything has its origin and is regarded as a constituent. Acting as the fate of the Western world and the will to will of the New Age man, technology is perceived as a tool of demythologizing the world of history and the Universe, while it in fact enchants man with its spell:

The bewitchment by technicity and its constantly self-surpassing progress are only one sign of this enchantment, by virtue of which everything presses forth into calculation, usage, breeding, manageability, and regulation. Even "taste" now becomes a matter for this regulation, and everything depends on a "good ambiance." The average becomes better and better, and by virtue of this bettering it secures its dominion always more irresistibly and more inconspicuously. (Heidegger, 1999, p. 87)

As a reflection of the being of reality, the New Age metaphysics inspired by Plato's teaching on ideas, in which a craftsman constantly repeats in his actions as realizations of what was already seen in the empire of ideas and only repeats what already is real:

This is, in fact, hidden behind the structure of the reality of production as a unique practical-poetical-theoretical activity of the New Age man. Since this new activity gives life to the man of the civil society, this civil opinion treats this activity as its known basis, which it is pragmatically trying to upgrade, forgetting that this foundation is only a historical form of managing life but not being itself. (Rodin, 1974, p. 70)

Marx has taught us that "Consciousness can never be anything else than conscious existence, and the existence of men is their actual life-process" (Marx, 2000, p. 13). Thus, we can conclude that technology as a way of managing life and shaping the community prefers the technical thought that from of it, the thought of technical life which ideologically analyses and explains its actions or makes it self-aware as the technical, metaphysical or cybernetic-ontological-ethical thought. Science as the paradigm of all productive powers is trying to find the single-sided world of labour. For this kind of science,

[...] there is no such thing in nature (physis) that could resist the construction and the creation, which would not be permanently established and has its existence (persistence) and constitution, its permanence and independence [...] For this science which has spread to the end of the era and is gathered at the

centre of the circle in a never-ending vortex, technology is not its mere application nor is, conversely, science its servant[...] but the light that the scientific view (theoria) is based on is the techno-logic, the condensed essence of managing everything for its operations and calculations. This science is in its essence techno-logical, i.e. it is a scientific technological technical science. (Sutlić, 1974, p. 137-138)

Science is itself as a reflection of the reality, which means that the complete consciousness of the technical society is within the scope of the scientific and technical desolation understood by man as creation and self-creation. Science goes hand in hand with cybernetic ontology in its aim to produce life in a planned organized manner and it becomes a tool of mindlessness and chaos of reality: "Every scientific operation in the reality of the modern technical world, which itself operates on the scientific principles, represents nothing else but a new form of realization, a new form of technical and scientific domination over the living world." (Rodin, 1974, p. 89).

The questions of science and technology are related, according to Heidegger, to the important question of philosophy as metaphysics – the question of being. Heidegger believes that technology is an instrument of satisfying technology-external aims. Its instrumentality does not show its true essence and truth. In his lectures on essence of technology, Heidegger claims that there is nothing technical in the essence of technology, warning us that it constitutes a historical way of managing life and that the historical life of man resides in its essence. Its true essence lies in the fact that technology as production allows something to come out of the shadows into the presence as something clearly visible. Production does not denote just manual and artistic creativity, but it also includes the spontaneous natural growth which does not require another entity to bring something into existence as it is in and of itself an act of creation. Thus, *poiesis* takes something out of the covert world into the overt and Heidegger regards this as an act of revealing. *Téchne* reveals that what is not produced by itself, which makes it the science of revelation i.e. the truth: "Technology is a way of revealing. If we give heed to this, then another whole realm for the essence of technology will open itself up to us. It is the realm of revealing, i.e., of truth." (Heidegger, 1977, p. 12). It challenges nature so that the energy hidden within the nature is liberated, transformed and stored so that it could be further distributed and switched – all of which is an act of revealing. All that is uncovered by challenging – unconcealment – which Heidegger regards as a state (object) over which man has no influence. The self-concealing as a state is enframing. Technology as revealing, challenging, establishing, delivering shows that:

Enframing means the gathering together of that setting-upon which sets upon man, i.e., challenges him forth, to reveal the real, in the mode of ordering, as standing-reserve. Enframing means that way of revealing which holds sway in the essence of modern technology and which is itself nothing technological. The word stellen [to set upon] in the name Ge-stell [Enframing] not only means challenging. At the same time it should preserve the suggestion of another Stellen from which it stems, namely, that producing and presenting [Her- und Dar-stellen] which, in the sense of poiesis, lets what presences come forth into unconcealment. (Heidegger, 1977, pp. 20-21)

The enframing, thus, shows the being in its unconcealed truth. Enframing as an act of showing the being is at the same time revelation of the reality as a state (object). Heidegger further argues that man is not the subject of concealment but it cannot occur without his participation in the act. However, enframing as a way of revealing the unconcealedness of being can block the path towards the original truth as the being is also hidden in the enframing in the revelation. Being as enframing needs man as man as the *shepherd of being* belongs to being by giving it the ability to act within history that is heading towards a philosophical shift towards the oblivion of being, i.e. the very truth of being. Man as the enframing of being is the one who expects this truth with his actions. Enframing represents a danger for man but also the salvation at the same time which is why:

To save is to fetch something home into its essence, in order to bring the essence for the first time into its genuine appearing. If the essence of technology, Enframing, is the extreme danger, and if there is truth in Hölderlin's words, then the rule of Enframing cannot exhaust itself solely in blocking all lighting-up of every revealing, all appearing of truth. Rather, precisely the essence of technology must harbor in itself the growth of the saving power. (Heidegger, 1977, p. 28)

Heidegger saw the technology as the destiny but also as part of the current age which contains the way out from *overcoming and surpassing* it. *Surpassing* technology and metaphysics requires a carefully created relationship between the technical mind and thought, i.e. the inevitability of technology in everyday life and paying attention that it does not completely take control over us. In the *Letter on Humanism*, Heidegger makes a *shift* in his thought and is looking for thought whose starting point is the technical reality and cybernetic ontology. However, this thought is not based in itself but is looking through preservation of the language of being and wandering in the *intermezzo* of worlds for the lost source which precedes the metaphysics of technology – *scientific work*, but is hidden in the non-metaphysical thought. The experience regards this kind of thought as historical thought or the thought of truth of being or purposeful thought. His aim is to produce it through non-metaphysical, poetic words which are not rational, but which does not imply their irrationality or anti-rationality.

Heidegger's *shift* in thought is the thought of metaphysics as nihilism and a transition from nothing to being. The fading of being is complete once philosophy is filled and dissolved in the world of science and technology, reaching its end. At that moment clarity is revealed (unconcealment and concealment), an occurrence of being which opens up a possibility of another and different thought beyond the empty talk of metaphysics and the rational thought of science and technology. As the fate of modern world and the basic form of will to will of the New Age subject in the world of fulfilled metaphysics, technology can be a hint or a foreplay to a different kind of world. This premonition that Heidegger had is becoming real today, but not as a shift in being-oriented thought. *Scientific work* as the paradigm of scientific history of *scientific humanism as naturalism* truly is an introduction into a new age of the world but not the post-metaphysical one in which man as the shepherd of being brings forth the essence of its language into historical time. This new world different from Heidegger's is not a part of being *above and beyond any man's disposal* but a world which is completed through establishment of partially or completely artificial entities, "instances that manage man like destiny, sending him signals and instructions as a way of persuading him to take part in the establishment of nothing" (Sutlić, 1974, p. 150) as the *scientific work* in the age of *scientific humanism as naturalism* sucks the life and thought with its actions, placing them onto a new ontological and anthropological scientific and technological basis. Can "life skip one of its eras, a period of shadow towards which the world is drifting, or will it depart from it once it starts to need life moving into its light?" (Sutlić, 1974, p. 150).

3.3 SCIENTIFIC HUMANISM AS NATURALISM

Scientific humanism as naturalism is our concept which stems from the scrutiny of Marx's understanding of real history in which *humanism and naturalism* of man are regarded as equal, and Sutlić's analysis of Marx's thoughts about the one and only science, the science of history as the *scientific history*. Marx's standpoints of *humanism and naturalism* as real history and Sutlić's analysis of this position as *scientific history* represent the ontological foundation of our understanding of the term *scientific humanism as naturalism*, i.e. as cyborgized historical reality. *Scientific humanism as naturalism* is not a witty thought but an attempt to follow these contrasted standpoints in a dialogue to reach that what was not exhausted in the Marx-Hegel dialogue and in Sutlić's philosophical reanalysis of basic terms in Marx's work, particularly in the concept of labour. Sutlić believes that the nihilistic situation requires the philosophical thought to be placed in the essence of time and be connected to the sources of its possibilities. This, in particular, refers to the historical determination of philosophy from Plato and Aristotle onwards and it also requires that the age of the philosopher in question be brought into a wider view of the entire Western history. What has so far remained hidden and was not reached by the *logos* in Marx's *humanism and naturalism* and Sutlić's *scientific work* as the space of scientific and technological dehumanization of man is the actual realization of the *scientific work* which is manifested in the scientific, technical and bio-scientific procedures of enhancing and reshaping the human body and mind, which gives human life a new inhuman foundation, in the name of fast scientific and technical evolution an cybernetic "humanization and naturalization" of historical space and time. These are, of course, the philosophical aspects of cyborgization of man and his reality in the age of *scientific humanism as naturalism* and possibility of thought.

We have previously shown Heidegger's idea that the view of being as presence that Plato and Aristotle advocate can be seen as the root of the metaphysical *oblivion of being* and its reduction to entity and ontological roots of possibilities of the current *techno-logic*. We have shown that Hegel tried to overcome this *nothingness* which stands in the essence of today's historical reality of the world by using the speculative-dialectal synthesis in the self-development of the self-aware world spirit which reached the absolute knowledge – truth and freedom – in its revelation at the end of history. In Hegel's history of the absolute – the absolute history – there is no room for the real man as a self-active and essential entity. In the spirit of Marx, Hegel's abstract self-aware man is trying to confirm himself in the otherness as such and regards this as his true existence. This means that the mind (man as a mental entity) is a non-mind in itself as the reality of the world is non-mental and estranged. The absolute as its own ontological movement needs man as a medium of his revelation in his history through philosophy. In this respect, one might argue that man as an intermediate and the medium of revelation of the absolute in history in a way participates in the absolute reaching of itself (in all three stages of the *Philosophy of the Spirit*), i.e. in its freedom, and it gives a particular supposition of subjectivity and power through the self-creation of the absolute. Sutlić believes that thinking as a path of realization and absolute reaching consciousness – self-awareness in philosophy whose interpreter is a man-philosopher, draws man into the centre of systems of its absolute power, as to be means to be established by the absolute. Since the absolute is revealed in history through man, and man exists by virtue of the absolute, one could observe the similarities and the differences between the absolute and man, but the rationalization of the overall system of the absolute and everything within in the purview of its ontological self-development and self-establishment positions man, at least for a moment, into eternal presence. This analogy of identity and differences between the spirit and man is also visible in Marx's understanding of labour, man and nature (identity and differences). The final consequence of the revelation of the absolute in history is its apparition, as it does not have its being within itself. With regards to speculative philosophy, Marx claims: "The object is therefore something negative, self-annulling – *a nullity*. This nullity of the object has not only a negative but a *positive* meaning for consciousness, since this nullity of the object is precisely the *self-confirmation* of the non-objectivity, of the *abstraction* of itself" (Marx, 1844, p. 70). For consciousness, something becomes real "insofar as the latter *knows* this *something*" (Marx, 1844, p. 70). It knows that

[...] object is only the semblance of an object, a piece of mystification, which in its essence, however, is nothing else but knowing itself, which has confronted itself with itself and hence has confronted itself with a nullity – a something which has no objectivity outside the knowing. Or: knowing knows that in relating itself to an object it is only outside itself – that it only externalizes itself; that it itself only appears to itself as an object – or that that which appears to it as an object is only itself. (Marx, 1844, p. 70)

In the critique of Hegel's *Logic*, or onto-Logic, Marx emphasizes that the "abstraction comprehending itself as abstraction knows itself to be nothing: it must abandon itself – abandon abstraction – and so it arrives at an entity which is its exact opposite – at *nature*. Thus, the entire logic is the demonstration that abstract thought is nothing in itself; that the absolute idea is nothing for itself; that only *nature* is something" (Marx, 1844, p. 73). For Marx, the transition of Hegel's *Logic* into the *Philosophy of Nature* is nothing but a transition from abstraction into observation – fulfilment with content. The being of this absolute does not come from within itself, as it would seem from Hegel's philosophy, but from the historical revelation, which does not mean that its absolute is established by history. Even though the absolute is not an immanence of the history itself, i.e. it is history-external, it determines it absolutely as the history is the history of the absolute. In contrast to Hegel who "fell into the illusion of conceiving the real as the product of thought concentrating itself, probing its own depths, and unfolding itself out of itself" (Marx, 1857, p. 34), Marx points out a concrete substance – the subject of history, not denying that the concrete totality of history as a meaningful totality is a product of conceptualization "but not in any way a product of the concept which thinks and generates itself outside or above observation and conception; a product, rather, of the working up of observation and conception into concepts" (Marx, 1857, p. 34-35). Consciousness as a conscious being is only a "theoretical shape of that of which the living shape is the real community, the social fabric, although at the present-day general consciousness is an abstraction from real life and as such confronts it with hostility" (Marx and Engels, 1988, p. 105). The reality of productive powers and production relations is "the real basis of what the philosophers have conceived as 'substance' and 'essence of man'" (Marx and Engels, 1998, pp. 61-62) and as estranged consciousness of an alienated human being confronted it to the real life as the superior power – the absolute knowledge of the absolute and the subject of world history. Thus, Marx sees in the historical reality as an essential non-truth the framework for possible reconciliation of essence and existence from which man and beings can regain their full sense, although "It is not enough that thought strive to actualize itself; actuality must itself strive toward thought." (Marx, 1977, p. 138). The real power and the real subject of history is

Where speculation ends -- in real life -- there real, positive science begins: the representation of the practical activity, of the practical process of development of men. Empty talk about consciousness ceases, and real knowledge has to take its place. When reality is depicted, philosophy as an independent branch of knowledge loses its medium. (Marx and Engels, 1998, p. 53)

Beyond Marx's expectations, philosophy is replaced by "*scientific work*, working science as a mytho-logic which has 'techno-logic' as its core: the modern prophecy, the "unconscious processing of everything that technology is, not as applied science but as its essence" (Sutlić, 1974, p. 146).

Understanding labour as the true essence and necessary condition for the existence of man and his alienation in the civil society lead Marx's thought towards the scientific portrayal of liberation of labour for the purpose of "uninterrupted development of productive powers of labour" – *scientific work* which leads towards true history, which Sutlić regards as the practice of *scientific history*. In contrast to the ontological overcoming of reality in the philosophical element of thought through the self-aware development of the absolute spirit of world history, Marx postulates *the one true science – the science of history*. However, the essence of this science has not reached the level of maturity and enlightenment in Marx's thought as he merely advocated the abolishment of Hegel's speculative metaphysics through realization of its content in the world reality. Marx describes the position of the *one and only science* in non-philosophical and empirical terms, as a medium in which man and nature are intertwined as well as the foundation for the possibilities of natural sciences and philosophy, as he is counting on the overtness of the object (man and nature) and the activity of the being that caused it all to happen: "natural science will lose its abstractly material – or rather, its idealistic – tendency, and will become the basis of human science, as it has already become – albeit in an estranged form – the basis of actual human life, and to assume one basis for life and a different basis for science is as a matter of course a lie. The nature which develops in human history – the genesis of human society – is man's real nature; hence nature as it develops through industry, even though in an estranged form, is true anthropological nature." (Marx and Engels, 1988, p. 110), *humanism = naturalism.*

We have shown that Marx starts from the practical assumptions that without man there is no historical world or philosophy. History and historicity cannot be tied to mere reproduction but should represent "a step away from what was given with the addition of new. It is not enough for something to be remembered to be declared historical. The criterion of historicality is contained within the new quality of survival" (Veljak, 2013, p. 604). The practical denotes the specific human activity which is used to "emphatically speaking, deliberately and freely produce and shape the human world" (Veljak, 2013, p. 606). This is how historical time and space are shown as the essential residence of the social man and philosophy. For man, to be means to be historical as there is no history without man nor is there human labour of history, which Marx explicitly states:

[...] the entire so-called history of the world is nothing but the creation of man through human labor, nothing but the emergence of nature for man, so he has the visible, irrefutable proof of his birth through himself, of his genesis. (Marx and Engels, 1988, p. 110)

Thus, "Generatio aequivoca is the only practical refutation of the theory of creation." (Marx and Engels, 1988, p. 112). In order to overcome the metaphysical thought (the absolute spirit) after its abolishment and realization in the reality of history, we believe Marx finds it crucial to understand history and the historical time as openness in which all beings appear and vanish, in addition to the concept of scientific work. Everything that is in history has a defined temporal weight for as long as the causes and conditions that caused it persist. Marx's relativization of everything that is historically provides us with a task of historically relativizing Hegel's absolute spirit of world history as the cause of eternal presence, just like Marx's absolute scientific work, thus opening time and space of history for human development. The uncritical acceptance of Hegel's concept of labour does not treat with philosophical consistency this openness to other relations in the production, qualitative change of productive powers and different consciousness of man based on it, as he sees the foundation for development of human potential in the uninterrupted quantitative development of productive powers of labour. Relativization of being in history has opened up the place of transphilosophical thought for him, but the non-philosophical postulation of the concept of labour, productive powers, scientific work as the foundation for man-appropriate development of society in which the humanism and naturalism of man are equivalent has at the same time cordoned off this path for him.

For Marx, the relation between man and nature is the relation of identity and difference of nature, man and his labour and is self-understandable as such. Nature on the one hand is the basic condition for the existence of man as it:

[...] provides labor with [the] means of life in the sense that labor cannot live without objects on which to operate, on the other hand, it also provides the means of life in the more restricted sense, i.e., the means for the physical subsistence of the worker himself. [...] Physically man lives only on these products of nature, whether they appear in the form of food, heating, clothes, a dwelling, etc. The universality of man appears in practice precisely in the universality which makes all nature his inorganic body – both inasmuch as nature is (1) his direct means of life, and (2) the material, the object, and the instrument of his life activity. (Marx and Engels, 1988, p. 75-76)

Marx concludes that "Man lives on nature – means that nature is his *body*, with which he must remain in continuous interchange if he is not to die. That man's physical and spiritual life is linked to nature means simply that nature is linked to itself, for man is a part of nature." (Marx and Engels, 1988, p. 76). Understanding and mastering nature is in the essence of man and his labour, so that it should reach consciousness and history. It is about-self available for historical work as it is by itself labour. "All things that labour frees from their immediate connection to their earthly whole are in nature found objects of labour. Man stands against the very natural substance as one natural power" (Sutlić, 1994, p. 139). Both man and nature reach itself in labour. Nature is the unconscious labour that becomes conscious nature through man (his labour).

In his explanation of the "self-mediated being of nature and of man" (Marx and Engels, 1988, p. 112), Marx employs Hegel's dialectic:

Don't think [about the progress of existence, I.G.], don't ask me, for as soon as you think and ask, your abstraction from the existence of nature and man has no meaning. Or are you such an egotist that you conceive everything as nothing, and yet want yourself to exist? [Marx tries to overcome this level of metaphysical speculation through real scientific evidence, I.G] Since the real existence of man and nature has become evident in practice, through sense experience, because man has thus become evident for man as the being of nature, and nature for man as the being of man, the question about an alien being [the creator, I.G], about a being above nature and man – a question which implies the admission of the unreality of nature and of man – has become impossible in practice. (Marx and Engels, 1988, p. 113)

At the same time, man introduces nature into history with his labour, turning it into a conscious historical nature. Thus, we can argue that:

The human aspect of nature exists only for social man; for only then does nature exist for him as a bond with man – as his existence for the other and the other's existence for him – and as the life-element of human reality. Only then does nature exist as the foundation of his own human existence. Only here has what is to him his natural existence become his human existence, and nature become man for him. Thus, society is the complete unity of man with nature – the true resurrection of nature – the consistent naturalism of man and the consistent humanism of nature. (Marx and Engels, 1988, p. 104)

Marx regards labour as an unproblematic, specific relationship of man towards everything that is and towards himself as man:

In creating an objective world by his practical activity, in working-up inorganic nature, man proves himself a conscious species being, i.e., as a being that treats the species as its own essential being, or that treats itself as a species being. [...] man produces universally [...] man reproduces the whole of nature [...]. (Marx and Engels, 1988, p. 76-77)

Nature formally and materially determines labour as such, as "The worker can create nothing without nature, without the *sensuous external world.* It is the material on which his labour is realized, in which it is active, from which, and by means of which it produces" (Marx, 1844, p. 29). Nature is the same as human activity, human activity is the same as nature. Marx argues that production is not creation but reproduction of the original nature as man uses his labour to repeat [change form] what nature creates and brings it to consciousness – the self-consciousness of man and nature. Since nature is reproduced in the historical reality through man as the labour being, Marx can only equate *humanism to naturalism* as to be natural means to work and, consequently, to be human. Man is natural when he uses his free labour to realize his naturality in human social relations because "[...] production is his active species-life. Through this production, nature appears as his work and his reality." (Marx and Engels, 1988, p. 77). For this reason

[...] the essence of man – the work process, does not only include nature as an object, but is natural itself, it is like nature and, one might argue, represents a paradigmatic instance of nature. In the work process, man is natural as he relates towards other natural powers as 'one true power' and also as he works as the very nature itself. (Sutlić, 1974, p. 11)

This means that man "can work only as Nature does, that is by changing the form of matter" (Marx, 1867, p. 31). Sutlić argues that

[...] the anthropological and cosmological connection, unity and difference, rest on ontological specificity of man of working as nature to be his 'essence', his 'nature', his 'condition of existence'. Labour is, thus, nothing else but developing the 'dormant potentials' in nature. In other words, nature in the sense of all living things (all living beings) and man (in the sense of a single being in this unity) is the nature which produces and the nature which works. (Sutlić, 1974, pp. 11-12)

To paraphrase Sutlić, we could claim that nature and man in Marx's understanding of essence and existence are the creative beings [self-creators, I.G.] as they owe their existence, their here-being to themselves. The process of self-creation of the historical nature and man (in the historical senses shown as self-birth) can be, Marx claims, dialectically shown and scientifically proven. Man is familiar with the process of self-creation and labour as the aware being which unites in itself the process of creation and the knowledge of the process gives him irrefutable scientific evidence.

In his analysis of the man-labour relationship, Marx claims that the historical man became what he is through his own labour. This leads us to the conclusion that Marx believes the subject of the entire process of becoming a man in historical reality is man's labour in the form of substance subject, or in Engel's words, that labour created man. In this view of labour as the absolute, man is, just like in Hegel's concept of the absolute world spirit, merely a moment and the function of the absolute through which it is realized in the reality of world and history. The identity and the difference between man and nature show to confronted attitudes about who is the real subject of world history in Marx's interpretation – is it merely labour or human labour? In this dichotomy, nature is clearly shown to be the substance subject. Sutlić questions two possibilities: 1) man as subject substance (self-creation through human labour), which gives man the status as the absolute divine creator of world history; 2. human labour is the substance subject, the absolute divinity in which man is only a moment – the medium of his realization in its reaching self-consciousness in world history (of nature and society). These perspectives on the relationship between labour, nature and man show the historical importance of human labour – *scientific work,* the substance subject which makes man a moment in its own self-creating movement that subsumes the entire historical reality in its alienation – nihilism. Thus, we are not dealing with man, but with his reduction to a worker in the world of work – a science worker.

We have emphasized that Marx did not scrutinize the concept of labour in its ontological essence. He takes labour as the real and historically unproblematic fact which, unscrutinised in this manner, becomes a central part of his dialectal portrayal of the development of reality and its change based on the free development of productive powers of labour. In the circle of Hegelian dialectic way of exploring and portraying reality, labour remains in the foreground of empirical terminology (primarily political economy) while its ontological status remains obscured. The essence of labour described in empirical terms remains hidden from philosophical thought:

So far therefore as labour is a creator of use value, is useful labour, it is a necessary condition, independent of all forms of society, for the existence of the human race; it is an eternal nature-imposed necessity, without which there can be no material exchanges between man and Nature, and therefore no life. (Marx, 1867, p. 31)

This kind of labour, Marx argues, belongs to man and man alone.

Subsumed under the empirical term, productive powers of labour etc., Marx sees only the humane aspect of labour as the necessary condition for existence, whose freedom in uninterrupted relations of the capital represents the essence of man and nature in world history. This liberated labour brings carries the face of true history, freedom, *humanism and naturalism* – the *empire of freedom.* Because of this, Marx finds it self-understandable that without the progress of science and the growth of productive powers based on it the existence of man may come into question. The key issue of their character and qualitative development remains outside of purview of Marx's thought. And just as labour is not scrutinized and is covered by the concept of productive powers of labour, so is the concept of nature self-understandable to him. Man and nature, with their similarities and differences, come into existence by themselves and have proof of that in the circular movement of the dialectic thought and natural sciences which "invaded and transformed human life all the more practically through the medium of industry; and has prepared human emancipation, although its immediate effect had to be the furthering of the dehumanization of man." (Marx and Engels, 1988, p. 110). Nature is merely a subject to human labour and man's task is to master nature and natural forces through labour.

Marx sees the dialectic contradictions of the civil society, which is governed by the relations of private ownership over the factors in the productive process, as an inhibiting factor for further development of productive powers and the foundation for alienation of man in his historical essence, existence and nature. In Hegelian terms, the real resolution of contradictions in civil reality can only be achieved when the capital as the other-being of labour is returned to itself. Since the dialectic of reality only applies to civil society, Marx uses the tripartite dialectic to complete the concepts from the economic domain and then destroy the entire epoch through open dialectic so that labour can become free upon its return from its other-being through abolishment of private ownership and be established in itself. The relations that hold between the *empire of necessity* and the *empire of freedom* are those of mutual interaction and not a tripartite dialectic as labour constitutes itself as mutual action of essence and existence. The alienated labour represents the foundation of another historical subject of realization and elevating the reality onto conceptual level – proletariat, which can "can thus only exist world-historically" (Marx & Engels, 1998, p. 57) as its revolutionary act abolishes private ownership as an estrange relation in the productive process and alienates entire history. This aim is expressed in the following claim: "Philosophy cannot be actualized without the abolition [Aufhebung] of the proletariat; the proletariat cannot be abolished without the actualization [Verwirklichung] of philosophy" (Marx, 1977, p. 142). Once the proletariat is abolished, the mutual relation of the two empires is founded on the liberated labour as the true subject of real history, now manifested in the unity of labour, man and nature, in their free and conscious activity (self-activity), object, means and products: "[...] individuals must appropriate the existing totality of productive forces, not only to achieve self-activity, but, also, merely to safeguard their very existence." (Marx and Engels, 1998, p. 96). Everything has been folded into one, into liberated labour as the substance subject of everything that is. Despite Marx's efforts and understanding of freedom, *humanism* and *naturalism*, this opens up a pathway for uninterrupted development of *scientific work*, which is established onto itself, is its own self-movement in self-consciousness and freedom. This labour has all the features of Hegel's absolute spirit as the substance subject and the absolute of world history. Marx describes the final historical outreach of the metaphysical understanding of labour as essence and existence of man with the relation of two empires – the *empire of freedom* and the *empire of necessity*. The *empire of necessity* as the biological foundation of existence of man founds the empire of freedom as a fundamental possibility for the true freedom of man. The *empire of freedom* always ontologically precedes the empire of necessity. In terms of Hegel's *Science of Logic*, which deals with the return to the basic, primal and true, the empire of freedom is the basis from which the empire of necessity arose as a direct correlate. Thus, the *scientific work,* which is shown as the concrete and the final truth of all beings, is reached at the end of their progress. For Marx, this relation (progress) leads to the history of liberated labour – *practice of labour as scientific history* in Sutlić's terms. The liberated labour in this respect is not established and not finished as such – it represents self-movement which establishes itself in itself as its own space of self-realization in that what is. In this eternal establishing of self in what is, it reaches its awareness about what is. We can say, "it [history of absolute labour, IG] is the riddle of history solved, and it knows itself to be this solution." (Marx and Engels, 1988, p. 102). In Marx's view, its knowledge and its science have an empirical character. From the historical experience of today, we can say that this knowledge is science which unites all sciences – cybernetics, as the ontology of *scientific humanism* as *naturalism* in the age of *scientific history*.

The true history as the history of the liberated labour and its empirical existence and the "full development of the forces of production has become the *condition of production*; and not that specific *conditions of production* are posited as a limit to the development of the productive forces" (Marx, 1857, p. 467), Marx describes with the following words:

This communism, as fully developed naturalism, equals humanism, and as fully developed humanism equals naturalism; it is the genuine resolution of the conflict between man and nature and between man and man – the true resolution of the strife between existence and essence, between objectification and self-confirmation, between freedom and necessity, between the individual and the species. Communism is the riddle of history solved, and it knows itself to be this solution. (Marx and Engels, 1988, pp. 102-103)

This time and space of history for real, free human development, development of human capabilities, for uninterrupted development of productive powers of labour, enables humans to meet their basic needs in a way which "leaves everyone surplus time for other activity." (Marx, 1857, p. 537). For time to become space for human development, Marx is counting on the application of science in the production, the objective power of knowledge in productive powers as *scientific work* to suppress the need for concrete individual labour of workers in the capitalist work time.

Free time – which is both idle time and time for higher activity – has naturally transformed its possessor into a different subject, and he then enters into the direct production process as this different subject. This process is then both discipline, as regards the human being in the process of becoming; and, at the same time, practice [Ausübung], experimental science, materially creative and objectifying science, as regards the human being who has become, in whose head exists the accumulated knowledge of society. (Marx, 1857, p. 631)

This is not a case of time and space for higher activity but time as space of higher activity. Labour in the described *empire of necessity* as space and time for securing the necessary conditions for the existence of man can only by free "only (1) when its social character is posited, (2) when it is of a scientific and at the same time general character, not merely human exertion as a specifically harnessed natural force, but exertion as subject, which appears in the production process not in a merely natural, spontaneous form, but as an activity regulating all the forces of nature." (Marx, 1857, p. 534).

Marx here understands knowledge – science as the general mind which directly becomes a productive power of labour. In other words, he sees the *scientific work* as the essence of man – the worker of science, as the condition and the possibility of existence of man's mastering of nature and its forces, as well as the possibility of opening space-time for essential labour, labour free from the natural necessity of securing one's existence, labour as self-creation, self-engenderment the freedom of man as a self-aware scientific worker in the *empire of freedom* in reality. Developing productive powers in the *empire of necessity* and freeing up time for human development are the material foundations of development of the *empire of freedom*. One might also say that in Marx's philosophy the *empire of freedom* always ontologically precedes the *empire of necessity* and enables it, but not as true ontological space-time. Synthetic activity of labour in the mutual relationship of the two empires brings Marx's understanding of *humanism* and *naturalism* of man and humanism of nature and naturalized world history to complete identity. In *scientific work*, in working science as a paradigmatic form of productive powers of labour, Marx sees the real subject substance which would ultimately solve the problems of speculative and metaphysical synthesis in a practical way as "a real problem of life, which *philosophy* could not solve precisely because it conceived this problem as merely a theoretical one." (Marx and Engels, 1988, p.109). One can see in the world today that the accumulated knowledge of society in the productive powers of labour requires an adequate worker – a scientific worker who is also a product of the accumulated knowledge of the society. That is the real foundation for the cyborgization of man and his reality of scientific *humanism* as *naturalism* in *scientific history*.

We have stated at the onset that our standpoint of understanding *scientific humanism* as *naturalism* in *scientific history* whose character establishes the actual realization of *scientific work* is not a philosophical whim. Instead, it represents the time in history in which the biological body and nature in it are transformed and enhanced with the help of science, thus making a man into an adequate scientific worker whose body and mind are integrated into scientific and technological facility of *scientific work* in the cyborgized historical time. We are witnessing the *humanism* and *naturalism* of the cybernetic metaphysics whose *abolishment of being* and its reduction to entity has forsaken man as a self-standing entity and his relation towards nature within itself, the purposeful life and nature as *fizis*, placing man's world onto a new ontological and anthropological basis. This *humanism* as *naturalism* is not place of origin of man's essence nor his life. It is the work-science-technology-cybernetics *scientific humanism* as *naturalism*.

This analysis of Hegel and Marx has shown that the nihilism of history as *scientific history* is trying to blend in with the total and optimal functioning of the world and the age of *scientific humanism* as *naturalism*. At the same time, the age of *scientific humanism* as *naturalism* attempts to neutralize and surpass this nihilism with its complete scientific and technological development – the *scientific work* which takes in its grasp the very medium of its revelation in history – the working man. Here, we are not dealing with the estranged historical world of man before the *abolition* and *realization* of philosophy as metaphysics in the *one true science*, the *science of history*, but with the time of the world of liberated labour which has overcome the limiting productive relations of the civil society in the mechanical, scientific, and technological machinery as the new form of securing life and a new shape of the historical community – the community of science technology and information of *scientific humanism* as *naturalism*. This new machinery of science and technology and the scientific, technical and information community of *scientific humanism* as *naturalism* which sprung from it establishes the appropriate working man, the scientific worker, which, like the machine as the means of labour, contains the accumulated knowledge of scientific and technological mind of social work. The self-realizing machinery of *scientific work* and its reality need a cyborg, a part man part machine organism, as the transition from the real life of the alienated man towards the scientific and technical reality of partially or completely artificial nonhuman beings. This represents the transition from the alienated biological life into death. The question of historical time remains open – will this scientific and technical reality of artificial beings even need the alienated man? This dilemma has been present since Marx's description of primary technification of capital: "For capital, the worker is not a condition of production, only work is. If it can make machines [cyborgs, IG] do it, or even water, air, so much the better." (Marx, 1857, p. 422).

Thus, the capital needs man's realized labour – physical and mental – the accumulated knowledge of society or the power of nature, and the entities which make up the basis of this new scientific and technological synthesis apart from the automated machines are beings established on the scientific and technical foundations of life – cyborgs and robots. They are pushing out the living labour of man from the means of production of scientific work. In the cyborgized reality, the relations take part not between humans but between new metaphysical characters – cyborgs, robots, avatars, apparitions and clones.

To characterize the *scientific work* in the age of *scientific humanism* as *naturalism* using Marx's metaphysical categories one has to understand it as a unity to equivalence of *theoria, praxis and poiesis*. This substance subject as the real being establishes everything what is in history. The entire world history of nature and man as the medium of its establishment are the never about-self completed objects of its permanently reshaping, enhancing and overcoming activity in the circle of self-overcoming its on nature. The history is the history of the absolute but it is most certainly not the absolute of history. Its knowledge – science is no longer collected in the traditional metaphysics but in the cybernetic ontology or the metaphysics as the absolute science of the *scientific work* which serves to organize, rationalize, actualize and informatize all that is in its purview and defines what, how and whether something will take place in the historical reality. In the circle of this infinite vortex, nothing is completed and everything is eternally reshaped, enhanced and overcome. Cybernetic ontology as the absolute science of the absolute *scientific work* is a science, not in the material sense but in the sense of being, a science of never completed approach to truth of the world constructed by itself in the endless manipulation of beings as raw materials, moments and functions of *scientific work*. In Marx's terms, this science is not the consciousness of the human reality but is the consciousness of the scientific reality itself, as established by the *scientific work*.

The operative scientific language of the *scientific work*, devoid of any essential real content, the world of man, nature, beings and history has been transformed into the communication language, the language of mathematical symbols that hide the estranged essence of the *scientific work* more than the metaphysics can reveal. When talking about the scientific and technical information community – *scientific humanism as naturalism* – the language of the ideology of scientific work intoxicates man with its promise of unlimited possibilities of development in real and imaginary space and time. Grandiose words of humanism, equality, ecology, rights of man, scientific evolution, ethical life, freedom, eternal life have all been adapted to the uninterrupted development of *scientific work*. The magical power of actions of *scientific work* in establishing and covering everything that is, the ideologized ethics, the media presentation of the future and the real possibilities convince the estranged man to take part in the constructive cooperation. It is indicative that social and special sciences are also organized as services of the techno-logic in securing the *scientific work*. The very scientific *humanism as naturalism* as the current era in the *scientific history* is spinning in the temporal circle of metaphysical presence in which the scientific worker as an enhanced human being, non-human, will lose every real possibility of mediation between nature, history, being and sense of existence. The enhanced and reshaped cyborg beings will be closed within themselves, in their own internal meaningless existence or will be a part of the oblivion of the world of illusions with their bodiless existence. The actions of the scientific work reshape and enhance the worker man, thus establishing him as the scientific worker compatible with his scientific and technological machinery of production and the cyborgized community. This man, scientific worker, represents the synthesis of his scientific and technical essence and existence, he is the realized concept and the conceptualized reality. This scientific, technological and information society, the society of *scientific humanism* as *naturalism* will become so "humanized" and so "naturalized" that it will no longer need man, not even as a means of self-reproduction nor as the interpreter of its own nihilistic history and senselessness.

We have shown that nothing as the metaphysical being is manifested as a new way in the historical time of scientific humanism as naturalism through man, its historical mediator and interpreter, revealing the nothingness of the sentient, material man and his entire material nature as the source of natural life. Today's levels of cyborgization of man in the historical reality still allow man to be to be in the service of the medium of revelation of being in history, even though that is not the point nor the final purpose of man's existence. Man is more than just the material sentient body or than just a being with a mind, which puts him apart from all other natural and historical beings. Instead, he is, in Heidegger's terms, the shepherd of being. It depends on this calling and the anthropological understanding of oneself as a gendered entity whether or now "we may see ourselves as the responsible authors of our own life history and recognize one another as persons of "equalbirth", that is of equal dignity." (Habermas, 2003, p. 29). This also implies that "moral community of free and equal subjects of human rights does not form a "kingdom of ends" in the noumenal beyond, but remains embedded in concrete forms of life and their ethos (Habermas, 2003, p. 37).

We hold that the destiny of man is still within his intermediate power, despite the omnipotence of *scientific work.* In his description of *scientific humanism* as *naturalism,* Habermas claims that:

[...] technologically assisted life-processes of the human organism, nano-engineers draw up visions of man and machine fused into a production plant subjected to autoregulated processes of supervision and renewal, permanent repair and upgrading. In this vision, self-replicating microrobots circulate in the human body, combining with organic matter in order, for instance, to stop ageing processes or to boost the functions of the cerebrum. Computer engineers, as well, have not been idle, contributing to this genre by drawing up the vision of future robots having become autonomous and evolving into machines which mark flesh-and-blood human beings as a model doomed to extinction. These superior intelligences are supposed to have overcome the flaws of human hardware. As to the software, which is modelled on our brains, they promise not only immortality, but unlimited perfection. (Habermas, 2003, p. 41)

From Habermas's description, it is clear that the current cybernetic tendencies of reshaping and enhancing not just man's body and mind and transfer of the mind onto a mechanical computer interface lead towards overcoming of man as a medium of revelation of *scientific work* in the age of *scientific humanism* as *naturalism.* This means that *nothingness* as being of the alienated world revealed through man as a mental entity will remain hidden, forgotten and will soon be interpreted only by the language of cybernetic ontology, which already represents the point of merger of essence and existence of cyborg beings. Man's mind is reduced to a technical computational mind in which the world can make any sense as world and the entities of the world are only the objects of its will for appetite. The unpredictable part of man can only be known in its truth from the power of true thought as it "transports the man of the future into that "between" in which he belongs to Being and yet remains a stranger amid that which is" (Heidegger, 1977, p. 136).

The world of *scientific humanism* as *naturalism* reduces the already alienated man onto his technical and anthropocentric essence, onto the scientific worker and this humanism cannot be the transmetaphysical "developed naturalism, equals humanism [...] the *genuine* resolution of the conflict between man and nature [...] between existence and essence [...] between freedom and necessity, between the individual and the species." (Marx and Engels, 1988, p. 103). In the *scientific humanism* as *naturalism,* "we are dealing with the primeval wanderings, the non-truths from which Marx tried to deliver us into true history, not just because it did not contain enough 'humanism', but because this fundamental, too 'humanistic' response – project was untrue at its core (which does not mean that it was incorrect with regards to 'real possibilities'), which makes it inhuman (despite its 'humanistic' nature') (Sutlić, 1972,

p. 275) and which makes it unnatural despite its claim of naturality of its progress. The tendencies of the absolute *scientific work* have already been established but the procedures of enhancing and reshaping man's body and mind show us that all consequences of the totality of its manifestation are nowhere near realized. Without a philosophical foothold from which we can observe and rethink his actual realizations, everything that wants to be achieved can still occur in its name.

We have emphasized that despite the current tendencies of enhancing and reshaping man's body and mind through cyborgization procedures in the absolute world of *scientific humanism as naturalism*, man-worker still stands with his illusory power and freedom as the beneficiary and the medium of creation of *scientific work*, like a narcissist drunk with his own power, perceiving himself as the absolute self-creating and free entity. This

[...] predicts the possibility that work as objectification (= equality and equivalence of the objectifying act, objectified object, means of objectification, the effect of objectification) in its essence, in appropriate and successful preparation of the prepared (téchne), which is the source of all acts (technical procedures), means (instruments of use), objects (raw materials) and effects (machines, facilities, 'synthetic goods') will reverse, decompose and convert into historical production of beings from events of being and time, which will provide a more pleasant enterprise to the essence of man than economy and economization. (Sutlić, 1974, p. 283)

Man must overcome this metaphysical trap, this narcissistic horizon of self-understanding and self-creation, so that his new primal spirit older than the world of scientific work can be released from the depths of metaphysics. Scrutinizing the historical time of *scientific history* in the age of *scientific humanism as naturalism* in its finite state and historical transience needs to take place in the name of biological life which is in its immediate change the first and the final reality, while the metaphysical realization of the living entity of man is the manifestation of this being. History stems from life and the values of life, and these are humanism = naturalism = freedom. Life is not a passing form towards something bigger – the empire of truth, goodness, holiness. Life is presence, permeation with nature, and the self-aware man recognizes his relation towards nature and other beings in which man sees himself – he is the shepherd of being. The history of metaphysic has shown that nothing can be accepted once and for all, as every new stop is equal to stepping out of the world or losing worldliness, or a sign that something else has overpowered man. If one wants to overcome the metaphysics of life, the will to power, then the inevitable final act in the post-metaphysical thought must be life being measured in relation to being itself – life as the highest value of being.

As the being in the age of *scientific humanism as naturalism, scientific work* is not the one, eternal and unchangeable entity, it is not the creator of historical time nor its absolute, though it is manifested like it. Historical time has given him the chance to appear in time and space but as a *one-time historical event*. It is its own history, but not the eternal history nor the time of history of man, which treats beings, and historical entities, according to Sutlić, as the *conductor of entities from beings,* carrying them into historical world. The time of history of man as the space for human development is not by history nor by *scientific work* but, in fact, the other way around. The history of man and the *scientific work* as the being of *scientific history* occurred in time and by time, which does not mean that time is being. The time of history of man is the true space for the possibility of a meaningful relationship between being, essence and entity of man.

Rethinking *humanism* as *naturalism,* which is not characterized as *scientific work* must "invoke a new world within the realm of possibilities of the old world, so that it fits that what never existed but will always be and will persist as the future in everything past, present and tomorrow" (Sutlić, 1972, p. 277). Equating the speculative concept of Hegel's humanism as well as Marx's realistic *humanism* as *naturalism* established by liberated labour, with freedom understood as self-realization of the absolute entity, situates philosophy further down the cul-de-sac of metaphysics. The essential life, the life of primal freedom of life, life turned towards being, life free of anthropocentricism and supreme beings, the absolute in history and other illusory and metaphysical apparitions is the highest possible value of *humanism* as *naturalism.* Thus "the extent to which man's *need* has become a *human* need; the extent to which, therefore, the *other* person as a person has become for him a need – the extent to which he in his individual existence is at the same time a social being." (Marx and Engels, 1988, p. 102). Biological life as the origin of vitality of man in history, life as the origin of freedom and care in the community does not require special metaphysics of life but the thought of life from the historical time in which man and entities come into being. Despite the totality of the absolute *scientific work,* which takes hold of life of man from its roots or, in Sutlić's words, the vitality of life, the life of man is always naturally, historically and spatially free. It is older than the concept of labour even though it tries to occupy it speculatively, philosophically, biologically and technologically, sucking the vitality out of it until it fades into oblivion. The time of history as the space upon which everything appears and disappears enables the life to meet its own origin in the depths of realized metaphysics and through it. It is only from there that man's resistance against the total philosophication of life can start and it is still up to man to decide whether he wants to remain a life-filled man or not. This last is a decision of the will for life before the completion of the initiated procedures of enhancing and reshaping man, mutual relations of men and the complete historical life on Earth. As the centre of movement of *scientific work,* the medium of his complete effective revelation, man is also his only true historical possibility of survival on the basis of *unavailability of life*. Sutlić's idea that life is older than the world of labour shows us that the primal, other and different, the old-new was in action throughout the entire history, but was covered with the metaphysics of philosophical tradition. We can hope that a historical equivalence of life of man and nature is possible, with all events that this relationship entails.

Humanism as naturalism will only be possible upon completion of the world of *scientific history*. The thought situated in the essence of *scientific humanism as naturalism* has a real chance of supporting the rush of life into the reality of *scientific humanism* as *naturalism* so that this reality could be, to paraphrase Sutlić, related to life and be shown as a function of life. Our starting point is the fact that life in most inexplicable manner is, it is the highest value of being and the completion of metaphysics is not its unconditional self-establishment, but its internal decomposition. However, this decomposition allows its final possibility to come into light – a change to become the metaphysics of life as something real that contains all articulations of metaphysics. Much alike Heidegger, we can argue that stepping over the essence of *scientific humanism* as *naturalism* can be conceptualized only from the *truth of being* that is in power in it, in order to open up a new future in which all entities and human beings stem from the being itself. In the cyborgized reality, thought is no longer, in Kantian terms, the supranatural element in the dual character of man's being, but is a goal for which we must fight with care and responsibility. The philosophy of wisdom as the talk of being can step over the horizon of the metaphysical and cybernetic thought as well as the metaphysics of life. This thought will flash over the dwindling source of life, flash with words clear of logical structure, concepts and metaphysical talk, nurturing the fundamental words of wisdom itself.

3.4 THE TRANSHUMANIST PERSPECTIVES

Jean-Luc Nancy argues that humanity is becoming an expression of endless denaturalization as a result of Platonic approach to science and technology [*scientific work*, I.G.]. Technology is shown as a step away from nature towards hominization. *Homo faber* – the author and the creator of the *homo sapiens* is a technician managing his own survival as a supraperceptible *homo animalis*. This movement – production – starts with the "man" and from the other side stems from his very nature. But this production gains a new direction in history. Instead of securing the survival of man as a faulty being, it produces new conditions for his existence in nature and history (Nancy, 2007, pp. 75-91). This is a process of complete denaturalization of man, human relations and nature. This way, the history is contrasted with technology [*scientific work*, I.G.], with "nature" in *scientific history*. The process of scientific and technological denaturalization of man is concisely expressed by the transhumanist standpoints.

The use of the term *transhumanism*, in today's sense of the term, stems from the works of Max Mora who, in conjunction with Tom Morrow, published the first issue of the *Extropy Magazine* in 1988 and founded the *Extropy Institute* with Morrow four years later. The *World Transhumanist Association* was founded in 1998 by Nick Bostrom and David Pearce and this association has a significant influence on public opinion. For the purpose of promoting their ideas, they published the *Transhumanist Declaration* which is given in its full form below:

(1) Humanity stands to be profoundly affected by science and technology in the future. We envision the possibility of broadening human potential by overcoming aging, cognitive shortcomings, involuntary suffering, and our confinement to planet Earth.
(2) We believe that humanity's potential is still mostly unrealized. There are possible scenarios that lead to wonderful and exceedingly worthwhile enhanced human conditions.
(3) We recognize that humanity faces serious risks, especially from the misuse of new technologies. There are possible realistic scenarios that lead to the loss of most, or even all, of what we hold valuable. Some of these scenarios are drastic, others are subtle. Although all progress is change, not all change is progress.
(4) Research effort needs to be invested into understanding these prospects. We need to carefully deliberate how best to reduce risks and expedite beneficial applications. We also need forums where people can constructively discuss what should be done, and a social order where responsible decisions can be implemented.
(5) Reduction of existential risks, and development of means for the preservation of life and health, the alleviation of grave suffering, and the improvement of human foresight and wisdom should be pursued as urgent priorities, and heavily funded.
(6) Policymaking ought to be guided by responsible and inclusive moral vision, taking seriously both opportunities and risks, respecting autonomy and individual rights, and showing solidarity with and concern for the interests and dignity of all people around the globe. We must also consider our moral responsibilities towards generations that will exist in the future.
(7) We advocate the well-being of all sentience, including humans, non-human animals, and any future artificial intellects, modified life forms, or other intelligences to which technological and scientific advance may give rise.
(8) We favor allowing individuals wide personal choice over how they enable their lives. This includes use of techniques that may be developed to assist memory, concentration, and mental energy; life

> extension therapies; reproductive choice technologies; cryonics procedures; and many other possible human modification and enhancement technologies. (Bostrom, 2005).

As the advocates of reshaping and transforming human beings, the transhumanists attempt to ethically justify the goals they stand for and hide their true intent in their justification regarding the rights of all individuals to make decisions about their own enhancements. According to their expectations, the technology will free man of his physical limitations, make him better, healthier, stronger, smarter and longer-lived, so that he is better adapted to requirements of time and space of *scientific humanism as naturalism*. They base their transhumanist beliefs and support for the acts of enhancing and reshaping of the human species on the principles of freedom and rights of every individual in choosing procedures and contents of enhancements and reshaping one's own body and mind and the ethical commitment of every individual to optimize their working capabilities. The overall character of *scientific work* gives way to the assumption that their actions will strive towards higher degree of uniformity and homogeneity of the world and man-cyborgs, which is why the enhanced cyborgs will be classified into other super-human, or sub-human in our view, categories. The transhumanists assume the primacy of rights of individuals over good and righteous life and even life itself.

John Harris argues that every individual with economic possibilities for realizing their personal enhancement or reshaping should not be bound by a legal or ethical principle of equality of all men as "Fairness [...] does not require that benefits should not be provided to any until they can be made available to all" (Harris, 2007, p. 28). In his book *Enhancing Evolution: The Ethical Case for Making Better People,* he proposes a positive moral obligation of every individual to enhance themselves as enhancements "make us better, not perhaps simply by curing or ameliorating our ills, but because they make us better people. Enhancements will be enhancements properly so-called if they make us better at doing some of the things we want to do, better at experiencing the world through all of the senses, better at assimilating and processing what we experience, better at remembering and understanding things, stronger, more competent, more of everything we want to be." (Harris, 2007, p. 2). Harris argues that the slow and selective Darwinian evolution can be replaced by a controlled scientific and technological evolution that will replace the natural selection by a "deliberately chosen process of selection" (Harris, 2007, p. 4) so that humans would evolve "into a better species altogether." (Harris, 2007, p. 5). The term "species" does not apply to all humans but just the ideal survival of the most powerful social group. The fact that an asymmetrical relationship will arise between enhanced and non-enhanced people in real life and in the moral domain does not seem to present an ethical issue for Harris.

In his justification of ethics of enhancement and reshaping, Harris avoids defining the means and content of medical treatments in relation to enhancement and reshaping procedures by putting forth a rhetorical question he knows no generally accepted answer can be given to: if "enhancement is departing from species-typical functioning in an upward trajectory" (Harris, 2009, p. 44), then it is difficult to determine where treatment ends and enhancement begins in all those cases where "These same powers that can repair and replace diseased or damaged tissue may, in a healthy individual, augment normal functioning. [...] Who would think this an unacceptable side effect of treatment?" (Harris, 2007, p. 33). In defending his case for enhancement and reshaping of the humankind, Harris foregoes the postulated ethical equivalence of treatment an enhancement so that his following question could reveal new, bigger possibilities – why not change the human structure and function if this will be a response to dangers in the environment (Harris, 2007). As an argument for this question, Harris takes the non-existence of a unified humanist agenda that would provide an answer to the question what man is. This void justifies the enhancement of this entity with questionable essence through directed scientific evolution, so that the entity may find itself on a higher level of natural evolution. Aware that enhancements and reshaping carry a certain risk for individual's health, Harris advocates the use of a scientific and technological revolution to create a better society, with individuals taking responsibility for their personal decisions and risks. The new biotechnology and the prosthetic procedures will soon be widespread within mankind and will be as natural as natural evolution. One might argue that Harris's, just like Savulescu's, ethics of enhancement as a moral obligation of individual will inevitably lead to an authoritative imperative that could have dire consequences on the future of mankind.

The other member of the transhumanist group is the aforementioned Julian Savulescu. In our opinion, Savulescu also avoids defining a clear difference between justified treatments of humans using new biotechnology and prosthetic technology, as enhancements in his view do not imply changing the traits of an individual but an absolute ethical value that implies creating assumptions for a longer and better life. He does not question what life is and whether life looks to increase its longevity and quality through enhancement using artificial technology or by unifying life with a pre-programmed genetic code. He emphasizes the necessity of enhancement by claiming it needs to become a moral obligation of every individual (Savulescu, 2007, pp. 131-155) as this will increase the mental capacity, speed and endurance of man in sports, and reduce negative social behaviour. He exposes his attitude on moral obligation of enhancement in a typical transhumanist manner with another rhetorical question – how come we find the social and environmental effects on the change of individual's behaviour which affects the change of biological traits to be ethically justified and desirable, while we regard the scientifically oriented methods ethically questionable (Savulescu, 2007, p. 522). Savulescu provides another argument in favour of his ethical standpoint: "If we accept the treatment and prevention of disease, we should accept enhancement" (Savulescu, 2007, p. 522). This simplified and engaged ethical standpoint leaves aside the real fact that the human organism represents a unified whole and that, unlike a machine, it does not consist of parts which can be removed and repaired without any major or unwanted consequences. His stance also disregards the fact that ethical aims of treatment are different from goals and aims of procedures for enhancing and reshaping mankind. Creating improved partially artificial beings made through integration of artificial intelligence into healthy human flesh or using new biotechnology to create artificial beings made up of organic matter ultimately lead towards meaningless denaturalization of human life. Interestingly enough, the entire ethical argumentation of the transhumanist is based on life – better, stronger, happier eternal life, which in all probability draws all meaning from the pre-programmed features. Savulescu believes that we should change the standpoint from health to enhancement of life and that its quality should become the priority, which is where technology can offer us the possibility of enhancing our lives (Savulescu, 2007). He believes "that to be human is to be better. Or, at least, to strive to be better. We should be here for a good time, not just a long time. Enhancement, far from being merely permissible, is something we should aspire to achieve." (Savulescu, 2007, p. 533). In response to his claims, we can say that in the final outcome of enhancement and reshaping we no longer see life – not better, not stronger nor happier life, as it becomes reduced into a single non-human formula, leading towards faceless, uniform, meaningless and lifeless world of *scientific work*. Instead of the actual historical and biological man, this is where scientific workers dwell – cyborgs, clones, robots, artificial non-human beings as real metaphysical characters established on new cybernetic, ontological and anthropological foundations of life.

The third representative of the transhumanist view that we shall cover here is Nicholas Agar who is, unlike Savulescu and Harris, not primarily concerned with ethical issues of enhancement. He simply believes that it is the basic political right of every healthy individual to make the best life decision with regards to enhancement. Agar is a supporter of the *pragmatic optimism* which is characterized by the faith in the power of new scientific possibilities in the domain of enhancement of man, which will serve to remove all subjective dilemmas and objective difficulties in their current use along with the established potential boundaries of use. *Pragmatic optimism* does not stem from the ethical principles but from the standpoint of safety of enhancements and it does not, according to Agar, cross the boundary of logic and metaphysics as it has no ambitions to play the role of the creator but to stay in the ever-increasing domain of biological possibilities of man.

Nick Bostrom, one of the founders of the transhumanist movement, advocates the right of parents to choose the genetic features of unborn babies, as this choice is justified by the natural errors in the blind evolutionary development. In addition, what makes someone human is not his or her genetic code but the created scientific, technical and social context. It is not clear from Bostrom's stance on genetic enhancement of man whether he perceives the new technical possibilities as a space of freedom for man that should be regulated accordingly or as an instance of self-empowerment of man who does not need self-limiting norms and behaving in accordance with them. We find it is justified to stand up for moral control over scientific and biotechnical manipulations of the human body so as to maintain the unavailability of the human being. We can agree with Bostrom in that man's historical development is affected by the level of achieved scientific and technological development and the form of the community based on it. However, that does not answer the question of potential *eccentric positionality*[3] of man in relation to his medial position at the very centre of the scientific and technological being in the age *scientific work*. This *eccentricity* would imbue him with the space-time of *humanism as naturalism* in which it is possible to reach the self-purpose of man without the *scientific humanism as naturalism e*stablished by the *scientific work* that Bostrom proposes.

The dialogue on transhumanism is complemented by Jonas's modified understanding of Aristotle's teleology of nature and life which represents the "the casual modus of nature or the immanent teleology" (Jonas, 1997, p. 65). This refers to a "teleological or a finalistic nature of life" (Jonas, 1997, p. 163), which is in its essence the "testament of life", its purpose and its nature. Unlike the "blind" Darwinian evolutionism which promotes fighting for survival as the fundamental meaning of life devoid of any purpose, Jonas believes that both the lower and the higher forms of life have within themselves a meaning and a purpose insofar they are imbued by spirituality which requires an ethical standpoint. The Darwinian evolutionism leaves the lives of lower and higher species without any internal purpose and treats them as the mere raw materials of evolutionary history, while man as the highest point of evolution represents a cosmic coincidence which must remain primitive in order to justify the vision of evolution (Jonas, 1997, p. 82). For Jonas, freedom is the internal purpose of life and the very exchange of matter is the "original freedom" which lies in the "dark rules of the primordial organic substance" (Jonas, 1997, p. 18). An organism's freedom also denotes its independence and causality in relation towards nature.

One can see that the transhumanists advocate the enhancement of the Darwinian understanding of evolution as a selective struggle for survival in which life has no inner purpose, thought or freedom of self-organization. This paves the way towards manipulation, enhancement and overpowering life according to the standards of science and technology. The Promethean prestige of technology in its destruction and reshaping of man, the non-human entities and human nature (society) is "reshaped by its advocates [or ideologists, I.G.] in an attempt to give its endless labour the most dignified and the highest purpose, i.e. to elevate something which arose as a means onto the level of purpose and impose it as the true determination of mankind" (Jonas, 1987, p. 30).

This *scientific history* is not the history of man but of cyborgs and of genetically modified beings established on a new inhuman foundation of life. These beings enhanced by the *scientific work* are scientific beings - scientific workers – non-human beings – apparitions. Simply put, they are partially or completely artificial beings which persist in the scientific history as eternal presence. We agree with Habermas that biotechnologization of the human nature will change the species-ethical self-understanding in such a profound way that we will soon be unable to understand ourselves as ethically free and morally equal beings oriented by norms and reasons.

If we understand the meaning of the prefix *trans* as 'going beyond something', and transhumanism as 'going beyond humanism', the question is – what is it that we are going over or transcending? We have shown that humanism and naturalism carry within themselves the ontological and anthropological possibility for the inhuman and unnatural, which is why the transhumanist stance is shown as the ontological version of thought of *scientific work* that justifies the scientific evolution of man from an ideological-political-ethical position of neoliberal capitalism. Truth be told, the transhumanist desire of man as a self-subject for enhancing and reshaping his own body and mind reflects the desire of the New Age man liberated from the revelation to use his knowledge and technology to secure the certainty of life and truth of what he knows in his own knowledge. Establishing everything that is includes the self-establishment of one's own body. However, in the age of *scientific humanism* as *naturalism*, the self-aware and the self-liberated subjects exists as a part of the *scientific work* and "everything that intends to be established, through representing objectification, as secured and hence as: in being, is binding for every man" (Heidegger, 1977, p. 151).

By dealing with the issues of bioscientific and scientific enhancement and reshaping of man, the transhumanist thought is at the brink of transhuman and posthuman sources of technical life and represents their direct reflection. For this philosophical view, the life of science and technology is, despite its claims for a more just and long-lived life, just a system of signs that should be constantly redefined in accordance with the progress of science and technology. As a distorted and false consciousness adequate to the being alienated by *scientific work*, this ideology is articulated:

[...] in such a manner which prevents us from seeing the real mechanisms which regulate the means by which the human world, human relations, means and modalities of life function and are reproduced. These are primarily economical or interest-oriented mechanisms, but they also encompass the mechanisms for reproducing self-serving power (regardless of the primacy of their economic, political or some other foundation). In this way, the ideology makes covert and opaque the mechanisms which regulate the manner in which the human life and world, human relations, and determinates and modalities of our social and individual life are reproduced. (Veljak, 2010, p. 596)

The effects of this "hidden ideology (i.e. the covert hiding of the reality of the deeply marked alienation and realization) is nihilism, a self-unaware nihilism as a social state of slaves sold to the consumer frenzy..." (Veljak, 2010, p. 603). The transhumanist ideology in its ultimate reach is shown as a plea for Marx's vision of humanism as naturalism – *scientific humanism as naturalism* whose real consequences he was unable to see in his own time.

The effects of biotechnology have an undeniable effect on self-understanding of the human body and human life, and they raise new ontological and anthropological questions about the possible biological completion of man or the end of his recent metaphysical and historical abilities. In this vein, futurologist Ray Kurzweil predicts that 2099 will mark the beginning of the end of humanity as we know it.

Overcoming the biological revolution is necessary, so Kurzweil concludes that the only thing left to do for us is the technological and scientific way of survival in the form of artificial intelligence. Life will become a different life as a spiritual game of immortals in an imaginary space. The scientific contributions of Kevin Warwick as an advocate of transhumanism for enhancing and reshaping of man through use of artificial intelligence are shown in the chapter 4.4.3 *Cyber-cyborg* and will not covered here.

In relation to the apologetic thought of the transhumanists which still has their body as their program, we support the thought which will be situated in the essence of *scientific work* and its current age – *scientific humanism as naturalism*, in the name of being and life older than *scientific work* and its metaphysic history. This thought will wait in its wanderings and travels through the depths of alienation of man, wait for an eventful flash to make a shift towards the primal humanism and naturalism of life and man. Man "cannot, of himself, abandon this destining of his modern essence or abolish it by fiat [...] only there where the consummation of the modern age attains the heedlessness that is its peculiar greatness is future history being prepared" (Heidegger, 1977, p. 153). Freedom cannot be fought without philosophical participation in the establishment of the *scientific work* as it is the deepest negation of freedom and life. There is still a chance for philosophical and practical action arising from the freedom on the new path that leads into the depths of nihilism of life.

REFERENCES

Anić, V. (2003). *Hrvatski enciklopedijski rječnik*. Novi Liber.

Berdyaev, N. (1934). *The Bourgeois Mind and other essays*. Sheed & Ward.

Bostrom, N. (2005). *A History of Transhumanist Thought*. Academic Press.

Dessauer, F. (1944). Filozofija tehnike. In Čovjek i tehnika. Kulturno-filozofski eseji. Matica hrvatska.

Ectors, L. (1971). Inteligencija, instinkt i kibernetika. In Zbornik radova. Izazov kibernetici. Centar za društvene djelatnosti omladine RK SOH.

Habermas, J. (2003). The Future of Human Nature. *Polity*.

Harris, J. (2007). *Enhancing evolution: The ethical case for making better people*. Princeton University Press.

Harris, J. (2009). Enhancements are a Moral Obligation. In J. Savulescu & N. Bostrom (Eds.), *Human enhancement* (pp. 131–154). Oxford University Press.

Heidegger, M. (1977). *The question concerning technology, and other essays*. Garland Pub.

Heidegger, M. (1999). *Contributions to philosophy: From enowning*. Indiana University Press.

Jonas, H. (1987). Warum die moderne Technik ein Gegenstand fur die Philosophie ist. In Technik, Medizin und Ethik, Zur Praxis des Prinzips Verantwortung. Surhkamp.

Jonas, H. (1997). *Das Princip Leben: Ansatze zu einer philosophischen Biologie*. Suhrkamp Verlag.

Marx, K. (1844). Economic & Philosophic. Progress Publishers.

Marx, K. (1857). *Grundrisse. Foundations of the Critique of Political Economy* (M. Nicolaus, Trans.). Marxists Internet Archive.

Marx, K. (1867). *Capital: A Critique of Political Economy* (S. Moore & E. Aveling, Trans.). Progress Publishers.

Marx, K. (1977). *Critique of Hegel's 'Philosophy of right* (J. J. O'Malley, Trans.). University Press.

Marx, K., & Engels, F. (1988). *Economic and philosophic manuscripts of 1844* (M. Milligan, Trans.). Prometheus Books.

Marx, K., & Engels, F. (1998). *The German ideology: Including Theses on Feuerbach and introduction to The critique of political economy*. Prometheus Books.

Nancy, J. L. (2007). *The Creation of the World, or, Globalization*. State University of New York Press.

Rodin, D. (1974). *Marxova misao zajednice*. Biblioteka velika ediceja ideja.

Šarić, L., & Čatić, I. (1998). Raznoznačnost naziva tehnika i tehnologija. *Mehanizacija Šumarstva, 23*(3–4), 157–162.

Savulescu, J. (2007). Genetic Interventions and the Ethics of Enhancement of Human Beings. In B. Steinbock (Ed.), *The Oxford Handbook of Bioethics* (pp. 517–528). Oxford University Press.

Selak, M. (2013). *Ljudska priroda i nova epoha*. Naklada Breza.

Sutlić, V. (1972). *Bit i suvremenost*. Izdavačko preduzeće Veselin Masleša.

Sutlić, V. (1974). *Praksa rada kao znanstvena povijest: Ogledi uz filozofijsko ustrojstvo Marxove misli*. Kulturni radnik.

Sutlić, V. (1994). *Uvod u povijesno mišljenje*. Demetra.

Veljak, L. (2010). Medij ideologije. *Filozofska Istraživanja, 120*(30/4), 595–604.

Veljak, L. (2013). Praktična i praktička filozofija. *Filozofska Istraživanja, 132*(33/4), 603–610.

Wiener, N. (1950). Cybernetics or control and communication in the animal and the machine (2nd ed.). MIT Press.

ENDNOTES

1 Source: Translated by F. M.

2 Cf. Šarić and Čatić (1998).

3 The term *eccentric positionality* stems from H. Plessner, who believes that man's center of survival has been shifted from its neutral position. Man is and is not in his own center as he is yet to make himself a man. This means that man is not a passive consumer of life but that it leads him towards fulfilment.

Chapter 3
The Age of the Cyborg

ABSTRACT

Following the historical and philosophical grounding, this chapter draws our attention to the question of what it means to be living in a cyborg era, presents the existential and ontological significance of cybernetics, and describes in detail the known forms of cyborgization. This chapter covers the key considerations including conceptualizations of cyborgs, replacement procedures, normalization procedures, improvement procedures, reshaping procedure, and types of cyborgs (animo cyborgs, homo cyborgs, cyber cyborg, and robo cyborgs). It also deals with targeted areas of human-cyborg enhancement, including bionic arms, bionic fingers, bionic legs, bionic ear, bionic eyes, bionic hearts, bionic skin, bionic tongues, bionic brains, and bionic clothes. The importance of computer interfaces and key applications of cyborg research are described throughout the subsection of this chapter, such as Kevin Warwick's and Stelarc cyborg projects, the popularisation of cyborgs in art, Steve Mann's wearable computers, augmented reality goggles, and robo-cyborgs.

INTRODUCTION

The procedures of enhancing man's biological capabilities through external extensions have been present in the history of mankind ever since the stone age. The idea of enhancing the human body originates in Descartes's thought whose starting points is a mechanical vision of the world, a vision according to which all natural phenomena can be explained using the principles of mechanics:

This is the way he reached the animal-machine hypothesis. According to this hypothesis, animals are only highly perfected mechanisms, comparable to watches and machines that existed in his time. This idea was supported by revelations from the domain of anatomy of that period, which perceived muscles and joints as systems of levers. However, there is no doubt that the crucial argument came from Harvey's (1628) discovery of blood circulation, as it proved that the heart, whose role was so vital in the life of man, was just a pump, i.e. a mechanism...He definitely saw the human body as a mechanism, but he accepted the existence of some kind of consciousness, some kind of immaterial soul, which was ascribed higher human activities, such as thought, volition, etc. Descartes's standpoint was re-accepted and brought to an

DOI: 10.4018/978-1-7998-9231-1.ch003

extreme by La Mettrie in his work Man a Machine (1784). In it, he claims that the whole man, not just his body, must be understood as a complicated machine, a physical manifestation of the activities of this machine. La Mettrie's major argument was that the activities of the soul, such as thought, consciousness etc., are conditioned by physiological states – brain injuries, consumption of alcohol or tranquilizers have an immediate effect on the psychological properties. (Vermeersch, 1971, p. 186).

Descartes' division of being into *res cogitans* and *res extensa* has led to a far-reaching division into the "thinking" and the "extended", that is, towards the unnatural division of life on its mental and corporeal part. In Jonas's view, the corporeal represents the central problem for ontology "The body which is alive and mortal, which has the world and belongs to the world as a part thereof, which can feel and be felt, whose external form is the organism and causality and the eternal form only the being and finality – is the memento of the still unresolved question of ontology what the being is..." (Jonas, 1987, p. 39). Furthermore, "without the body and its elementary experience, without this starting point of our most extensive and general extrapolation into the wholeness of the reality, one could not gain any insight into the power and the activity in the world, nor the active connections between all things and, thus, no understanding of nature in general (Jonas, 1987, p. 46). The task of philosophy is to "regain the internal dimension of the organic by reading the biological text" (Jonas, 1987, p. 9), thus overcoming the division into *res cogitans* and *res extensa*. This way, man himself, according to Jonas, is returned to the unity of the remaining living world and encompasses the phenomena of life as a whole.

For Jonas, the organic is "spiritualized" in all its essence and at all levels of life, and the spiritual is in its essence aimed at the organic. Both are intertwined in the "mystery of the living body" (Jonas, 1987, p. 9) – the organism. The lowest forms of organic life are imbued with spirit and the highest form of the spiritual aspect of man remains a part of the organic. This perception is reflected not only in anthropology but also in biology, as the being of man and the very phenomenon of life are observed in a new way in the unity of the natural (biology) and the spiritual (philosophy). A new life and new living beings represent an inseparable fusion of the spiritual and the corporeal. These two substances cannot be observed in isolation for the purpose of any kind of scientific and technical research, nor can they be thought of as separate ontological or biological entities.

However, despite Jonas's attempts, *scientific work* as the being in the era of *scientific history* reshapes, enhances, and denaturalizes the inadequate, transient, and mortal body of man through the means of science and technology. The technological procedures are used to integrate the organic and inorganic nature, man is changed and reshaped into a partially or completely artificial being. Somewhere among us are the intelligent machines capable of showcasing the activities we once held to be exclusively human. Prolonging human life and the yearning for immortality are reflected in the perfection of biological organs through technology or biotechnology, and through acts of establishing life on a new non-biological basis which transcends the metaphysical human nature as the unity of natural and supernatural in a mortal body. These attempts to reach eternity as the immortality of man carry within them the superhuman-nonhuman nature and devalue life as a whole. Artificial intelligence and technology create the conditions for a completely artificial life of the inhuman, whose origins do not lie in our known biological historical reality. The tendency for immortality and eternity of non-human exist outside and beyond the space-time of possible human history.

Man has a body because he lives within its boundaries, aware of its finiteness and possibilities. Plessner highlights the crucial difference between our organic 'equipment' that we bestow upon ourselves through various enhancements and the living body in which we are one with the nature. Marx argues that man lives of nature as it is represented by his body with whom he must remain in a permanent process if he wants to stay alive. Man represents the nature's connection with itself, not only in the organic sense but through man's essential activity. This is why human life can be perceived as a self-aware life worthy of human beings. The other living beings on Earth have no awareness of their life and its finiteness and they live like immortals. Fukuyama also warns that man's emotional nature cannot be reduced to an artificial basis as human nature represents the unity of mind and feelings within the boundaries of the human body.

4.1 TRUE AND ONTOLOGICAL SIGNIFICANCE OF CYBERNETICS

A new basic science which defines sciences and governs them = cybernetics.

Heidegger

Norbert Wiener's 1948 book *Cybernetics: Or Control and Communication in the Animal and the Machine* marked the beginning of a new scientific paradigm which would raise cybernetics onto the level of an ontological science and through it affect the scientific foundation of all sciences (natural or social), enable the automatization, digitalization and informatization of all productive powers of labour, change the character of production itself and push the modern world in to a cyborgized post-modern scientific and historical reality. As an analogy with Hegel's determination "what is rational, is actual, and what is actual, is rational" (Hegel, 1991, p. 29), the information code as a means of the scientific and technological mind is the transcendental substance subject of establishment of the simulacrum[1] as a world of digital images, mass media, virtual world of the network and the augmented reality. This ontologizing, reality-based essence of cybernetics as the *techno-logic* of the *scientific work* which reached the unity of its theoretical/poietic/practical self-activity is in stark contrast with Wiener's and Marx's expectations that the liberated, onto-self established *scientific work* guided by the scientific mind and the ethical principles will be in the service of *humanism = naturalism*, freedom and truth, as the instrument of all human purposes. Unaware of his misconception, Wiener claims:

There are, however, hopeful signs on the horizon. Since the publication of the first edition of this book, I have participated in two big meetings with representatives of business management, and I have been delighted to see that awareness on the part of a great many of those present of the social dangers of our new technology and the social obligations of those responsible for management to see that the new modalities are used for the benefit of man, for increasing his leisure and enriching his spiritual life, rather than merely for profits and the worship of the machine as a new brazen calf. (Wiener, 1990, p. 162)

The term cybernetics comes from the Greek word *kybernáo* 'to steer (a ship), to manage', while the word *kibernetiké* (téhne) denotes the art of managing or steering a ship. It is the science of general laws of control and communication processes, means of receiving and using information in organized systems – machines, living organisms and their mutual relationships with one another. The term was first used by Plato (427-347 BC) in his *Dialogues*:

Cybernetics delivers us from the dangers, not only of the soul, but also of life and opulence. It is a wise and humble science, it does not boast by achieving something magnificent, as the helmsman knows that as he unloads his passengers he did not leave them better than they were at the moment of boarding, both in body and in soul. (Moles, 1971, p. 58)

In recent times, Ampere was the first to use the word *kybernetes* in his classification of sciences, describing it as a science of control (Boulanger, 1971, p. 19). Its goal is to discover and formalize means of communication and to establish communication-based control. The crucial features of cybernetics are: system theory, information theory and management theory. Control is the basic and the crucial part of cybernetics, a system is the object of its study and information is the substrate of management. Numerous neologisms related to the concept of cybernetics stem from the cybernetic organization and production of reality, such as *cyber space*, *cyborg* and *cyberpunk*, etc.

Its current meaning "cybernetics owes to a group of mathematicians and physiologists from the Harvard University – Wiener, Rosenblueth, Warren, and McCulloch. This group was later complemented by Neumann, Lorente de Nó and Goldstein. It would be fair to say that cybernetics was derived from neurophysiology and is related to mathematical physics (Laborit, 1971, p. 159). The US mathematician Norbert Wiener is regarded as the founder of cybernetics. In his 1948 book *Cybernetics: Or Control and Communication in the Animal and the Machine*, he laid out the foundations of the synthetic development of science within the framework of cybernetics. In his deliberations about the fundamental goals of cybernetics, Wiener starts from the accomplishments of the modern physics, physiology, statistics and mathematics, but also the need for cooperation between the scientific disciplines, so that one could study the scientific truth in a synthetic manner and reduce the entropy in the world. The era of great men like Albert Einstein, Werner Heisenberg, and Max Planck is followed by a revolution in physics, initiated by, in Wiener's opinion, Ludwig Boltzmann and Willard Gibbs. Thanks to them, "physics now no longer claims to deal with what will always happen, but rather with what will happen with an overwhelming probability" (Wiener, 1990, p. 10). This practically implies "that the rigid Newtonian basis has been discarded or modified, and the Gibbsian contingency now stands in its complete nakedness as the full basis of physics" (Wiener, 1990, p. 10). Gibbs discusses the matters of entropy – the likelihood of all worlds which contain a possible answer for the entropy of our world, which means that Gibbs believes that the final destination for all of us is chaos and/or death. As "entropy increases, the universe, and all closed systems in the universe tend naturally to deteriorate and lose their distinctiveness, to move from the least to the most probable state, from a state of organization and differentiation in which distinctions and forms exist, to a state of chaos and sameness" (Wiener, 1990, p. 12). It is most likely that the universe will end up in chaos and that man is faced with his personal death and eradication from earth. Our task is, according to Wiener, to utilize cybernetics to stand up against the complete chaos and establish the organization on Earth and thus fight the likelihood of death.

How should cybernetics accomplish this aim? The key instruments of cybernetics are information and means of communication used to establish interaction among the systems, between man and machine, machine and man, or between machine and machine. Wiener argues:

To me, personally, the fact that the signal in its intermediate stages has gone through a machine rather than through a person is irrelevant and does not in any case greatly change my relation to the signal. Thus, the theory of control in engineering, whether human or animal or mechanical, is a chapter in the theory of messages. Naturally there are detailed differences in messages and in problems of control, not only between a living organism and a machine, but within each narrower class of beings. It is the purpose of Cybernetics to develop a language and techniques that will enable us indeed to attack the problem of control and communication in general, but also to find the proper repertory of ideas and techniques to classify their particular manifestations under certain concepts. (Wiener, 1990, p. 16-17)

The concept of the message stems from Wiener's idea that language is not an exclusively human attribute and that a man-machine and machine-machine dialogue can be established through coded symbols. Semantically non-trivial information for humans and machines are those pieces of information which stimulate a particular activity – the one which activates the organism.

By receiving and using the information from his environment, man tries to adapt to the vagaries of reality and effectively maintain himself in it. The state through which all living beings defy entropy is called homeostasis and it represents the attempt of all human beings to preserve the functional structure in their living organisms. What matters in homeostasis is not the fact of the matter that the organism consists of (as it changes every day) but the continuity of the process and the memory of continuity of this process. Wiener thus concludes that the body is more form than matter, and as form it can be transferred over distance as a message. To remain active, we have to preserve the continuity of the process and be well-informed, as "communication and control belong to the essence of man's inner life, even as they belong to his life in society. (Wiener, 1990, p. 18).

Man communicates with his environment with his senses and they represent the primary source of information upon which he his decisions are based. The same applies to the automatic machines which have receptors for receiving and processing information. Since the automatic machines need vast amounts of data, they operate on the principle of a predefined program of possible functions. Controlling the automatic machine using a real program and not expected actions is called feedback and it requires the use of sensors as the machine's senses for communicating with its environment. This establishes order and the real activity of the machine which is in contrast with the entropy and the chaos of the environment.

Today, it is well-known that every human cell functions as a cybernetic machine – it has its own memory, program, operational units and the command vector. In this regard, man is a message on its own. Verschraeghen (1971, p. 273) argues that we can reach this conclusion by comparing the body with the embryo, which has been long regarded as a chromosomal message transferred by the genes. However, the 'information' transmitted by the genes is a complete program which normally leads to the 'formation' of a perfect individual. Man established in this way no longer lives in the duality of his/her gender identity as man and woman, as he/she has been simply transformed into a sign. This leads to the crucial question of cybernetics – where does the animate begin and the inanimate end? Wiener's key hypothesis is that "the physical functioning of the living individual and the operation of some of the newer communication machines are precisely parallel in their analogous attempts to control entropy through feedback" (1990, p. 26).

From the standpoint of cybernetics, the question whether the machine is alive or not is purely semantic as both the machine and man oppose the ever-growing stream of entropy. This means that both the machine and life belong to the same paradigm. This does not entail that "the specific physical, chemical, and spiritual processes of life as we ordinarily know it are the same as those of life-imitating machines (Wiener, 1990, p. 32). Wiener (1990) argues that the main difference between man and machine that exists today lies in the fact that man utilizes internal models which he can use to form the image of the outside world. These models allow him to imagine the situations which had not come to pass in nature (imagining forms) and also recognize the inbound stimuli as forms (recognizing forms). What makes the machine alive is the "ability to make decisions it can produce around it a local zone of organization in a world whose general tendency is to run down" (Wiener, 1990, p. 34). According to him, the existence of these local islands (synthesis of man and machine) which oppose entropy allow us to conclude that progress does indeed exist.

The quantum theory shows that progress can oppose entropy of nature and the universe by dedicating a small amount of information necessary for transfer of information, messages and orders. Since we are "shipwrecked passengers on a doomed planet" (Wiener, 1990, p. 40), and cybernetics provides us with the ability to steer, we have two options at our disposal – the first one is to put our trust in the unlimited capabilities of science and technology that will enable us to control our historical reality, nature and society, which is just a different way of formulating *scientific humanism as naturalism*. Wiener believed in the possibilities of cybernetics and its application for preserving life. The second option is a meaningful and reasonable acceptance and moral participation in the game of the scientific mind and its metaphysical apparitions (with all the dangers which come with it) as they provide the possibility for, in Heidegger's terms, "the favor of the turning about of the oblivion of Being into the truth of Being" (Heidegger, 1977, p. 44). Man as the *shepherd of being* is the one who anticipates this truth with his mental and moral participation.

The ontological nature of cybernetics as the science and technology defying entropy within the scientific and historical reality is manifested in the current reshaping and enhancement of the man-worker into a scientific worker-cyborg, a partially artificial non-human entity, an apparition which defies entropy in the cyborgized information and communication reality of *scientific humanism as naturalism*. It achieves this goal by drawing all life from man with its procedures and establishing this new life as an organization with a transient material structure trying to preserve the continuity of organic processes by memorizing information which reflect the continuity of its relationship with its environment. This possibility of enhancing and reshaping both body and mind was inadvertently and jubilantly announced by Wiener in 1948 in his romantic humanism (in fact, *scientific humanism as naturalism*): "We have modified our environment so radically that we must now modify ourselves in order to exist in this new environment. We can no longer live in the old one. Progress imposes not only new possibilities for the future but new restrictions" (Wiener, 1990, p. 46). He makes an even more determined claim on this matter:

Those who would organize us according to permanent individual functions and permanent individual restrictions condemn the human race to move at much less than half-steam. They throw away nearly all our human possibilities and by limiting the modes in which we may adapt ourselves to future contingencies, they reduce our chances for a reasonably long existence on this earth. (Wiener, 1990, p. 52)

For the sake of scientific integrity, we have to remark that that in some parts of his work, Wiener also argues that "man's advantage over the rest of nature is that he has the physiological and hence intellectual equipment to adapt himself to radical changes in his environment. The human species is strong only insofar as it takes advantage of the innate adaptive, learning faculties that its physiological structure makes possible" (Wiener, 1990, p. 58). However, this is concealed with the overall movement of the cybernetic logic and we believe it serves more as a role model of a perfect natural organism that can utilize its physiology to resist entropy and can serve as a model for building a cybernetic machine. These ambitions are also reflected in his following stance: "Theoretically, if we could build a machine whose mechanical structure duplicated human physiology, then we could have a machine whose intellectual capacities would duplicate those of human beings (Wiener, 1990, p. 57)".

We can conclude by saying that cybernetics as ontology or metaphysics of the end of the modern and postmodern historical reality has enabled the current realization of the essence of *scientific work* in which man as a participant of self-activity of *scientific work* is enhanced, reshaped and established as a scientific worker which defies the entropy of the Universe by preserving the process continuity of its organism as a partly artificial, non-human entity. The goal of this mechanical and physiological life is the exchange of information with the environment so that through "defiance of Gods" (Wiener, 1990, p. 184) life could be preserved as "an island here and now in a dying world" (Wiener, 1990, p. 95).

Wiener believes that man should sacrifice his body and mind to defy the entropy of reality in symbiosis with his mechanical counterpart. In this process, the material life of man is dematerialized from his organic substance so that it could be translated into the digital code and information that could be transferred over large distances with minimal energy used. Man's senses are replicated by sensors which react to the environmental stimuli, and man's mind is stationed in the software of the program controlling the continuity of processes of the mechanical machine in the perspective of the mechanical being. More than that, cybernetics has enabled the simulation of the historical reality and has left a thin dividing line between the real and the unreal. As the power and strength of establishing the world on the principles of scientific work, cybernetics superimposes itself over everything without a valid cause. To encourage us to accept the benefits of the cybernetic progress with ethical determination and judicial foundation, Wiener claims: "The *Monkey's Paw* of skin and bone is quite as deadly as anything cast out of steel and iron [...] The hour is very late, and the choice of good and evil knocks at our door" (Wiener, 1990, p. 186).

Against Wiener's claims, we have to remark that *scientific work* with its procedures for enhancing and reshaping man's body and mind uses the foundations of science and technology to ontologize the life and essence of man as a scientific worker and establishes life on a new, artificial anthropological basis. With this radical cut in the scientific and technical overcoming of all limitations that stem from the traditional understanding of unity of human nature, the *scientific work* allows the *empire of freedom* to prevail over the *empire of necessity*.

The perfected body and the cyborgized social relations are understood not just as planned and controlled media of defying the entropy of the Universe, but also as a victory of freedom of life (from its natural limitations) in the eternal persistence, beyond the limitations (entropy) of the *empire of necessity* and its historical space-time. This represents the ontological realization of Marx's attempts to equate *humanism and naturalism*. The empire of freedom is manifested as the true empire of *scientific humanism = naturalism* without man and biological life. As an opposition to the cybernetic *empire of freedom* in the *scientific history*, man has a possibility to act as the protector of being and time to expect a historical event of being, the essence of man and being in the openness of historical time. This historical event and the very history in time presuppose man as their possibility, as reflected in Heidegger's 'being-in-world' which manifests its freedom as the 'co-being' in the historical community of equal human beings. Expecting the historical event of being presupposes the philosophical journey towards being, as well as appropriate moral action on man's part.

4.2 THE REALITY OF CYBORGS

In this part, we shall describe the existing cyborgs and cyborgized social relations of our historical reality. Apart from scrutinizing the ontology of cyborgs, our intention is to stimulate a wide-ranging intersubjective discussion on procedures and goals of procedures of reshaping and enhancing human body and mind which should result in a consensual set of ethical norms and moral procedures and legal solutions. We want to emphasize the unquestionable benefits that the scientific and technological achievements bring within the domain of treatment and prevention of diseases, but we also want to express caution about the dangers which stem from the essence of the ontological movement of *scientific work*. In hope of a possible historical event of thought, we advocate the ethically agreed upon norms and moral actions. This exploration of reality and ontology of cyborgs consists of two interrelated parts. In the first part, we shall present a segment of a long-term research activity on cyborgization of living beings, as well as those that are yet to come in the near future both in the real and virtual world. Within this part, a cybernetic analysis of man as a system is also provided. In the second part, we focus our attention to the cyborgized reality, virtual and augmented reality.

4.2.1 The Term Cyborg

The term *cyborg* represents a blend of the phrase *(cyb)ernetic (org)anism*, which is used to denote the connection between a biological organism and a mechanical machine, i.e. a self-regulated man-machine system and an exogenously augmented organizational complex that unconsciously functions as an integrated homeostatic system (Featherstone and Burrows, 1995, p. 35). An ontological denotation for a cyborg is *hybris*, i.e. a monster which has no essential centre, which makes it an essential-less, home-less, alienated and non-human being. We can perceive it as a virtual *hybris* of man in the cybernetic space time.

The term cyborg was coined by scientists Manfred Clynes and Nathan Kline in their 1960 paper *Cyborgs and Space* (Clynes and Kline 1960). Their paper analysed the assumptions about the ways for adapting humans to survive in space and the possibility of manned space flights, claiming that "Altering man's bodily functions to meet the requirements of extra-terrestrial environments would be more logical than providing an earthly environment for him in space" (Clynes and Kline, 1995, p. 26).

The scientists assumed that man's adaptation to space without space suits could be possible with the aid of implants which would enable his enhanced body to persist in unfavourable conditions. According to their ideas, implanting a machine part into a living organism would enhance and modify the functions of the organism, which would function as a cybernetic self-regulatory man-machine system. For this reason, Clynes and Kline propose the term *cyborg* as a name for any "exogenously extended organizational complex functioning as an integrated homeostatic system" (Clynes and Kline, 1995, p. 27) (Figure 1).

4.2.2 Historical Strands of Emergence of the Term Cyborg

The term cyborg stems from two historical strands. In the first historical strand, the term cyborg stems from the domain of cybernetics. The connection between the human nature and the machines can be traced way back to mechanical metaphors from ancient literary texts. The Indian national epic Mahabharata, written around 300-500 B.C., portrays a machine with the form of a lion. In his *Discourse on the Method*, Descartes argues that animals are nothing more than highly perfected machines, unlike man who is a being of thought. In the 18th century, La Mettrie, compares man to a mechanical contraption by claiming "the human body is a clock" (La Mettrie, 1954, p. 141). What makes cyborgs different from these mechanistic projections is the in-built system of information. Haraway argues that cyborgs "are information machines. They're embedded with circular causal systems, autonomous control mechanisms, information processing - automatons with built-in autonomy" (Kunzru, 1997).

An article published in the New York Times on May 22nd, 1960 claims: "The Cyborg deliberately incorporates exogenous components extending the self-regulatory control function of the organism in order to adapt it to new environments."[2] In their 1960 article *Cyborgs and Space*, Clynes and Kline

Figure 1. A projection of a cyborg man

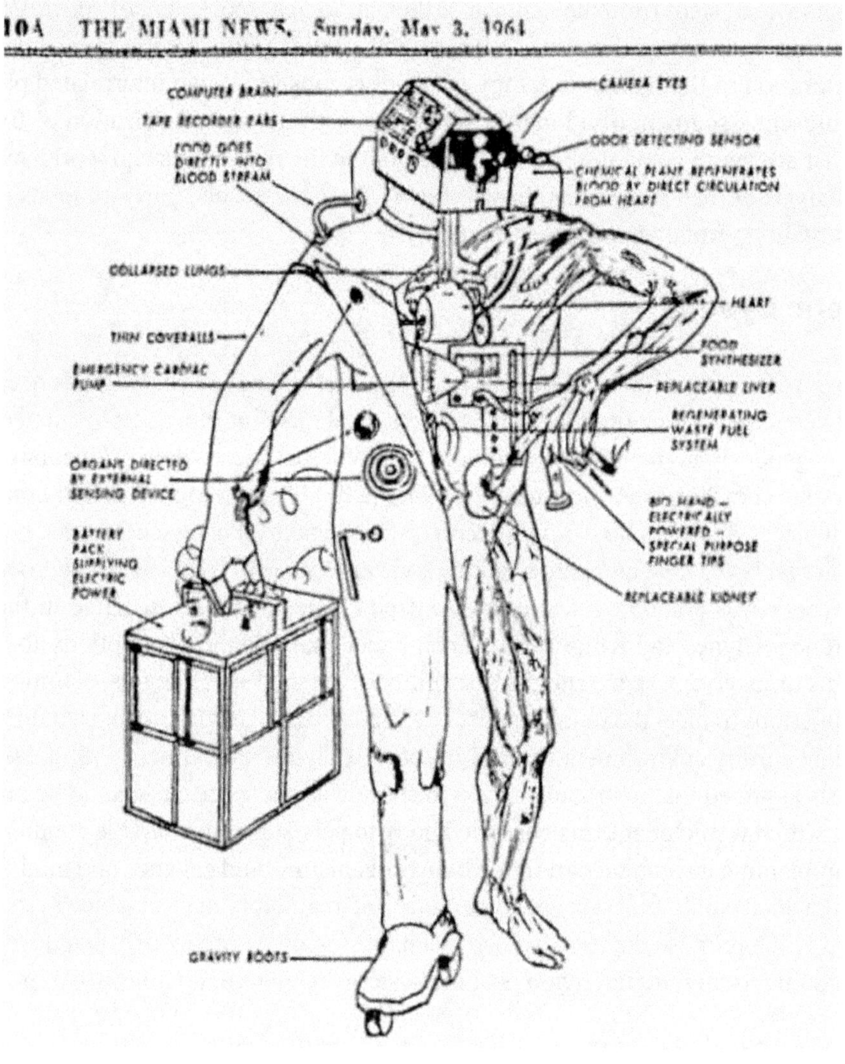

announce their new stance about the means of survival of human beings in the inhospitable extra-terrestrial environment and found the concept which would "bring about the biological changes which might be necessary in man's homeostatic mechanisms to allow him to live in space qua natura" (Clynes and Kline, 1995, p. 27). In their view, the cyborg is the solution for the frail human body in a perilous environment as the modified human body would be able to withstand all the dangers of being in space. In his 1966 sci-fi collection *Other Worlds than Ours*, Maxwell Cade (1966) describes the cyborg as an animal-machine combination of the future. In *Future Shock* (1970), A. Toffler describes an advanced combination of man and machine called the cyborg, claiming it is much closer than most people think.

The second historical strand is derived from the science fiction literature published before and after the World War II. In 1818, Mary Shelley's *Frankenstein* describes a type of cyborg made of biological and chemical elements. Frankenstein is a metaphor for the increasing social anxiety over the growing development of science and technology (Shelley, 1988). In 1843, Edgar Allan Poe's *The Man That Was*

Used Up (Almeida, 2010, pp. 163-173) describes a man with a large prosthesis. In 1908, Jean de la Hire wrote his *L'Homme Qui Peut Vivre Dans L'eau* (The Man Who Could Live Underwater) in which he describes the man-machine Nyctalope, which has artificial organs and enhanced mental capabilities (Zehr, 2011). In his novel *The Comet Doom* (1928), Edmond Hamilton described astronauts as a mixture of organic and machine parts (Stableford, 2006, p. 115). In 1923, Edwin Vincent Odle's novel *The Clockwork Man* describes the man of the future whose body and mind are regulated by a clock mechanism embedded into his head (Stableford, 2006, p. 115). Catherine Lucille Moore's story *No Woman Born* (1944) describes a dancer called Deirdre whose body was burned and brain was transferred into a faceless, flexible mechanical body (Flanagan and Booth, 2002, p. 301). A similar story with a lead female character was written in 1973 by James Tiptree (Flanagan and Booth, 2002, p. 301). In his 1984 novel *Money*, Martin Amis describes himself as a robot, android and cyborg. The most famous literary version is the creation of the homunculus in Goethe's *Faust II* (Goethe, 2008). According to Vermeersch (1971), these newer instances of science fiction works based on mythology bring us perceptibly closer to cybernetics as creation of man was taken beyond the artificial or alchemic and into the domain of robots and automata (Vermeersch, 1971, p. 184). Cyborgs are also present in the film industry, where humans are portrayed as mixtures of organic and machine parts with capabilities which by far supersede those of humans (Frankenstein 1931, Terminator 1984, Terminator 2: Judgement Day 1991, Blade Runner 1982, Matrix 1999, Heatseeker 1995).

An essay *The Cyborg Manifesto* (1980) by Donna Harraway utilizes the trope of a cyborg to explicate its meaning as a product of science and technology. Haraway defines a cyborg as "a cybernetic organism, a hybrid of machine and organism, a creature of social reality as well as a creature of fiction[...] By the late twentieth century, our time, a mythic time, we are all chimeras, theorized and fabricated hybrids of machine and organism—in short, cyborgs" (Haraway, 2016, p. 150). The existence of cyborgs indicates the time of new ontology in which the natural identity and organic integrity of the human body is no longer discussed. The symbiosis of the natural and the artificial allows man to live his life in a half-natural, half-artificial reality. The cyborg is our new human ontology. The significant influence of science and technology creates and establishes new forms of human beings. In this new reality of modern life, the relationship between human beings and technology has become so intimate that it is no longer possible to make a clear distinction between man and machine due to their ever-increasing merger. For Haraway, to be a cyborg does not mean to be just technologically augmented with technology and prosthetics in a human body:

[...] the cyborg, as I use that term, does not refer to all kinds of artifactual, machinic relationships with human beings. Both the human and the artifactual have specific histories [...] The cyborg is intimately involved in specific histories of militarization, of specific research projects with ties to psychiatry and communications theory, behavioral research and psychopharmacological research, theories of information and information processing. (Haraway, 2000, pp. 128-129)

Machines from the end of 20th century erase the boundary between the natural and the artificial life as well as the difference between the spirit and the body. Haraway describes an image of a mutating body which is merged with "scary new networks" (Haraway, 2016, pp. 28) which destroy "old hierarchical dominations" (Haraway, 2016, p. 28) and produce new forms of power and control which lead towards a world of a dangerous "deadly game" and blasphemy (Haraway, 2016, p. 28). Today, real cyborgs are walking among us. All individuals with artificial organs or limbs (e.g. a pace-maker), everyone who gets vaccinated to resist a disease or consumes medicine to improve their mood (psychopharmacology) is technically a cyborg (Gray et al., 1995, p. 3). The cyborg represents the fascination with artificial life. Weibel claims that artificial life can also refer to life with no natural substances, such as computer-simulated systems that grow and develop (Weibel acc. to Gržinić, 1988, p. 200). It is the creation of animate organisms using inanimate elements that can carry information to receive life.

4.3 THE PROCEDURES AND THE GOALS OF CYBORGIZATION OF MAN

The procedures of cyborgization of man are listed according to goals and methods used. All procedures lead to the creation of cyborgs.

1. Replacement procedures: Physical replacement of lost limbs and organs with partial or complete restoration of their functions (*homo* cyborg).
2. Normalization procedures: Use of implantation to establish functions human species-specific functions that the body is missing (*homo* cyborg).
3. Improvement procedures: Expanding the existing functions of the organism – creation of man of science and technology (*cyber/digital* cyborg).
4. Reshaping procedures: Adding organs or functions not characteristic for human species with the aim of creating a man of science and technology who will be able to persist in an otherwise unfavourable natural and historical environment (*cyber/digital* cyborg) (Gray et al., 1995, p. 3).

The cyborgization procedures can be perceived from two perspectives. On the one hand, the implants help the human body restore lost functions and normal functionality. On the other hand, they pave the way for the possibility of cyborgization at a particular stage, with the aim of creating enhanced, partially or completely artificial beings in which the dividing line between the biological and the artificial will be very slim. The replacement and normalization procedures fall under the heading of reconstructive cyborgization which implies replacement and restoration of lost or damaged bodily functions and organs. This type of implantation of artificial technology into a biological organism has allowed the injured people to regain their state of health (e.g. pacemakers, artificial hips, bionic limbs, artificial cochlea, etc.). Enhancement and reshaping procedures belong to the domain of enhancing cyborgization, the goal of which is to expand the existing and adding new functions that man does not intrinsically possess (e.g. chip implantations), which necessarily changes the biological structure of man with the aim of creating a man of science and technology with superhuman capabilities. The enhancing cyborgization procedures will inevitably lead to the creation of artificial beings and future robotic life forms – androids and robots. These could suppress the biological man in the current reality, due to needs of the already initiated cyborgization of the entire historical and natural reality of man.

4.3.1 Replacement Procedures

Cyborgization as a procedure of replacing limbs and organs that have been put out of function as a result of a disease or an accident and their complete or partial restoration represents a medical procedure that aims to restore the health of the patient.

1. Complete replacements of organs and limbs without restoring functions – these are the most primitive organ replacement procedures without restoring their functions (e.g. leg or eyeball prostheses which only simulate the lost part of the body but are not functional).
2. Partial replacement of organs and functions is performed using prostheses that can replace parts of bone structures. Replacement functions are partial and are achieved via mechanisms activated through the use of remaining muscle groups (e.g. ordinary leg or hand prostheses).
3. complete restoration of organs and functions is a procedure of restoring functions which requires the nerve and muscle structure related to the organ are intact (e.g. artificial heart valves, artificial eye lenses).
4. reconstruction of organs and function with interventions into the peripheral and autonomous nervous systems denotes all those procedures that are used to affect the nervous system and replace or upgrade muscular or neurological structure involved in the regulation and control of a particular organ (e.g. bionic prostheses, artificial hips, pace-makers).
5. reconstruction of organs and functions by acting upon the structures of the central nervous system is based on the use of implants that are embedded into a particular part of the central nervous system and which restore the lost function (e.g. air muscles used as aid with muscular dystrophy).

4.3.2 Normalization Procedures

Cyborgization as a normalization procedure refers to medical implantation which aims to restore the missing bodily functions that are characteristic or at least desirable for a human organism. This, of course, is a form of treatment for a disease-ridden organism. These procedures can be divided into two groups:

1. procedures for normalization of physical features – procedures used to return the patient's organism, which does not possess certain physical features characteristic for a human organism, into its normal state (e.g. vision restoration chips for persons blind from birth).
2. procedures for normalization of mental features – procedures used to remedy the inadequate or lost mental features and functions of patients lost due to a disease, such as emotions, memory, volition (e.g. electric brain stimulation).

4.3.3 Enhancement Procedures

Cyborgization as a procedure for enhancing a healthy man is a scientific experiment used to create an improved version of man. It is manifested through two types of procedures:

1. procedures for enhancement of physical features – these denote the creation of a cyborg man with artificial organs and limbs which would be more durable and resistant to disease and atrophy of bone, muscle, and other types of tissue. Chip implantation would expand the physical functions of particular organs (e.g. eyesight restoration chips could be used to extend the functionality of vision so that we are able to see all parts of the electromagnetic spectre – infrared or ultraviolet. This kind of chip would enable us to see at night, improve the resolution of our vision through zooming capabilities, etc.).
2. procedures for enhancement of mental features – these would allow us to expand the mental capacities of memory, improve perception etc. (e.g. memory chips).

4.3.4 Reshaping Procedures

Cyborgization as a procedure of scientific reshaping of a healthy human individual implies adding organs or functions not typical for human species, with the aim of allowing it to persist in unfavourable conditions. It is manifested in two different aspects:

1. procedures for reshaping the physical features – these denote adding those functions or traits which are not characteristic of human bodies. For instance, these include increasing physical strength using the exoskeletons; chip implantations that allow us to see the ultraviolet or the infrared parts of the spectrum; adding neural infrastructure to transmit new types of senses, e.g. ultrasonic signals etc. It remains an open question whether the human organism can accept and decipher these received signals, i.e. whether man can perceive the world around him using the senses that are not a part of his natural biological toolkit, whether he can remain a human being in that process or whether he becomes a non-human artificial being.
2. procedures for reshaping the mental features – these denote reshaping the mental features in such a way that man is given structures and functions that he does not possess in its original form. Through implantation of different kinds of chips into the central nervous system, cybernetics steps into the domain of the organism as a homeostatic system which it tries to empower in its battle against entropy and it also changes man's relationship towards the world, both sense-wise and thought-wise. A particularly notable aspect of change of mental capabilities of man is life in virtual space or the world of human mind apps in man's robot descendants.

All of these procedures, regardless of whether they are treatments or acts of enhancing or reshaping healthy persons, carry within themselves two messages and two real possibilities. On the one hand, they are life-saving solutions for the diseased. On the other hand, they serve to further deepen the inequality among human beings and create partially or completely artificial beings with superhuman capabilities.

In the remainder of this book, we will show that there already exist nanotechnical and biotechnical implants, genetic therapies and forms of AI which help the diseased overcome their health-related difficulties and bodily impairments. All these procedures and implants from prostheses, artificial ears, BCI interfaces, 3D implants, artificial organs and genetically-manipulated therapies, are overwhelmingly costly and therefore unavailable for most patients.

The standards that the medicine of the 20th century has established in terms of generally equal possibilities of primary healthcare for all people are slowly turning into an elitist project available only to the wealthiest. Through enhancements and reshaping of the human body with technical procedures and genetic manipulation, the aim of extending the lifespan of human body with the ultimate goal of eternal life in a computer interface thus becomes an ideological opium for the poor and a real foundation for the survival of the wealthy. With these assumptions in mind, it is necessary to uphold the Hippocratic Oath about the equality of all patients and the preservation of dignity of human life.

4.4 TYPES OF CYBORGS

Cyborg bodies can be analysed in terms of their cybernetic foundation and functionalities with regards to the type of cyborgized entity and the technical accessories used. Using this methodology, we have reached the typology of cyborgs in Figure 2 (Greguric, 2010):

1. ***animo* cyborg** (animal cyborgs)
2. ***homo* cyborg** (biomedical cyborgs)
3. ***cyber* cyborg** (cyber/digital cyborgs)
4. ***robo* cyborg** (roboticized cyborgs) [3]

Figure 2. Types of cyborgs

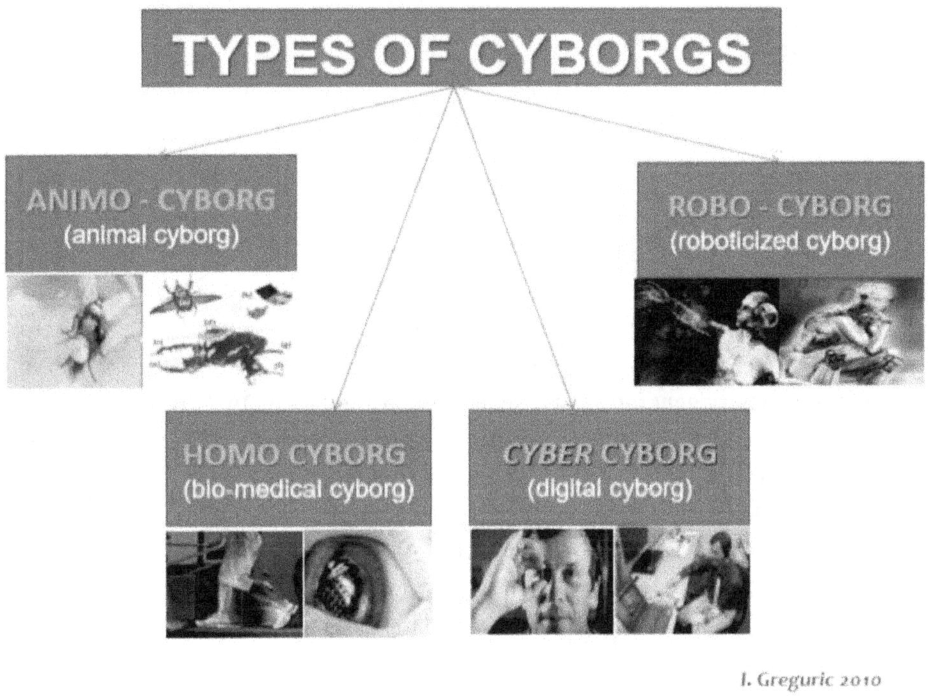

4.4.1 Animo Cyborg

The emergence of the *animo* cyborg is seemingly at odds with the philosophy of cybernetics, whose main goal is enhancement of the human organism for the purpose in the fight against entropy. But, as we shall see, cyborgization is also used in other practical purposes, which confirms our assumption that the essence of *scientific work* is not in the preservation of life of man as a mental being but that *scientific work* controls history and nature towards its own purposes.

The animo cyborg is an animal cyborg, a combination of an animal body and technology, remotely controlled using sophisticated implants and wireless network. It is believed that the first *animo* cyborg was a white lab rate made in the late 1950s as a part of the experimental program in the national hospital in Rockland, New York (Clynes and Kline, 1960, p. 30). The rat (Figure 3) was implanted with an osmotic pump which injected a finely controlled dose of chemicals, which caused different physiological changes in the rat's body and resulted in his behaviour (Clynes and Kline, 1960, p. 27, Clark, 2003, p. 15).

Figure 3. The first animo cyborg[4]

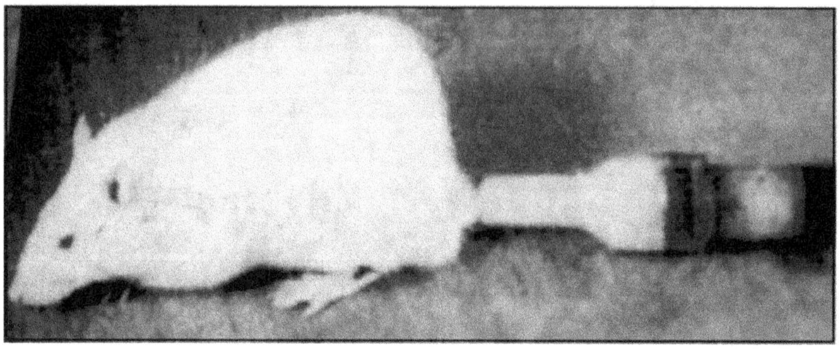

4.4.1.1 Cyborg Insects

Cyborg insects are a product of the military industrial complex and are used to collect strategically important information. They are created by growing an insect body around implanted computer parts. They are regarded as one of the most ambitious projects of the US *Department of Defense*.

The projects started in 2007 as a part of the *Defense Advanced Research Projects Agency* (DARPA) with the creation of the cyborg generation of insects. According to the DARPA project, the insects which have a modified body structure and micro-electromechanical systems (MEMS) built into them will be controlled via mechanical sensory activation or a similar mechanism (Page, 2007). Scientists are hoping that they will be able to harness the energy that insects emit during movement to power the internal MEMS. We will also mention other examples of cyborg insects which were implanted with electrodes, microphones and video cameras with the aim of performing intelligence-gathering tasks. According to Sato et al. (2009), these include: moth cyborgs, used in the fight against terrorism; cyborg bugs used to record monitor conversations; and cyborg crickets used in rescue missions (Figure 4):[5]

Moths are enhanced with computer chips implanted in the cocoon of the moth, which enables the control over their nervous system. The assumption is that it will be possible to have insects land in terrorist camps, transmit videos and other information to the US army using the system which its creators regard as reliable feedback between the tissue and the machine. The cyborg moth project has started in 2007 and it represents just one of the technical novelties that will be used for military purposes (CDC/JS, 2007).

The project *OrthopterNets* was started in 2009 and it deals with a network of bugs that send messages to each other by flapping their wings (Galić, 2009). Using minuscule biochemical sensors, they can sense people, nerve gases, explosives etc. Implanting electrodes into the brains and bodies of bugs to control their wing muscles has allowed transfer of messages over very large distances, which are then collected by the army. After the research phase, the project was experimentally used not only for military purposes but also for search and rescue purposes for injured and captured people in devasted buildings.

In 2009, a team of researchers from the *University of California*, Berkeley lead by Hirotaki Sato have created the first flying bug under complete wireless control via laptop (CDC/JS, 2009). This aim of this project was to create the first cyborg bug used to record conversations. The military agency in charge of research and development of new technologies has worked for two years on developing cybernetic organisms that contain both natural organic materials and artificial technological parts. During the cocoon stage, a bug was implanted with microbatteries, had its nervous system connected to the micro-battery, microcontroller with the transceiver (both a receiver and a transmitter). With the help of brain implants, the cyborg insect was issued commands to take off, fly, land and turn, which was performed using stimulated stronger flaps of left or right wing. The end goal of this research was to create cyborg insects which could fly 100 metres to their destination and remain there until they are issued the next command. During the experimental stage of creating this cyborg insect, three types of stag beetles from Cameroon were used. The smallest one was 2 cm long and the largest was 20 cm long (N.J.D., 2009).

N. Sharkley, an international expert for artificial intelligence, claims that there had been past attempts to control bugs like cockroaches. He doubts that this strand of research could yield any long-term success as bugs would have to be implanted with a GPS system and a camera, which is too cumbersome for a single bug. Berkeley researchers suggest that a cyborg insect might serve as a model for miniature air vehicles and as messengers in inaccessible areas. As we can notice, the US army was the first to create a cyborg generation of insects with the aim of creating an army of spies that no one could control but them (Singer, 2009). Flies, bees, bugs, rats and other animals are implanted with various spy devices and attempts are made to control them remotely. After several successful test on hamsters and sharks, cyborg insects have a particular advantage as other animals cannot pick up their scent. Leading researchers are convinced that spy insects will be virtually undetectable when they are completely equipped with technology as they will be physically identical to other insects.

4.4.1.2 Other Animal Cyborgs

Apart from cyborg insects, other cyborgized animals also exist. In the following examples of animal cyborgs, we can notice medically justified procedures as the aim of their cyborgization is not enhancement and/or military use as is the case with cyborg insects, but reconstructive cyborgization used to return animals into their state of health. Oscar (Fig. 5) was the first cat with bionic paws (Servant, 2016).

After an incident in which he lost his back legs, implants covered with a layer of hydroxyapatite (substance used to stimulate bone and skin growth over metal) were embedded directly into its bone. After four months, the can could evenly distribute his weight across all four legs and even run. It was a revolutionary case of implantation using the so-called ITAP technique, whose usage in humans is currently being tested. The development of this technology will forever change the future of orthopaedics as it allows implants in which bone and skin are fully grafted.

The second example of reconstructive cyborgization among animals is the Belgian shepherd Storm, which was given a bionic paw. After the amputation of its paw, which was attacked by an aggressive type of tumour, it was implanted with perfected prosthetic paws made of metal which enable the skin to be grafted with the metal, and this allowed the dog to walk again (Krmpotić, 2007). The third example of reconstructive cyborgization of animals is the first bionic goose in the world called Betty (Telegraph, 2009). After it was found with a broken leg, it was implanted with steel needles, bolts and screws which allowed her to move freely.

Figure 4. Cyborg insects

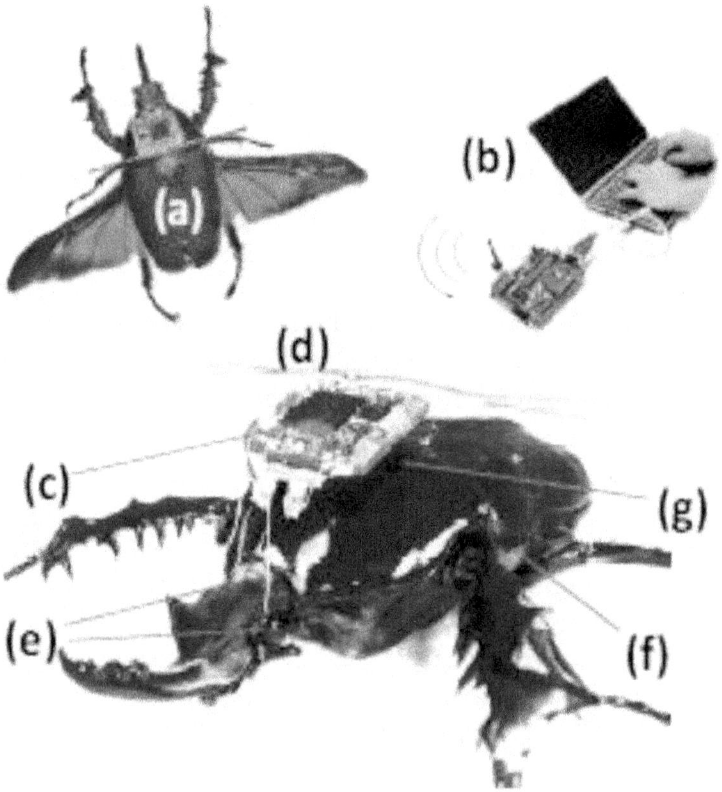

4.4.2 Homo Cyborg

Under this term, we shall analyse the biomedical cyborgs, i.e. the acts of implanting artificial technology into the human body with the aim of replacing lost organs or functions and removing the imperfections of the human body.

We should preface this by saying the bionics is a scientific domain of studying mechanical systems functioning as living organisms or parts thereof (Fig. 6). Bionics studies the regulatory (control) mechanisms in living systems, applies them in the electrotechnical systems and deals with constructing machines whose functions imitate the work of biological systems.[6] The term bionics was first used by Hans Ostreicher to denote the creation of new technical devices using the structural principles of biological organisms (Scott, 1987, p. 232).

Bionic prostheses, i.e. artificial replacements for parts of the body are becoming more and more sophisticated and truer to their models. Artificial limbs can nowadays be controlled at will as the preserved electrical signal from the muscle of the amputated limb with the in-build detectors stimulate the prostheses. Experimental exoskeletons are under development, which should enable the paraplegics to walk. Jarrold Pertovsky from the Wright State University, Dayton (Scott, 1987, p. 232) has conducted successful experiments in the field of re-establishing the functions of paralysed limbs using a microprocessor for artificial stimulation of motor nerves in limbs.

Figure 5. Animal cyborgs

By analysing the patterns of impulses and muscle activity which produce particular bodily movements, these data can be translated into a computer program, which can stimulate the paralyzed limb. In addition to an ever-growing and persistent need to help the disabled, this experiment wants to confirm the cybernetic assumption that there is absolutely no boundary between biological and mechanical systems.

We shall also list some positive examples which include the restoration of a lost bodily function to the injured persons. *Homo* cyborg (biomedical cyborg) is a man whose lost or damaged biological organs and functions are replaced or regenerated using bionic implants (Fig. 7).

Reconstructive cyborgization of humans leads to healthy and normal functions which the organism inherently possesses (EPFL, n. d.). As we shall see later, almost every single part of the body will soon be replaceable with a bionic implant, except for the artificial brain, which is expected in the near future (Paravić, 2010).

Figure 6. Bionics[7]

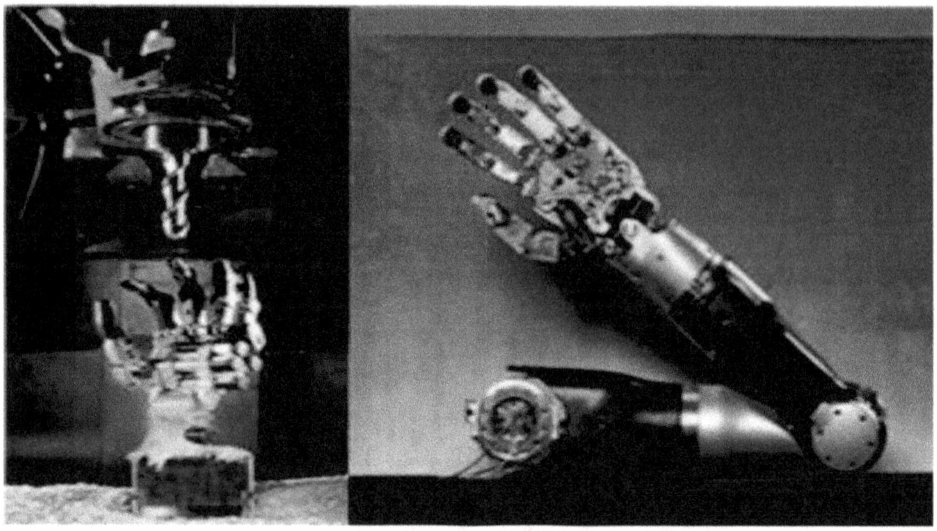

Figure 7. Homo cyborg (Fischman 2010)

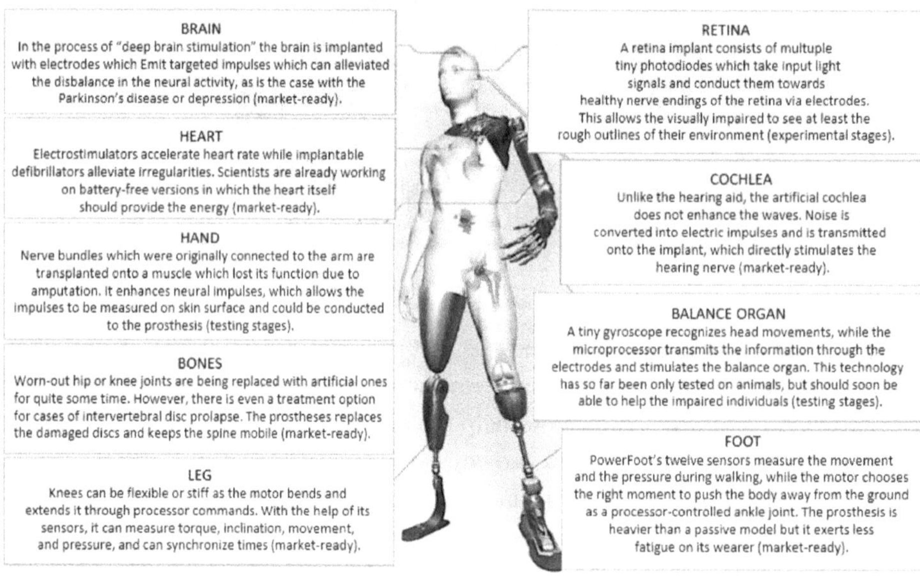

4.4.2.1 Bionic Arm

The bionic arm is a prosthesis which enhances the electronic signals in the muscles to move the electric motors, which in turn move the joints of the hand and fingers, thus enabling the sense of touch to be received and transmitted. The patient imagines a movement, and the muscle activity in the rest of the hand is deciphered by a computer and is transformed into actual movement. A prototype of the first bionic arm was transplanted in 2002 at the *Rehabilitation Institute of Chicago*.

During the press conference in Washington in 2006, the first bionic humans were presented to the global audience– Claudia Mitchell and Jesse Sullivan (Fig. 8), who both lost their hands in a traffic accident (Mucić, 2008). The surgeons cut the nerves that previously connected the hand muscles with the pectoral muscle, rerouting the signals that travel from the brain via the pectoral muscles.

By connecting the nerve endings with miniature electro-sensors, the patients' sense of touch was restored. Sensors were placed on parts of the pectoral muscle to read the brain activity transformed into a movement and convert it into electronic impulses for the processor in the mechanical arm. The patient operates the bionic arm by envisaging a hand movement and the muscle activity in the rest of the amputated arm is deciphered by a computer on its back and it causes an actual movement and even sensory stimulation (Rudež, 2006).

Figure 8. The first bionic humans[8]

The bionic arm is battery-operated and it has six engines which allow the fist to be opened and closed, and the elbow and the wrist to be bent (Fischman, 2010). While the natural biological arm can perform 22 different movements, the bionic arm so far only has four movements: shoulder rotation, elbow bending, fist twisting and fist opening/closing.

The father of the bionic arm is Todd Kuiken from the Rehabilitation Institute of Chicago (Kuikken, 2014), which is where the first bionic arms were merged with the amputated limbs of patients. Kuiken's idea was that nerves in the remaining part of the amputated limb could still transmit the signals from the brain and that the computer in the prosthesis could operate the electromotors for moving the limbs. The first step in realization of this idea was gathering the remainders of the main nerves from the patients' arms. But the problem emerged when the nerves were supposed to be connected with the computer cables. To amplify the nerve signals, he used the muscles that emitted electrical discharge during contraction, which were strong enough to be detected using skin electrodes. In a complex surgery, a surgeon rerouted these nerves from the motor cortex into the brain into different areas of the upper parts of the arm. Using this method, Kuiken developed a technique of targeted muscle reinervation, used to control the bionic limb and the remaining functional nerves of the amputated limb.[9]

4.4.2.2 Bionic Fingers

Bionic fingers represent a huge step forward in normalization of the locomotory system of the human body. With the help of micro-electric sensors, bionic fingers detect signals from the remaining muscles in the patient's hand, which enables them to move their fingers. They are covered with artificial skin that increasingly resembles human skin.[10]

The first bionic fingers *ProDigitis* (Figure 9[11]) were created by the scientists from West Lothian (Quick, 2008), the creators of the I-Limb hand (Fildes, 2009). Is made of light plastic, and is moved with the help of five motors located between the root and the joint of fingers and the sensors which receive the signals sent by the brain. An often-mentioned example of successful implementation is the case of a former concert pianist Maria A. Iglesias who lost her fingers due to septic shock. With the help of bionic fingers, Maria can once again play the piano and perform her everyday chores (Fildes, 2009). The scientists at the University of South Carolina have developed a *BioTac* sensor which imitates the tip of a human finger. The finger has a soft, flexible, vibration-sensitive skin, similar to human skin. The area beneath the skin is covered by a liquid which serves as a conductor, while its middle part contains the core for transfer of information. Apart from recognizing different materials with a 95% certainty, the finger also recognizes the direction and the temperature. So far, it has been tested on a robot, but the plan is to implant it onto patients to restore their sense of touch (SynTouch, n.d.).

Figure 9. Bionic fingers

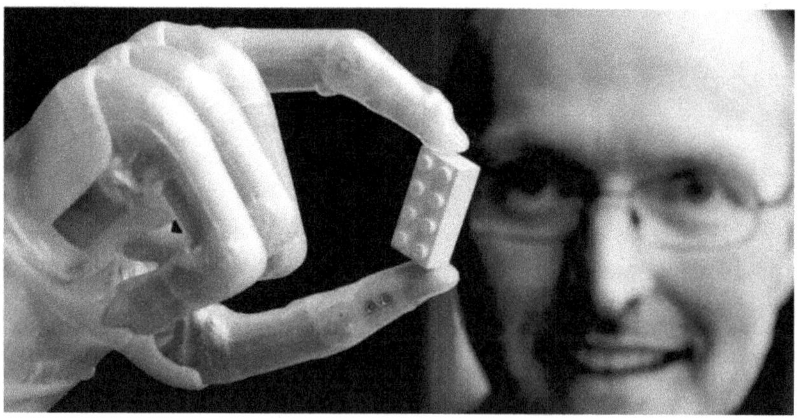

4.4.2.3 Bionic Legs

Bionic legs are constructed as electric limbs moved with the help of electric springs, which replace the lost biological legs. The successfulness of artificial legs is confirmed by several cases. The first example is that of colonel Greg Gadson (Fig. 10) who regained his former balance and full leg mobility with computerized *PowerKnees* electric knees. Motorized springs of the electric ankle function like real legs, save energy and reduce joint problems. US Army has developed bionic legs Bleex, which could be used by the infantry, firefighters, and emergency personnel in extreme missions which require movement over challenging terrains (Yang, 2004).

The most well-renowned example is the case of Oscar Pistorius, the South African athlete with artificial legs *Cheetah*. Due to congenital leg deformation, his legs were amputated during his childhood and replaced with prosthetic ones. The bionic prostheses are built in such a way that when his leg module comes into contact with the ground, the joints absorb 97% of the energy and store it to make the mobility as efficient as possible. The higher the speed, the more energy the joints store. The glutes and the thighs ensure the overall movement. Thanks to his artificial legs, Pistorius spends less energy on movement than a healthy individual.[12]

Figure 10. Bionic legs[13]

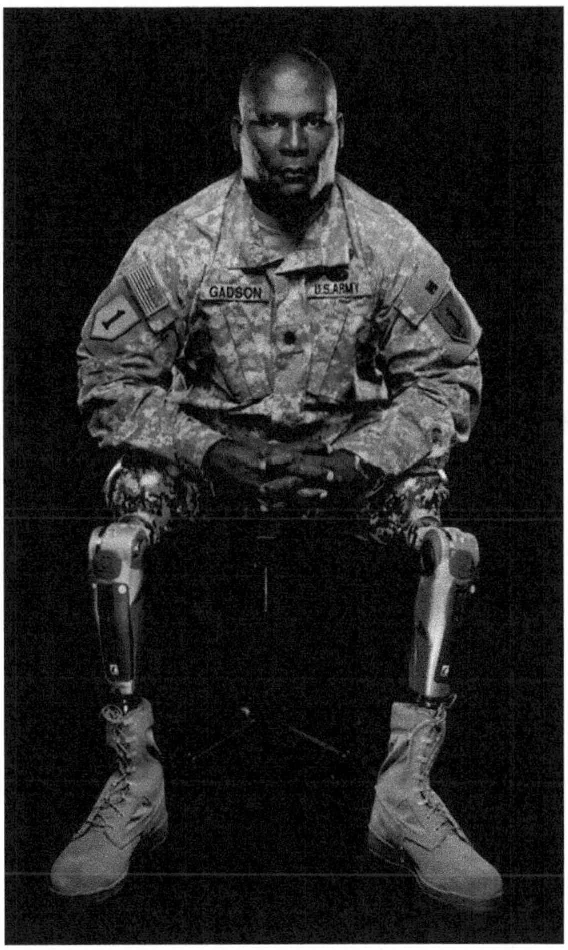

4.4.2.4 Bionic ear

The first bionic ear appeared in 1969. The bionic ear (a cochlear implant) consists of a digital microprocessor, which sends the signal into the brain and enables the deaf person to regain hearing (Fig. 11). An often-mentioned example of use is the case of Aiden Kenny, the boy who regained hearing with the help of a bionic ear. He was surgically implanted with two implants and 22 electrodes inside his ear, which collect sounds via microphone and transform them into signals so that the hearing nerves could recognize them. By bridging the non-functional parts of the ear, these implants transmit electronic signals (sounds) to hearing nerves. Several months after the surgery, Aiden was able to once again hear his own voice (Fischman, 2010).

Figure 11. Bionic ear[14]

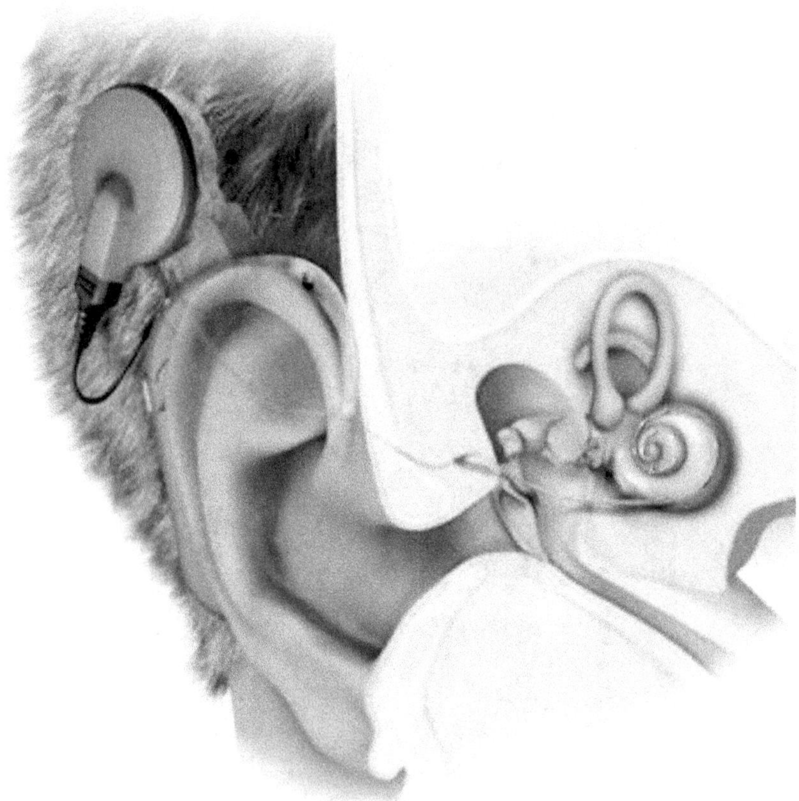

The scientists from the Princeton University have created a bionic ear which can hear the frequencies not available to the biological ear. This is a new approach to integration of biological and electronics materials. With the help of a 3D printer, nanoparticle cells were printed and a miniature spiral antenna with a cartilage was created. The recent attempts have faced the problem of interfacing a two-dimensional electronic component with the tissue. Project team leader Michael McAlpine claims that the goal of the project is not treatment via creation of replacement ears for the deaf and the persons without ears but to answer the question whether an average human being can be given a supernatural hearing power (Agencija VLM, 2013).

The artificial cochlea:

[…] has revolutionized the field of audiology, giving back the gift of hearing to the deaf. These artificial cochleas work by connecting electronic hardware with biological "wetware," that is, neurons. The cochlear implant has several components. A microphone is placed outside the ear. It receives sound waves, processes them, and transmits the signals by radio to the implant that is surgically placed inside the ear. The implant receives the radio messages and converts them into electrical currents that are sent down electrodes in the ear. The cochlea recognizes these electrical impulses and sends them on to the brain. These implants can use up to twenty-four electrodes and can process half a dozen frequencies, enough to recognize the human voice. Already, 150,000 people worldwide have had cochlear implants. (Kaku, 2011, p. 64)

4.4.2.5 Bionic Eye

Bionic eye represents a merger of an electronic device and an optical nerve of the human eye which enables a partial restoration of lost sight (Fig. 12). The functionality of this implant depends on a miniature chip with 60 electrodes which is surgically implanted behind the retina, on the back side of the eyeball. A damaged nerve is fixated using an ultrathin wire used to transmit the light and image into the brain where these are processed (Hanlon, 2007). The special glasses that the patients wear contain a tiny camera with a receiver which send images to the chip behind the retina. Stimulating different fields creates a contrastive image, which enables a person to recognize light and silhouettes. The implant requires no spare parts as the chip is implanted only once and lasts the whole lifetime while the battery is situated inside the mini-computer which has a rechargeable battery like a mobile phone.

Figure 12. Bionic eye (Fischman, 2010)

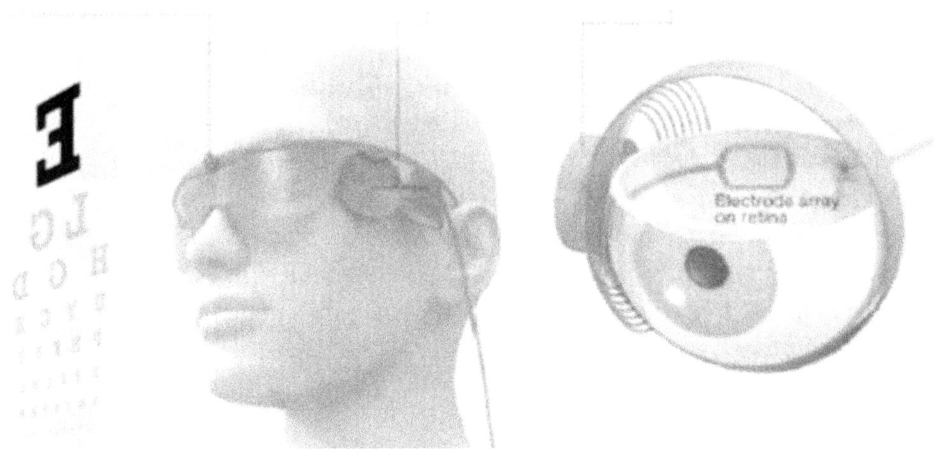

A very illustrative example is the case of Ann Lewis, a 79-year-old blind woman who regained partial sight after a surgical procedure. She was able to see the silhouettes of trees using a tiny camera which communicates with her vision nerve. The hardware built into the eyeball communicates with the computer and transmits the visual images towards her brain. The light signal from the environment is transformed into an electric one and is processed as a contrastive image through the visual nerve. This allows the completely blind persons to see objects and contours. British surgeon Lyndon da Cruz, one of the leading experts in the domain of development and implantation of bionic eyes, has conducted the first implantation procedures for blind patients in the Moorfields Eye Hospital in England. The operations were performed as a part of the *Argus II* project for restoring basic sight to patients that had gone blind due to macular degeneration or retinis pigmentosis (McKeegan, 2008).

Argus II bionic eye consists of a processor that the patient wears like a cell phone, a silicon chip built into the eye to behave like a retina, and glasses. The chip is connected to the video camera and a transmitter located in the glasses. The camera records the surrounding images that are translated into electronic signals and they are transferred to the chip via wireless signal. The bionic eye is efficient with patients who had normal eyesight as their brain knows how to process visual information. The visual nerve has to be at least partially functional or else the data will not be wire-processable. The researchers are trying to create a camera the size of a pea that could be built into the eyeball, replacing the natural tissue with artificial technology. Da Cruz believes that as much as 70% of blind people will be able to see regain their sight, probably in colour (Trifunović, 2009).

Several strands of researchers are searching for the methods to help the blind:

[…] by creating artificial vision, connecting a camera to the human brain. One method is to directly insert the silicon chip into the retina of the person and attach the chip to the retina's neurons. Another is to connect the chip to a special cable that is connected to the back of the skull, where the brain processes vision. These groups, for the first time in history, have been able to restore a degree of sight to the blind. Patients have been able to see up to 50 pixels lighting up before them. Eventually, scientists should be able to scale this up so that they can see thousands of pixels. (Kaku, 2011, p. 64)

By 2012,

[…] thirty patients have had artificial retinas with up to sixty electrodes. But the Department of Energy's Artificial Retina Project, based at the University of Southern California, is already planning a new system with more than 200 electrodes. A 1,000-electrode device is also being studied (but if too many electrodes are packed onto the chip, it could cause overheating of the retina). In this system, a miniature camera mounted on a blind person's eyeglasses takes pictures and sends them wirelessly to a microprocessor, worn on a belt, that relays the information to the chip placed directly on the retina. This chip sends tiny pulses directly into the retinal nerves that are still active, thereby bypassing defective retinal cells. (Kaku, 2011, p. 64)

4.4.2.6 Bionic Heart

During the 1950s[15], the era of successful implantation of pace-makers has begun. This device helps the patients with a dysfunctional heart rhythm by stimulating the regular heart rhythm via electric impulses. It consists of the body, pulse generator containing the battery and the electronic system components, and the electrodes that stretch from the pacemaker to the top of heart or some of its structures. This is where the electrode comes into contact with the heart tissue and sends its own electric impulses.

Artificial heart (Hevesi, 2012) *AbioCor* (Fig. 13) is made of titanium and plastic and it weighs only 900 grams. In 2001, *AbioCor* was implanted for the first time to its recipient Robert Tools (Grubin, 2007). The maximum life expectancy of the machine is 18 months, but this patient only lived for 151 more days. The company *Abiomed* is working on *AbioCore* II, a bionic heart which should last up to five years.[16]

Figure 13. Bionic heart (Guirguis, 2013).

4.4.2.7 Bionic Skin and Tissue

Artificial limbs could soon have skin sensitive to both touch and temperature. Carbon nanotubes the width of 1/10000 of human hair are the most effective conductors of heat and electricity currently known to man. The plan is to have them distributed in soft plastic skin to collect the sensory data via active endings of non-dead nerves in the body and differentiate the temperature from the pressure via sensors. Nanotechnology will also serve to create the ingredients for water-proof skin that will be laser-formed to look like real skin. E-leather is made of ultra-thin and flexible electronic material that could give superhuman powers to its users as in Fig. 14 (Someya, 2013). Around 10 years ago, researcher Takao Someya and his team started exploring the possibility of creating artificial skin for robots that would enable them to feel (Someya, 2013). Bioengineers at the *Harvard University* are working on creating artificial tissue. They have connected neurons, heart cells, muscle, and blood vessels with nano-wires and transistors. The tissue is one half living cells, one half electronics. The experiments have shown that living cells act normally and nano-wires and transistors acts as a network of sensors that connect them with the computer. The tissue is developed around a three-dimensional scaffolding made of collagen which makes up the connective tissue of animals. Apart from creating bionic tissue of rats, the scientists have managed to create a 1.5-centimetre bionic human blood vessel (Anthony, 2012).

Figure 14. Bionic skin

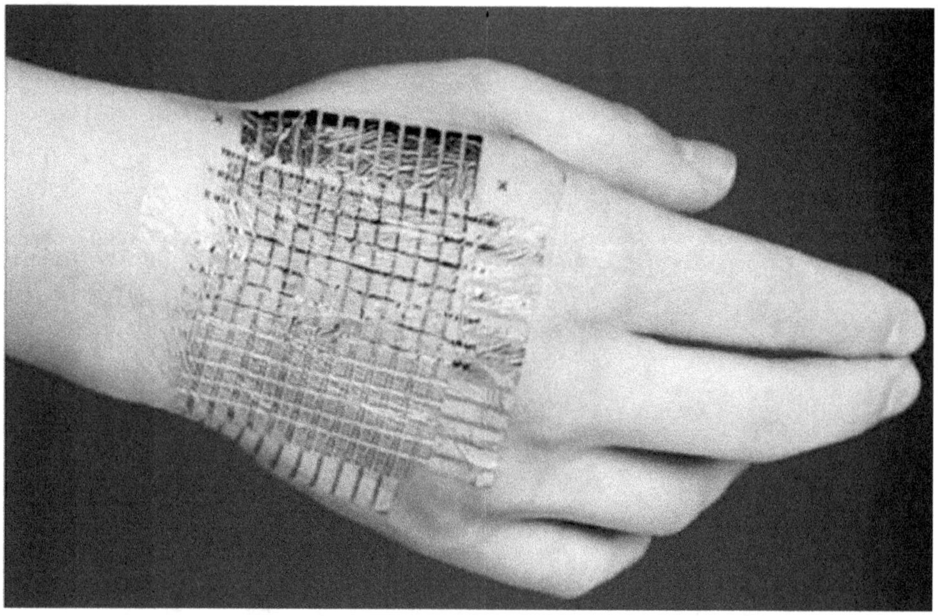

4.4.2.8 Bionic Tongue

In the nearby future, bionic tongue could also become available, according to the scientists at the *Medical University Luebeck*. The bionic tongue would be made out of muscles of the larynx and be connected with the source which emits the muscular signals similar to the pace-maker, and it has so far been tested on pigs (Rudež, 2008). A group of scientists headed by professor Shogo Minagi at the *University in Okayama*, Japan have developed the first artificial tongue prosthesis which enables the oral cancer patients to talk (Fig. 15).

4.4.2.9 Bionic Brain

Human brain is certainly the biggest challenge in the creation of bionic organs. During the last few years, several patients were given artificial brain implants. The most famous one is the *BrainGate*, a 2-square-milimetre chip embedded into the brain cortex. The scientists working on the project called the *Brain Gate* are trying to wire-connect the motor cortex of immobile patients directly with a computer so that they could lift remote objects with their mind. Researchers are also working on an artificial hippocampus, the part of the brain in which the memories are stored, so that it could be implanted into people who lost their memory.[17]

Figure 15. Bionic tongue (MYLE Team, 2016)

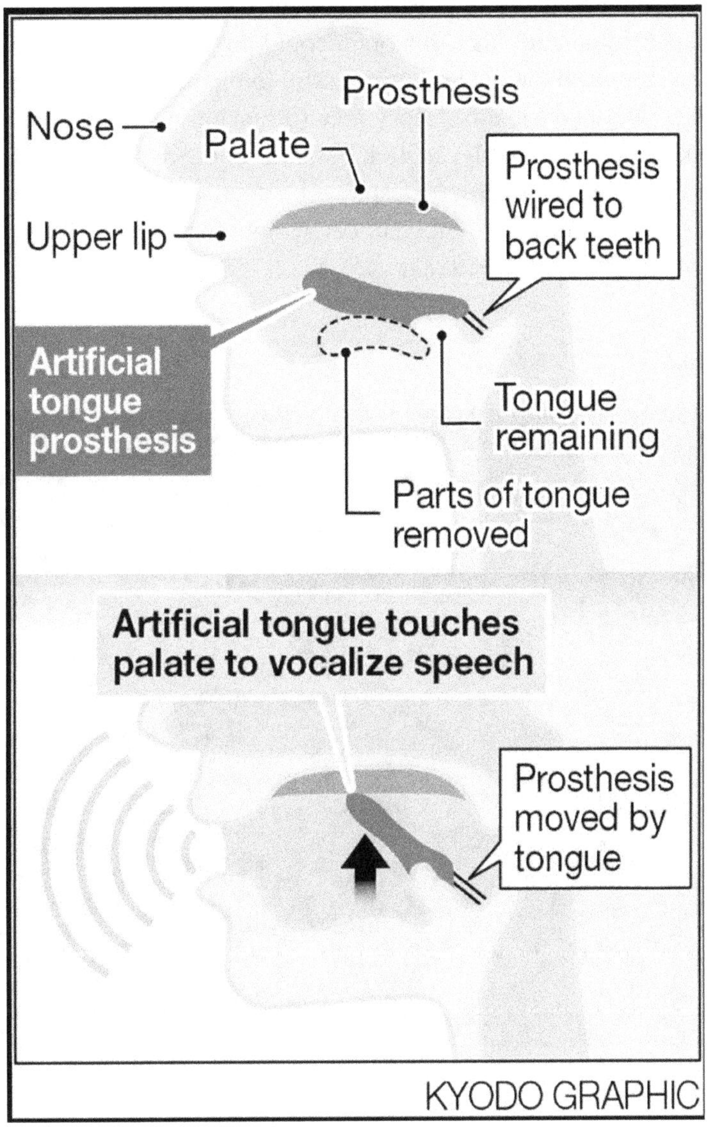

Henry Markram, the head of the *Blue Brain Project* (Fig. 16)[18] which was started in 2005 has announced that a complete and fully functional artificial human brain will be made in 2018. The project was initiated with the aim of investigating brain diseases and dysfunctionalities, finding a cure for mental diseases and uncovering the secrets of the brain.[19] The team of researchers that has been working on this revolutionary experiment for the past 15 years has focused on the neocortex as the most important part of the human brain. To describe the model of the brain and to discover the rules that underlie communication and the ways in which the parts of the brain are connected, members of the team have created a three-dimensional model on a super-computer. Markram explains that a single computer is required to conduct all calculations performed by a single neuron, which means that a machine with 10,000 processors is required for the entire workload (Fildes, 2009).

Scientists claim that the artificial brain will be made of silicone, copper and gold and will have the ability to develop thoughts independently of external input and will have the ability to feel. This raises concerns in the scientific community that this brain could do more harm than good, which opens up numerous ethical dilemmas. Will the potential successful implementation of this project live up to the vision of Mary Shelley who published her story about Frankenstein some hundred years ago on the very shores of the Lake Geneva where the artificial brain is being developed? The story ends with the tragic death of the scientist and his monster. Despite all potential of technology, we should learn from this example that knowledge comes with inherent danger and that striving for goals above the natural boundaries is a road to misery (Shelley, 1988, p. 45).

Figure 16. Bionic brain

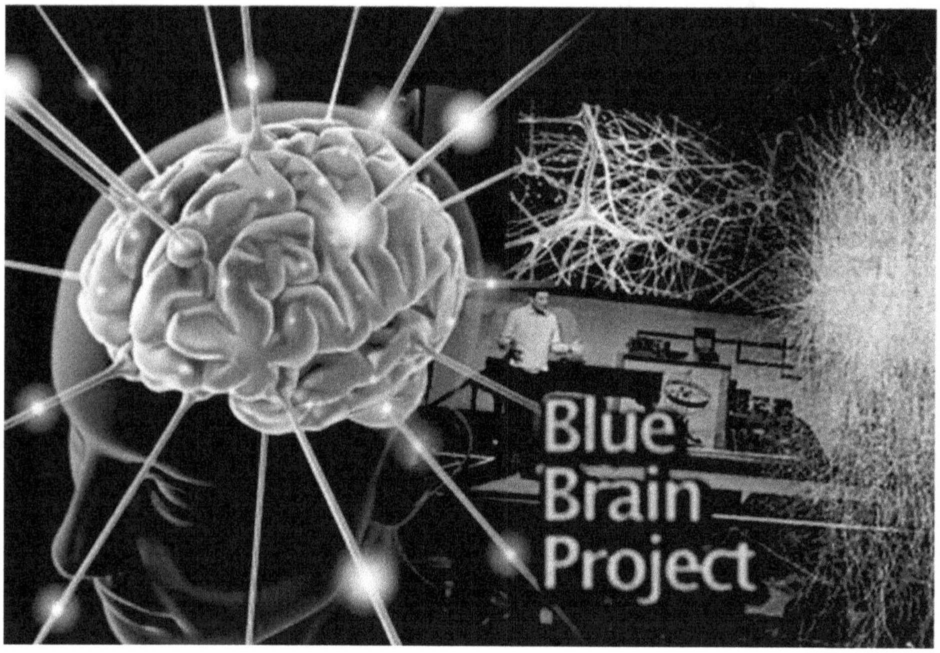

4.4.2.10 Bionic Clothes

Bionic HAL (*Hybrid Assistive Limb*) are bionic clothes, a moveable exoskeleton which helps the disabled by increasing their strength and replacing the injured limbs. The exoskeleton weighs over 23 kilograms and enables its user to move freely and easily lift objects ten times heavier than those that he/she could lift with his own muscles. It has motors equipped with 5-hour batteries functioning as its joints to increase its power. Skin-level sensors detect electric activity which the brain sends through the nervous system when making an order. Through the sensory neuron impulses, the exoskeleton receives the order to move and is moved a fraction of a second before the human muscles. There is also the possibility for the sensors which detect neuron signals on the user's skin to be adapted to track virtual signals and eventually protect the diseased person or call for someone to help him/her (Artemiadis, 2014).

HAL bionic suit is already available in Japan and its creator Yusjiyuki Sankai does not want HAL to be used for military purposes. His users are persons with movement difficulties, persons who suffered a heart attack or a stroke, and industry workers as in fig. 17 (Shelby, 2012).

Another example of this is the Walking Assist Device (Ikeuchi et al. 2009, pp. 4073–4079). This bionic device supports the entire weight of the body and makes walking and especially bending and crouching less tiring. It resembles a bicycle seat which supports the lower part of the body and facilitates movement. The legs of the device are attached to the shoes, which reduces the pressure on the muscles, bones and ligaments (Hanlon, 2013).

Figure 17. Bionic clothes

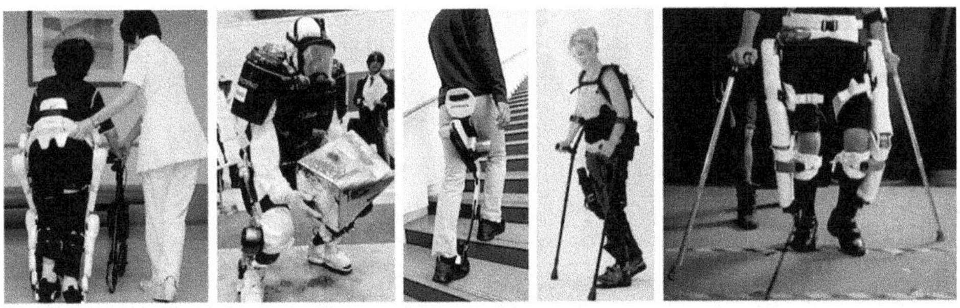

HAL in healthcare · HAL anti-radiation suit · Honda Walk Assist · Ekso Bionic · NASA's Exoskeleton

4.4.3 Cyber-Cyborg

A *homo cyborg* is a biomedical cyborg in which replacement and normalization procedures are used to return the disease-ridden body into the state equal to that of healthy persons. A *cyber cyborg* or a digital cyborg represents the enhancement of the human body via chips and other computer-connectable devices to increase human capabilities. As Harraway argues, "Modern machines are quintessentially microelectronic devices: they are everywhere and they are invisible" (Harraway, 2016, pp. 12-13).

Enhancing human beings is defined as "intervention that improves a condition that we view as a normal function or feature of members of our species" (Daniels, 2000, pp. 309-322). The definition of enhancement is also provided by *The President's Council on Bioethics*, which regards enhancements as acts of perfecting or extending particular traits, capabilities or activities of human beings.[20] Cyborg technology allows scientists to intervene into evolution via technical improvements and extension of human capabilities. Redesigning humans is the path towards the *empire of freedom – scientific humanism as naturalism* and it questions the essence, the existence and the very biological life of man. To answer whether implanting a chip into the brain is comparable to the aids that have been used so far, scientists say that the computer implant in the brain which grants access to a computer radically surpasses the level of functioning of the human brain, thus changing its biological function (Allhoff et al., 2009, pp. 8-9). It is expected that in the next 20 years the newer interface will not only have an extended dynamic range of senses but it will also increase memory and enable cyborg thought for invisible communication with others. As an advocate for transhumanistic enhancements of man, Savulescu argues that biology and medicine should be used to enhance human life, not only to prevent and treat diseases. The sense of enhancement implies securing a longer and better lifespan in comparison to today's understanding

(Savulescu, 2007, p. 517). Not only is it allowed to enhance human beings, it should also become a moral commitment (Savulescu, 2007, p. 528). To justify his standpoint, he provides the ethical basis for using biotechnology with this aim in several examples. According to him, one of the most widespread examples

The beauty industry is testimony to the attraction of enhancement. Body art, such as painting and tattooing, and body modification, such as piercing, have, since time began, represented ways in which humans have attempted to express their creativity, values, and symbolic attachments through changing their bodies. Modern professional sport is often said to be corrupted by widespread use of performance-enhancing drugs, such as human erythropoietin, anabolic steroids, and growth hormone. However, some effective performance enhancements are permitted in sport, such as the use of caffeine, glutamine, and creatine in diets, salbutamol, hypoxic air tents, and altitude training. Many people attempt to improve their cognitive powers through the use of nicotine, caffeine, and drugs like Ritalin and Modavigil. (Savulescu, 2007, pp. 517-518)

An even more radical example is Savulescu's advocation for using aborted or artificially created foetuses to prolong human life. It is assumed that the lifespan could be extended to 120 years through replacement of tissue and organs (Savulescu, 2007, p. 518). Thus, the transhumanist do not see any harm in unlimited use of biotechnology (Bostrom and Roache, 2007). As an opponent of enhancements, Fukuyama (2002, pp. 168-172) sees them as a drastic change in human nature and dignity.

The use of technology for enhancements and reshaping changes the functions and surpasses the capabilities of particular body parts, thus changing the living being as a whole entity. In addition to the division between the so-called natural and artificial enhancements, there is also a division according to the technology used, which has become increasingly applicable. These types of technologies come from four different scientific domains – Nano-, Bio-, Info- and Cogni-sciences (Spohrer 2002: 103). In addition, there is also the distinction between the natural and the artificial, i.e. internal and external technologies for human enhancements and several subcategories thereof (Spohrer, 2002, p. 103):

1. Body- and environment-external. This category encompasses new materials, new aids, different contexts of action (e.g. virtual reality), new subjects (e.g. Ray Kurzweil's virtual alter ego Ramona);
2. External to body and person with new mediators (tools and accessories) as a subcategory;
3. Body-internal – temporary technology which implies orally digestible enhancement methods (e.g. wisdom pills and existing pharmaceutical behaviour control drugs);
4. Body-internal – permanent technologies, which include new organs (new senses and nerve endings), new skills (new use of existing senses and nerve endings) (Spohrer, 2002, pp. 104-112).

4.4.3.1 The Brain-Computer Interface

The patient who is using an external brain-computer interface (Wolpaw, 2012) is not considered an enhanced cyber cyborg as the interface replaces lost functions instead of expanding the existing set. The open question remains whether the use of the brain-computer interface will be restricted to treatment methods. If this method were to be used on a healthy individual, then we would be facing an enhancement – a cyber cyborg.

The brain-computer interface (BCI) is a system through which the human brain communicates directly with an external device (Figure 18). It is used by paralyzed patients as it enables them to use their thoughts to communicate with a computer and manipulate the electronically controlled artificial limbs (Pandika, 2013). This external device consists of silicon chip circuits which are used to transmit or receive signals from and towards the brain, and these are then used as a medium of sensations and limb movement. The brain-computer interface has thus far been successful in repairing the damaged sight, movement and hearing, and it allows the surrounding objects to be controlled using thought, for instance turning the TV on and off, writing on a computer without using the keyboard, etc. The BCI system's success stems from the fact that the brain can adapt to the brain-computer interface and treats it as a natural part of the body. In this type of application, the BCI has an immeasurable positive impact on the state of affected individuals, but in cases where this device was used to enhance an individual's brain functions, the aim would not be treatment but enhancement.

Figure 18. Brain-computer interface (Guger et al., 2001)

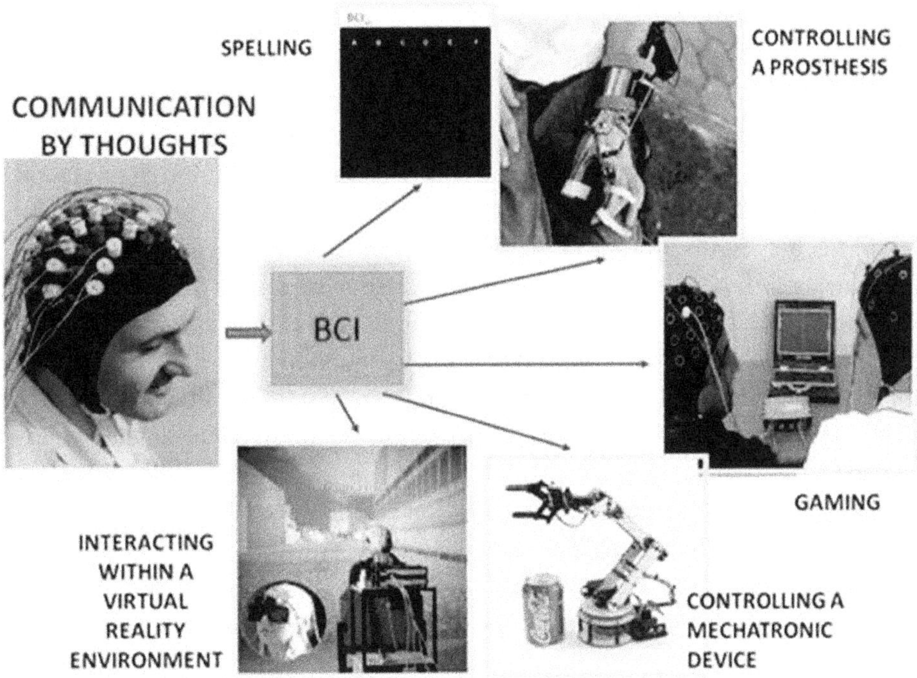

Roy E. Bakay and Phillip R. Kennedy, the neuroscientists at the Emory University in Atlanta have invented the Neurotrophic electrode which helps the paralyzed patients who lost the ability to speak due to spinal cord injury, fracture, or Lou Gehrig's disease. With the help of Neurotrophic, the BCI enables the brain to communicate with movement as it records the electric activity in the brain and translates it into orders, e.g. moving a cursors, moving the electric chair or turning on the appliances (Kennedy and Bakay, 1998, pp. 1707-1711). The clinical centre *Charité* in Berlin in cooperation with the *Fraunhofer Institute for Computer Architecture and Software Technology* (First) has developed a mental writing machine (Wolpaw, 2012). A cap with electrodes is mounted onto the patient's head and it receives the brain waves in the form of an electroencephalogram whose signals are used for control. Using the brain waves, the patient moves the cursor through the letters and they form a word. By focusing on images and words, the patient can choose which letter he/she wants to write (Vaughan et al., 2003, pp. 94-109).

BCI interfaces can be invasive, non-invasive and partially invasive (Anupama et al., 2012, pp. 740-742). Invasive BCIs are based on direct implantation of electrodes into the grey matter of the brain, which enables eyesight restoration via camera and movement of the limbs using brain activity simulation. Even though brain surgery is a high-risk procedure, these interfaces have the best results. Direct brain implants have been used since 1978 to treat congenital blindness. One of the first scientists to use the interface to eyesight restoration was W. Dobelle. The researchers at the Emory University in Atlanta used the invasive system in 1998 with patients who suffered from the Locked-in syndrome. In 2004, a paraplegic patient was able to send an e-mail and play video games using his thoughts thanks to a chip implanted into his brain. By implanting neurotrophic electrodes into the part of the brain where speech is generated it is possible to create sound using one's thoughts. When the patient is trying to speak, the electrode collects the brain signals and sends them via wireless link to a computer which transforms those signals into a sound. The whole process takes around 50 milliseconds, which is comparable to normal speech (Tutton, 2009). To treat the severe cases of depression, a device similar to a pace-maker is implanted into a patient's brain and it helps the patient via deep brain stimulation. Intervention is also possible with patients with severe brain damage who have minimal level of consciousness. After the pace-maker is implanted into the brain, the patient was able to communicate again: "The language of the brain is electricity. This is why we use electric signals to encourage the brain to use its own language" (Glas Amerike, 2007). This method is used to treat the patients with the Parkinson's disease and is regarded as one of the biggest discoveries in the field of depression treatment.

The crucial question is whether the BCIs will be limited to treatment of neurological states or they will be used to modify behaviour. What is the next step in this development? Will it be the enhancement of human capabilities to the extent that we will be able to communicate via thought? One of the future BCI programs is enhancement of human capabilities and mind control. In this way, the BCI would allow the patients with retina implants to recognize the contours in the infrared area or telepathic conversations between two persons. A Near-field communication (NFC) chip (1.3 mm in size) was needle-implanted onto the surface of the brain to collect the electric brain cell activity (Telegraph, 2009). The brain cell activity is recorded via the receiver signal that is implanted onto the skull. The advantage of this telepathic chip lies in the use of wireless technology which reduces the risk of potential infection and increases user mobility. In addition to being used to help the patients with speech impediments or those with dysfunctional nerve system motor skills, it is also believed that the chip could also be used for telepathic communication (Metro Portal, 2010). This would enable the enhancement of the healthy individuals as well, since this would allow them to control the lights, computers, and other devices using their thoughts.

Partially invasive BCIs are embedded into the skull but not within the brain. Impulse strength of this type of device is lower, but so is the risk of scar tissue formation in the brain. These devices use the electrocorticography (ECoG) technology which measures the electric activity of the brain. The ECoG technology was first used by E. Leuthardt and D. Moran from the *Washington University* in St. Louis in 2004.

Non-invasive BCIs are based on EEG recorded at the surface of the head. Their advantage stems from the fact that the patients are not exposed to the risk of brain surgery, but they have limited communication channel capacity. Unlike the invasive systems, the non-invasive ones have the lowest clarity of signal when it comes to brain communication because the skull deforms the electromagnetic wave signal. This device type has been successful in treating paralyzed patients as it moves implanted muscles and develops communication with the outside world. One of the most popular devices from this category is the EEG or electroencephalograph. The experiments with the technology that reads the signals of the human brain have begun with rats during the 1990s. Nicolelis and his colleagues developed the brain-computer interface which deciphers the brain activity of night monkeys and uses the device to reproduce the movement of the monkeys in the robotic arm. By year 2000, the device was able to detect brain signals of night monkeys when controlling the robot arms or reaching for food.

The brain-computer interface operates in real-time and can be controlled via the Internet. Its manufacturer *NeuroSky* has developed a prototype of an interface that directly connects the brain of a man with a computer. The interface collects the analogue electric brain waves that are converted into digital signals and sent into the controls of gaming consoles, computers or medical devices. Their goal is to find an effective method for brain treatment. BCI systems will be further developed for enhancing the capabilities in the virtual world online. Florin Popescu and his team of scientists have developed a procedure that allows brain waves and eye movements to be tracked in order to control robots. Within the *Myoplant* project, the *Fraunhofer Institute for Biomedical Technology* and its partners are developing the smallest type of electrodes. When these electrodes are implanted into a severed arm, a prosthetic arm can be controlled just like a real arm and the patient can even feel the touch. The biggest obstacle for the BCI technology is the lack of modality sensor which would enable a sure and accurate access to brain signals. Use of these sensors should expand the range of communicative functions that can be provided by the BCI system. The scientists are also developing a telepathy chip. In 2016, the *Defense Advanced Research Projects Agency* (DARPA) announced that it is developing a "mind reading device" within the *Neural Engineering Systems Design* (NESD) program. This human brain implant could enable the computer to have a direct communication with the brain and would enable a faster exchange of information. So far, all attempts to achieve this were based on the interpretation of electric brain signals, but the main problem of binary computers is that they only recognize a part of the logical state (1s and 0s), while the neurotransmitters (which activate neurons which transmit information via electric signals) enable the transmission of more complex states and interactions. The goal of this project is to process these pieces of information with over a million neurons at the same time and to develop a feedback system which would stimulate at least 100,000 brain neurons connected with the sight, hearing, touch, pressure and movement senses. Apart from the military application, this technology could be used to restore eyesight to the blind as the information could be delivered into the brain via implants (Jutarnji list, 2016). Jens Clausen from the *Institute of Ethic and History of Medicine* believes that the BCI interfaces "pose ethical challenges, but these are conceptually similar to those that bioethicists have addressed for other realms of therapy" (Clausen, 2009, p. 1081). In fact, he argues that the bioethics is well-prepared for solving the issues that may arise with the use of BCI technology. Expectations of effectiveness and value of the BCI technology plays a major role in analysing the ethical questions. When using this technology, several elements have to be taken into consideration: providing timely notification for the patient about the potential risks of the procedure, the side-effects and the effectiveness of therapy; respecting the concept of informed consent of patients who are unable to make decisions on their own[21]; defining the personal responsibility of physician/researcher (e.g. who is responsible for improper implementation of a neruoprosthesis), defining the future development of the BCI technologies – reading one's thoughts, mind and behaviour control, selective enhancement and social stratification. For every suggestion for further BCI research, it is crucial to measure the possible dangers and difficulties against the expected individual wellbeing and the overall benefit for the society.[22]

4.4.3.2 Kevin Warwick's Cybernetic Projects[23]

Kevin Warwick from the *University of Reading* in Great Britain earned the title of the first enhanced *cyber* cyborg due to his experiments with nervous system implants, RFID chips (Fig. 19) and a robotic arm (Michael Journal, 2013). For Warwick, a cyborg is a man-machine whose capabilities can be expanded beyond the traditional boundaries (Warwick, 2004, p. 61). This type of *cyber* cyborg includes enhancement of a healthy individual as

Implant technology (technology implanted directly into the human body) opens up all sorts of possibilities in terms of enhancing people's capabilities. Technically, we are in a position where we are just beginning to get grips with the fact that electro-chemical signals on the human nervous system can be transmitted to and received from a computer, via a fairly straightforward implant in a person's body. (Warwick 199: 286-287)

Like Hans Moravec, Warwick believes people should be united with the technology to become "machines themselves" (Warwick, 2004, p. 70). This is the main idea of the Cyborg 1.0 experiment by Warwick and his team at the *Department of Cybernetics*. In 1988, Warwick was implanted with an RFID (radio-frequency identification) silicone transponder chip in his forearm, which served as a receiver and a transmitter. "The chip, housed in a 23 mm long glass cylinder, was inserted under local anaesthetic" (Warwick, 1999, p. 281). The chip posed a health hazard for Warwick:

Attempts to sterilize the chip were unsatisfactory; he tried using boiling water but the chip exploded, and acid also did not work. Eventually, he heated the chip to 80 degrees in an oven. He had to be on antibiotics and there was always the possibility that it could have smashed and 'there would have been glass and wire all over the place'. (Warwick, 1999, p. 281)

The implantation of the chip allowed Warwick to communicate with a computer. A network antenna transmitted the radio waves to the computer programmed to react to Warwick's movements, recognizing his location within the building in which the experiment was being conducted (Fig 19) (Warwick, 2003).

Antennas allowed him to control the doors, lights, heater and other computers via transmitter without lifting a finger. With this experiment, Warwick tested the boundaries of the human body for accepting implants and the ability to transfer information from the implant, i.e. chip signal, to nearby objects connected to a computer. Warwick wanted to see how the implant would change his reaction both when being a cyborg with an implant and a human being when the implant was removed (Warwick, 2004, p. 175).

The following experiment conducted in 2002 involved a new implanted chip which would transmit the signals between the nervous system and the computer. In this *Cyborg 2.0* project, Warwick's nervous system was implanted with a neural interface that gained access to his nerve signals. The neural chip controlled the robotic arm that mimicked the movement of Warwick's hand by catching the nerve signals in his own arm via computer. Warwick (2003) managed to establish a remote electronic connection with a robotic arm via the internet (Fig. 20). The robotic arm experiment established a two-way electronic link between Warwick's nervous system in New York and Reading, which confirmed that:

Figure 19. RFID chip (Warwick, 2003)

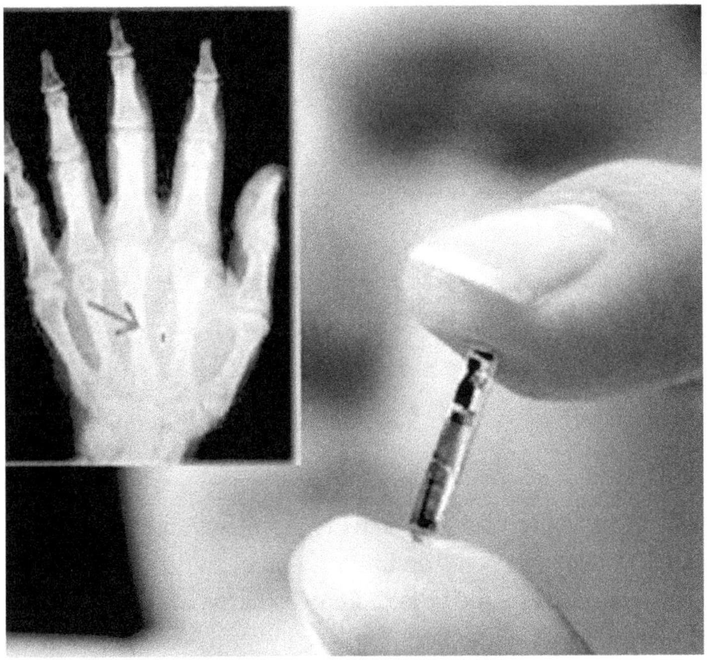

[...] signals from my brain could be transmitted around the world, via the internet, to operate a piece of technology. It really was a super-human power, meaning that, as a cyborg, your physical powers controlled directly from your brain are not restricted to your immediate body's capabilities. In essence, your cyborg body extends as far as you have an electronic connection. With the internet this means that your body extends, as a network, around the world. (Warwick, 2004, p. 258)

In a subsequent experiment, Warwick used a computer to connect his nervous system with his wife's nervous system and managed to activate a movement via the internet:

I felt a much stronger link as if we were coupled together in the same way as Siamese twins – there was a link there but we were two separate beings. When it was taken out, he says, he felt like he was losing a friend" (Warwick, 1999, p. 283). The experiment established the first electronic communication between the nervous systems of two people, i.e. pain, physical emotion and sexual arousal (Warwick, 2004, p. 285). This allowed for the brain signals to be transmitted "directly from brain to brain and bring about some form of thought communication by way of brain implants. (Warwick, 2004, p. 282)

Warwick's experiment indicates the possibility of brain communication between individuals. Once a human brain is connected to a computer and linked into a network with other connected individuals, there will be no isolated individuals; we will be a society of [enhanced cyber-, not: I.G:] cyborgs (Infonomia, 2011).

Figure 20. Kevin Warwick's nervous system live on the Internet (Warwick, 2003).

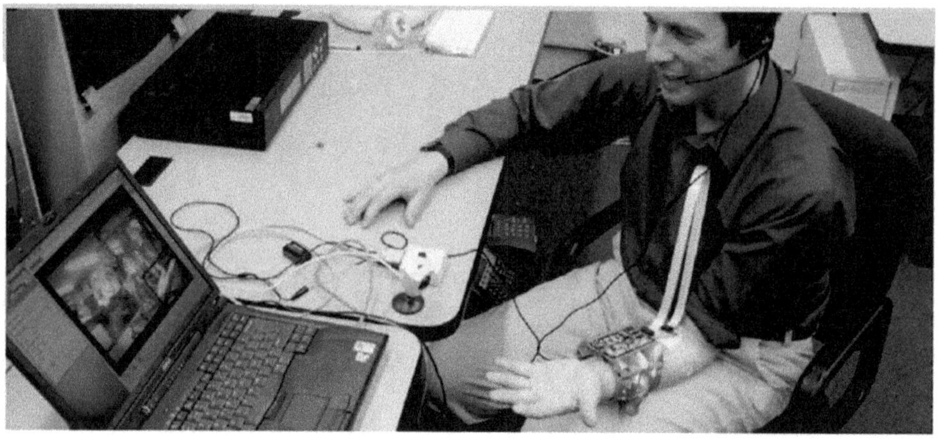

Warwick was unable to get a permission for his experiments due to ethical and legal limitations, which is why he conducted the experiments on himself and his wife. But, he emphasizes that there are numerous other technical issues that should be explored: "What would be the effects if different electrical currents were played down on to my nervous system?[…] would we discover any signals or feeling that we would not be able to understand […]?" (Warwick, 2004, p. 175). Warwick wants to "harness human thought to control computers – to make a computer able to read his mind […] and develop computers able to download information into a human brain – […] telepathy – but for real" (Warwick, 1999, p. 282). According to Warwick, his wife Irene claimed: "if people were going to be able to start to read other people's thoughts, she wanted to be the first to read mine" (Warwick, 1999, p. 282). For Warwick (2004), this also raises the question another question: "[…] would I be able to move her muscles simply by moving my muscles, and vice versa? If so, this could open up possibilities for remote-controlling patients with a nervous-system problem" (Warwick, 2004, p. 138).

Warwick has also conducted other experiments the human nervous system:

1. ultrasonic extrasensory input;
2. controlling robots via human nervous system;
3. remote controlling nerves via internet;
4. interacting with the network in a cybernetic building;
5. mimicking the arm movement with the 'third' artificial arm;
6. interactive jewellery.

The accomplishments of his experiments have opened up rather worrisome possibilities:

In the future it would mean that by connections to the nervous system and ultimately the brain, technology could be operated and controlled via the internet from just about anywhere in the world, merely by thinking about it. Not only that, but it would be possible to control a person's movements and actions at a remote location, by selecting and sending signals across the internet from a computer. (Warwick, 2004, p. 260).

This is why Warwick aims to examine the "signals loosely associated with physical emotions, particularly anger, shock and excitement. An overall aim was to record signals from my nervous system when I was happy; then, if my mood changed a day or so later, play the happy signal back down on my nervous system…" (Warwick, 2004, p. 133). This allowed for the feelings to be recorded, such as the pain caused by a clear electric signal on his way from a part of the body to the brain, or the feeling of happiness that could be reloaded when we are not feeling happy. Human brain correlates the feeling of pain with a particular part of our body and by modifying the stored signal, we could attempt to relocate the feeling into another part of the body: "Because of the way electronic signals can be used to change the electro-chemical balance of the human brain, it is possible that numerous medical problems could be tackled in this way. […] Will it be possible to remove a headache, not by taking an aspirin, but by inserting an electronic signal?" (Warwick, 2004, pp. 294-295). The possibility of sending electronic signals into the nervous system would have an effect of changing the mind of a treated person. Warwick sees this recreation of original feelings in the positive application of the e-medicine that would be able to treat the mental problems of the patients (Warwick, 2004, p. 134).

Warwick's standpoint is simple – human beings have very limited senses. To be precise, they have five of them, each of which is meaningless in and of itself. What dogs, cats and bats have in common is that they surpass human sensory capabilities. But, why should man be happy with the genetic equivalent, why not use the technology for extending the senses and gaining supersonic hearing or bionic vision: "With the human collection of senses so limited, our brains do not ordinarily process signals such as X-rays, ultraviolet, infrared or ultrasound" (Warwick, 2004, p. 132). The question is "Would I feel the ultrasound? Would my brain cope with it?" (Warwick, 2004, p. 132). Warwick believes that with the aid of "direct brain implants, it will be possible to upgrade the intellectual capabilities of humans by turning them into cyborgs. Extra memory, clip-on maths abilities and multi-dimensional processing […] Not just ultrasonic, but infrared, X-ray and more. However, thought communication is the biggest – yet most likely – step of all. It is certainly the direction in which I want my own research to head in the future" (Warwick, 2004, pp. 293-294).

When asked what would happen to people when they merged with the machines, Warwick responds believes that machines might become more important than other humans – those that become cyborgs will be one step ahead of others and just like humans have regarded themselves as the most valuable form of life, the cyborgs will probably look down upon humans who had not evolved (Infonomia, 2011). Can humanity reach a consensus to develop the boundaries of development or will those decision be made autonomously by the *scientific work* as the being of history and the interest groups that control the scientific and technological development and globalization?

4.4.3.3 Warwick's Vision of Cyborgs in Year 2050

In his book *I, Cyborg*, Warwick writes a chapter which he believes should not be read before January 1st, 2050. In it, he argues that by 2050, Earth will be completely ruled by (enhanced) cyborgs, technology-augmented human beings. Their development will not be marked by physical enhancements, such as bionic arms and legs, but by mental enhancements. The cyborgs should be able to communicate with one another using the signals of thought and their intellectual capacities: "Once connected in this way to a computer, via the Internet, communication could take place between people carrying implants, by means of thought processes alone. No more need for telephones or faxes, we will simply be able to think to each other" (Warwick, 1999, p. 287). The brains will be connected via radio to the global computer network controlling the entire system: "Linking human thought to computers could change the whole concept of language by making speech no longer necessary. People could eventually just think what they wanted to say and transmit it straight into someone else's mind. Vast information databases, like the Internet, could be plugged directly into the human brain." (Warwick, 1999, p. 282). All communication would take place over the network, from brain to brain; the use of telephones, television and other media would be superfluous and unnecessary as it would be possible to download information into an individual human brain. Through the use of electronic signals, it will be possible to completely remove the pain from our body as well as to download certain information and even achieve virtual sexual intercourse (Warwick, 2004, p. 302). Increased popularization and social acceptance of brain implants would result in new usage of mobile phones that could be embedded inside the head.

The potential benefits of the implant technology include:

1. Re-education of the brain and spinal cord-precise and varied stimulation patterns can be applied to central and peripheral nervous tissue; 2. Prevention of spinal deformity; 3. Treatment of intractable neurogenic and other pain; 4. Assisting bladder emptying; 5. Improving bowel function; 6. Treatment of spasticity; 7. Improvement of respiratory function – assisting coughing and breathing; 8. Reduction of cardiovascular maleffects; 9. Prevention of pressure sores – for example by providing sensory feedback from denervated areas; 10. Improvement and restoration of sexual function; 11. Improved mobility; 12. Improved capability in transfers and activities of daily living, especially through improved hand, upper limb and truncal control. (Warwick, 2004, p. 306-307)

After humans harness the powers of thought communication, Warwick believes that speech as a means of communication will die out. Its only use would lie in cases of potential network issues. The advantages of this system, apart from telepathic communication, would be crime prevention. If someone wanted to commit a crime, the network would detect this intention and could prevent the action (Warwick, 2004, p. 300). However, there is also a negative aspect of the use of this network, interaction and control over every nervous system. This machine Big Brother-like level of control would certainly abolish any form of privacy (Warwick, 2004, p. 302).

As a result of super-intelligent brains, enhanced cyborgs would have the ability to think in an infinite array of dimensions, completely discovering a new theory of the Universe and new forms of energy conversion. Better understanding of the nature of light and discovering the possibilities of traveling faster than the speed of light would enable us to visit distant planets and galaxies. A *Slimplant* chip would allow us to control the metabolism and would give diet instructions to every individual, depending on their bodily needs. Every *Slimplant* would be connected to the network, which would enable the download of genetic instructions in real time (Warwick, 2004, p. 301). The implants could also allow the control of numerous bodily functions in real team, such as heart rate, blood pressure and body temperature. By 2030, the implant records would be connected into the *Genetwork*, a trademark medical network projected for interactive stimulation and body control, which would expand the lifespan. The enhanced cyborgs that would be connected via implants to *Genetwork* could expand their lifespan up to 30 years (Warwick, 2004, p. 301).

According to Warwick, there are numerous positive reasons for using implantation technology. He believes that with the help of this technology we will be able to create a much stronger connection between a computer and the human nervous system:

So, we should be capable, before long, of sending signals to and from the human body in order to operate and interact with computers without the need for a keyboard or mouse. In way various emotional signals can also be transmitted. [...] In the future we could have memories of events that we have not witnessed, and mathematical abilities that far surpass anything of today. (Warwick, 1999, p. 287)

Despite this huge optimism, Warwick is aware of the fact that:

A human brain is a stand-alone entity, guaranteeing a unique human identity. But link a human brain via the Internet to other brains, both human and machine, and what of the individual then? (Warwick, 1999, p. 287)

Warwick believes that cyborgs will be given morals, values and ethics appropriate for cyborg life. Their intelligence would be part human, part machine-like. The machine side would include collective intelligence shared with other cyborgs. The machines will be able to learn on their own and Warwick has "already created in his laboratory machines capable of finding an electricity socket and plugging themselves in to recharge" (Warwick, 1999, p. 283). Machines with learning capabilities could become better than humans in every aspect and take over the world:

The human race [...] which has dominated this planet for perhaps 10,000 years, will see its domination come to an end. We will have created a being that will displace us. The human race will be as outmoded as the dinosaurs. (Warwick, 1999, p. 283)

What else to say about Warwick's transhumanistic ethics of cyborgs? In our understanding of enhancement and reshaping of man, cyborg as the scientific worker lives in the age of *scientific humanism as naturalism* that is established and organized according to the principles of science and technology and needs no ethics as a corrective factor. We could say that his values will be his program. If in the age of *scientific humanism as naturalism* we are dealing with ethics in general, then the ideological science stems from the self-movement of the *scientific work*, and it enchants people with its promises, luring them further into the mechanism of its self-reproduction. This ethics paralyses any possible moral activity that would lead in the same direction as life and man.

4.4.3.4 Improved Cyborg in Arts – Stelarc

Australian cybernetic performance artist Stelios Arcadiou, also known as Stelarc, with numerous enhancements of his body does not fall exclusively to the domain of cyber cyborgs, as he uses both computer-connected and computer-independent implants. When it comes to enhancements and expanding the human capabilities, he is described in the chapter on *cyber* cyborgs because he uses sophisticated medical instruments, prostheses, robotics, virtual reality and the internet to create his art (K. D., 2009). In his cybernetic performances, he tries to surpass the limitations of the human body through the experience of human-nonhuman symbiosis of man and the machine. Stelarc experiments with the new life of a transhuman body. The artist himself believes our bodies our outdated as we are reaching the end of traditional human philosophy and physiology.

Stelarc uses several methods for connecting his body with the artificial technology that we divide into five different approaches to increasing bodily capabilities:

1. technology connected to the body (third arm);
2. technology built into the body (stomach sculpture);
3. technology expanding body externally (exoskeleton);
4. extension of body into the network via technology (virtual arm);
5. connecting the body onto the network via technology (movatar, ping body, parasite).

In addition to giving his biological body non-existent organs, these interfaces and devices increase and expand the human form and functions. His most renowned works are: *third ear, third hand, virtual arm, stomach sculpture, exoskeleton, movatar, seaside suspension*, etc.

His performances *Seaside suspension* (1981) and *City suspension* (1985) depict different technologies of bending, piercing and opening the body as the flesh with the aim of adapting the body for future transhuman existence. In order to achieve that, Stelarc connects his body with the digital technology, pneumatic prosthetics and virtual reality.

Third hand (1982) is an artificial mechanical arm attached to Stelarc's right arm (Stelarc, 2021). The hand has the capability of independent movement and is activated via muscle contraction in his abdomen and legs, which allows his real arms to be moved independently. After several years of use, Stelarc can move his third arm intuitively without any need for focusing his muscles on moving it, as shown in Fig. 21.

Figure 21. Stelarc: Third hand (Stelarc, 2021)

Movatar (2010) is the concept of a virtual body or an avatar imbued with artificial intelligence which makes it partially autonomous and operative (Stelarc, 2021). By taking over Stelarc's physical avatar body via internet-connected device, he can act in the real world. When Stelarc puts on the device and logs into the avatar, he becomes a host for the intelligent virtual entity that expresses the movements of avatar. The avatar takes over the body and acts in the physical world.

Movatar uses the *Motion capture* recording technology used in the optical *Motion Capturing:* reflection and pulsating LED diodes. Cameras record the movement of the marker on Stelarc's body that moves the body. Systems with one or two cameras are used to record face movements, the movement of the entire body requires from 4 to 32 cameras, as seen in Fig. 22.

Figure 22. Stelarc: Movatar (Zylinska, 2002, Stelarc 2021)

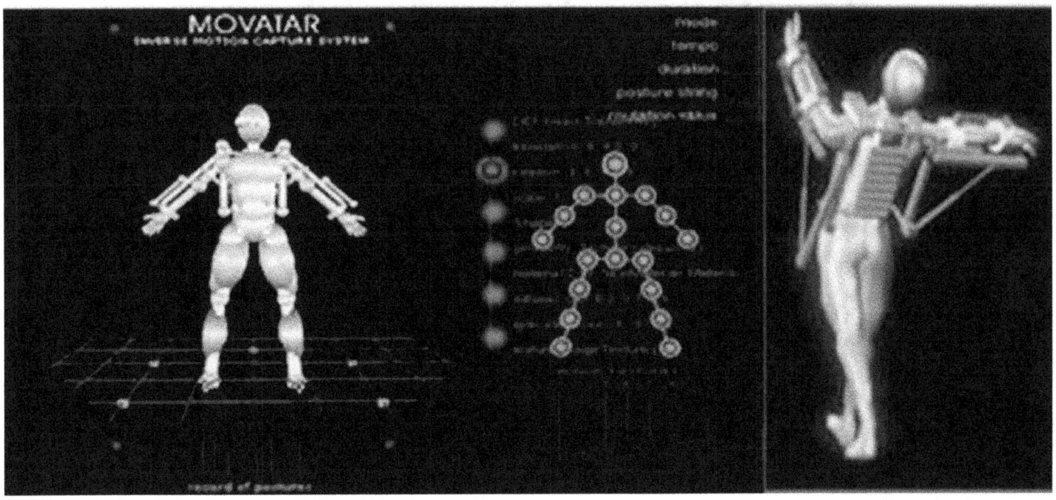

The additional *Third ear* (Zylinska, 2009) is a flexible prosthesis made of soft tissue and flexible cartilage which represents the expansion of the artist's body and resembles the natural ear in both shape and structure but not function, as seen in Fig. 23 (Zylinska 2009, Stelarc, 2021). It was grown in a laboratory from real human cartilage and was transplanted inside Stelarc's left forearm. The artist wanted to use the ear to makes his shows more attractive, but has caused numerous public controversies. The original idea was to have a balloon implanted under his skin, which would get gradually filled in the course of several months. During that period, a bubble was formed under his skin and the ear-shaped piece was implanted into it. The surgeon was supposed to sew the skin over the ear and place it onto Stelarc's head, behind his normal ear (Stelarc, 2021).

Stelarc was unable to get the adequate professional help for this project and decided to use a different approach to construct this ear. In cooperation with Oron Catts and Ionat Zurr of *Tissue Culture & Art* (Zylinska, 2002), he grew a replica of his ear from his own bone marrow cells. After the ear was fully grown, it was implanted beneath his skin on his arm. The prosthesis is a partially living form – part constructed and part grown. In the future, he plans to place a microphone into his third ear and enable the sound to be transmitted via the internet. He also plans to grow an earlobe from stem cells, but it is not yet known whether it will be implanted into his arm as a fourth ear or whether it will replace the hard cartilage of his current ear. The ear would not have the hearing function but would be used for emitting sound.

Figure 23. Stelarc: Third ear

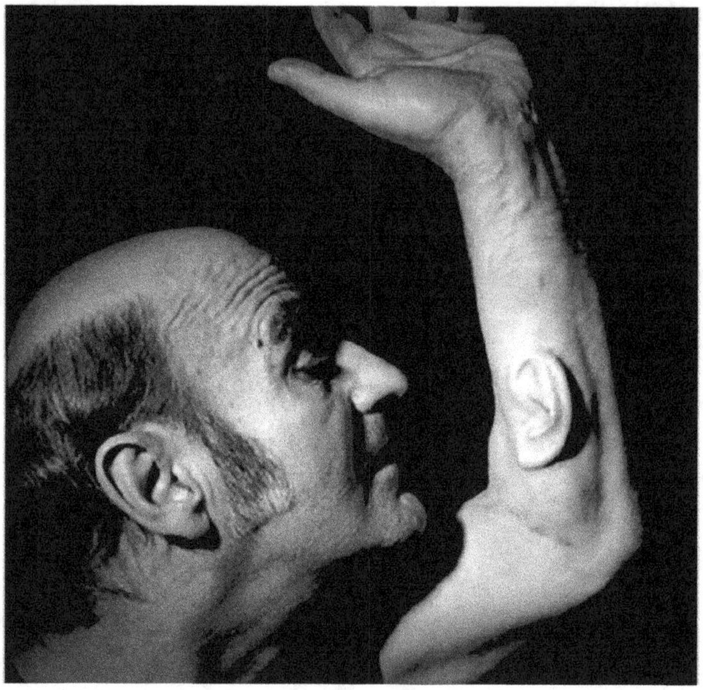

By connecting to a modem or a laptop, it would have the ability to emit audio and increase the sound that the actual ear can her. This additional ear would represent a type of internet antenna which telematically and acoustically increases the hearing possibilities. The project questions the idea of integrity of the body and confronts the growing potential of manipulating living systems. The *Tissue Culture & Art* project deals with ethical issues as well as question of perception about how the living tissue can grow and function outside the body.

The Hexapod (2002) is a hybrid human-machine architecture which consists of six robotic legs which can be controlled by Stelarc. It resembles an insect but its dynamic locomotion allows it to move like a dog. Robotic locomotion and control are guided by body movements and the rotation of the torso (Zylinska, 2002). *The Hexapod* weighs about 250 kg and moves with 5-metre legs. The muscle machine as a flexible and stable mechanism is moved by a combination of an electric and pneumatic system. When Stelarc's single leg is lifted, three mechanical legs are moved forward after the other three legs stop, as shown in Fig. 24 (Stelarc, 2021).

Stelarc's body stands on the ground in the machine's base, which includes the lower exoskeleton body connected to a robot. The coder on the side wrists allows Stelarc to control the movements of the machine and to change the speed at which he wants to traverse. The walking mode can be selective and is activated by hand gesture. The movement system has a sensory speedometer that delivers the data about Stelarc's movement and directs the machine's work. When the machine starts to move, it is impossible to ask who is in charge, Stelarc or the machine, as they become completely integrated and move as one.

Figure 24. Stelarc: The Hexapod

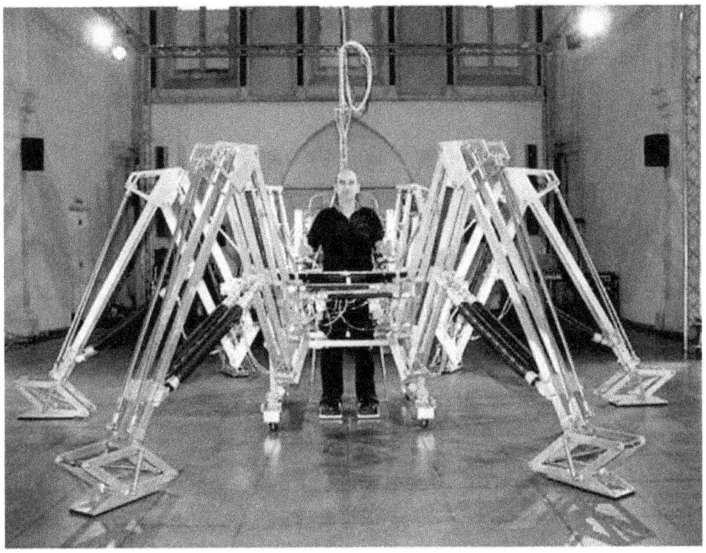

4.4.3.5 Steve Mann's Portable Computers

Steve Mann suggested the idea for the development of transportable information technology in 1970, and in 1994 he invented the *Wearable Wireless Webcam* (Mann, 1997). The movable helmet with a video screen and a camera allow him to stay connected to the cyberspace wherever he goes. The camera posts pictures online, allows him to check his e-mail, browse the internet or work on a computer while doing his job, thus granting him the ability to choose whether to look at the world with his own eyes or through his screen (Groothuis, 1999, p. 123).

Wearable Computing (Mann, 1997) refers to computers that are as inobtrusive as the clothes we wear and can be used on the move without using your hands (Fig. 25). This facilitates the use of this technology as it is always turned on and ready to work (Mann, 1997). The project is testing a new form of interaction between man and the machine, or man and the computer, and it serves to increase the intellectual capabilities, increase or improve the senses. Experts like Philip F. Binkley at the *University of Ohio*, Walter Frontera, David Standaert and Joel Stein at Harvard (all of whom are physicians, not engineers) agree that the use of clothing that allows continuous and uninterrupted tracking of physiological signals is vital for improving the diagnostic possibilities and treatment possibilities for cardiovascular and neural diseases as well as patient rehabilitation (Wenker, 2002).

It is important for the devices to record and retain as many physiological signals as possible over longer periods of time, as they are crucial for gaining a deeper understanding of the development and outcomes of the diseases and for revealing new physiological signs that can warn the physician about the need for an intervention.

Figure 25. Steve Mann – Wearable computing

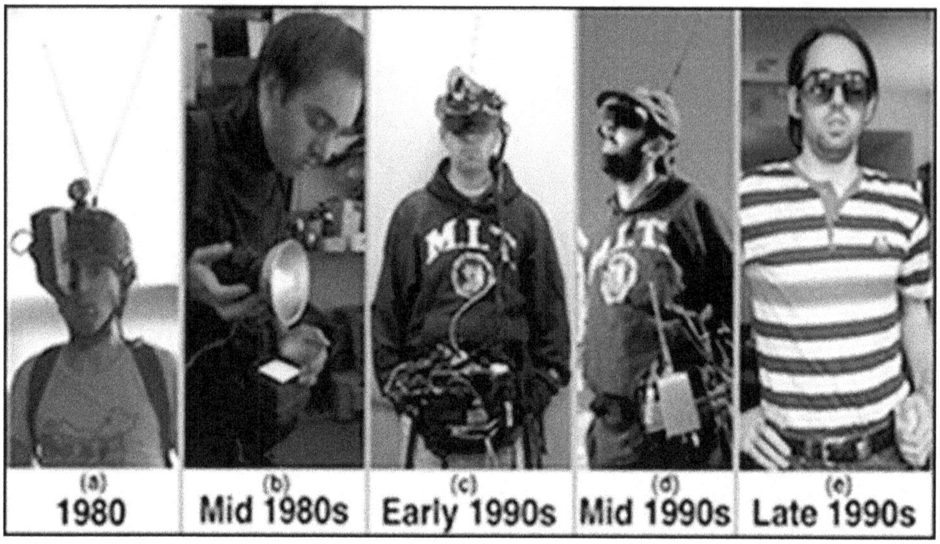

4.4.3.6 The Eyeborg Projects

The purpose of presenting the *Eyeborg* projects in this section is two-fold. On the one hand, it replaces the shortcomings of the human body that are a result of diseases or accidents, and on the other hand it expands the human capabilities outside its natural domain, thus creating a cyber cyborg. Neil Harbisson is a young painter who could only see the black and white tones due to a photoreceptor syndrome called achromatopsia. After implanting the *Eyeborg,* which transforms 360 colour tones into different sounds, Harbisson can hear and identify 36 different colours, from the red with the lowest frequency to purple with the highest frequency, as in Fig. 26 (Yasenchak, 2013, Sisley, 2016). He can even hear the colours that the human eye is unable to see, like infrared and ultraviolet, and can even paint the people's voices in colour. He can send the voices directly from his mobile phone to a chip in his skull via Wi-Fi. In the future, he plans to become a composer and learn to differentiate the sounds for the whole spectre of 360 colours (Marković, 2007).

Eyeborg I is a device which consists of a digital camera that is mounted onto the head of a user and reads the colours in the immediate vicinity with the help of a chip built into the skull. The camera is connected to a transportable computer which transforms the frequencies of colour light waves into frequencies of sound waves. Every colour has its own light frequency that is read by a sensor and transformed into a sound. The computer sends the sound of every colour into a device similar to an MP3 player, located in the user's ear. The creator of the Eyeborg, cybernetic expert Adam Montadon, hopes that this technology will help humans with impaired vision. The Time magazine declared the Eyeborg camera one of the best 50 inventions in 2009 (Time, 2009).

Eyeborg II or the "spy eye" is a miniature wireless web camera built into the eye with the aim of recording everything in immediate vicinity (Spence & Jaworski, n.d.) . It consists of a medical camera used for endoscopy, a battery, and a wireless transmitter. The creator of this device Rob Spence is a Canadian director and an engineer from the *Massachusetts Institute of Technology*. The aim of the *Eyeborg II* project was not to return sight to the blind, but to explore the boundaries of privacy and raise awareness

Figure 26. Neil Harbisson – Eyeborg

about the clandestine surveillance cameras. Rob Spence lost his eye in an accident when he was young. In 2008, he implanted a miniature wireless camera into his eye which is moved by the eye muscles just like a real eye. The miniature camera has a battery and a wireless transmitter mounted into a tiny plate. The physicians around the world have been working on creating an eye with an in-built camera which would enable the blind or the vision impaired to regain their sight.

4.4.3.7 Augmented Reality Goggles

Google Glass is an external addition to the human body which augments its ability to perceive the world. This augmented reality technology has been available for the past few years. *Google Glass* (Fig. 27) is a small computer without a screen and a keyboard that is connected to the internet via smartphone device. It is an Android-based device in the form of reading glasses which displays the data directly in the user's field of vision. It has a 5-megapixel camera, the capability of recording and HD video, and a memory of 16 GB. It is controlled by touch, voice and head movement and can even receive phone calls.

When fully functional, *Google Glass* shows the events around us in real time, augmenting our reality by connecting the physical spaces with the information and images from the computer databases. Solutions like this will change the way we communicate and take away the privacy as everyone will know everything about each other, depending on the information available in the database. In 2016, *Epson* announced the third generation of augmented reality glasses (Večernji list, 2009). The Moverio BT-300 model has a quad-core Intel Atom processor and an Android OS 5.1., which means that more power allows it to generate 3D content. The device is able to see and contextualize the user's environment through the front 5-megapixel HD camera, and is able to draw the content using the sensors. Silicon has replaced glass as the base layer, which enables a more precise display (Večernji list, 2009).

Figure 27. Google Glass (Russell, 2014)

4.4.4 Robo-Cyborg

A robo-cyborg (Fig. 28) is the project of the future, and it is a combination of a human mind and a machine body. Hans Moravec believes that humans will be merged with robots by the 22nd century. This merger will take place:

[…] by undergoing a brain operation that replaces each neuron of our brain with a transistor inside a robot. The operation starts when we lie beside a robot body without a brain. A robotic surgeon takes every cluster of gray matter in our brain, duplicates it transistor by transistor, connects the neurons to the transistors, and puts the transistors into the empty robot skull. As each cluster of neurons is duplicated in the robot, it is discarded. We are fully conscious as this delicate operation takes place. Part of our brain is inside our old body, but the other part is now made of transistors inside our new robot body. After the operation is over, our brain has been entirely transferred into the body of a robot. Not only do we have a robotic body, we have also the benefits of a robot: immortality in superhuman bodies that are perfect in appearance. […] In the ultimate scenario, we discard our clumsy bodies entirely and eventually evolve into pure software programs that encode our personalities. We "download" our entire personalities into a computer. If someone presses a button with your name on it, then the computer behaves as if you are inside its memory, since it has encoded all your personality quirks inside its circuits. We become immortal, but spend our time trapped inside a computer, interacting with other "people" (that is, other software programs) in some gigantic cyberspace/virtual reality. Our bodily existence will be discarded, replaced by the motion of electrons in this gigantic computer. In this picture, our ultimate destiny is to wind up as lines of code in this vast computer program, with all the apparent sensations of physical bodies dancing in a virtual paradise. We will share deep thoughts with other lines of computer code, living out this grand illusion. We have great, heroic exploits conquering new worlds, oblivious to the fact that we are just electrons dancing inside some computer. Until, of course, someone hits the off button. (Kaku, 2011, p. 67)

At the MIT, living tissue is being used in a new generation of robots which will be composed of both human and machine materials. One way for man to merge with robots without rejecting his body is for us to control the robots with our minds.

Figure 28. Robo-cyborg[24]

REFERENCES

Agencija, V. L. M. (2013). *Stvorili bioničko super uho pomoću 3D printera*. Poslovni.hr. https://www.poslovni.hr/sci-tech/stvorili-bionicko-super-uho-pomocu-3d-printera-242153

Agnew, S. P., Schultz, A. E., Dumanian, G. A., & Kuiken, T. A. (2012). Targeted Reinnervation in the Transfemoral Amputee: A Preliminary Study of Surgical Technique. *Plastic and Reconstructive Surgery*, *129*(1), 187–194. doi:10.1097/PRS.0b013e3182268d0d PMID:22186509

Allhoff, F., Lin, P., Moor, J., & Weckert, J. (2010). Ethics of Human Enhancement: 25 Questions & Answers. *Studies in Ethics, Law, and Technology*, *4*(1), 20121004. Advance online publication. doi:10.2202/1941-6008.1110

Almeida, D. (2010). The Dismembered Body: Poe's 'The Man That Was Used Up' and Cindy Sherman's Prosthetic Compositions. *The Edgar Allan Poe Review*, *11*(1), 163–173.

Amerike, G. (2007). *Dubinska stimulacija nudi nadu kod pacijenata sa povredom mozga*. https://www.glasamerike.net/a/a-34-2007-08-26-voa6-87090677/765078.html

Anupama, H. S., Cauvery, N. K., & Lingaraju, G. M. (2012). Brain Computer Interface and its Types – a study. *International Journal of Advances in Engineering and Technology*, *3*(2), 739–745.

Artemiadis, P. (2014). *Neuro-robotics: From brain machine interfaces to rehabilitation robotics.* Academic Press.

Bostrom, N., & Roache, R. (2007). Ethical Issues in Human Enhancement. In J. Ryberg (Ed.), *New Waves in Applied Ethics*. Palgrave Macmillan.

Boulanger, G. R. (1971). Što je kibernetika? In *Zbornik radova. Izazov kibernetici*. Centar za društvene djelatnosti omladine RK SOH.

CDC/JS. (2007). *Moljci kiborzi protiv terorista*. Portal Blic. https://www.blic.rs/vesti/svet/moljci-kiborzi-protiv-terorista/r2hrwkx

CDC/JS. (2009). *Američka vojska stvorila bube-kiborge za prisluškivanje*. Portal Blic. Rs. https://www.blic.rs/slobodno-vreme/zanimljivosti/americka-vojska-stvorila-bube-kiborge-za-prisluskivanje-video/hllf1qd

Clark, A. (2003). *Natural born cyborgs: mind, technologies, and the future of human intelligence*. Oxford University Press.

Clausen, J. (2009). Man, machine and in between. *Nature*, *457*(7233), 1080–1081. doi:10.1038/4571080a PMID:19242454

Clynes, M. E., & Kline, N. S. (1960). Cyborgs and Space. *Astronautics*, 26–76.

Clynes, M. E., & Kline, N. S. (1995). Cyborgs and Space. In C. H. Gray (Ed.), *Cyborg Handbook* (pp. 29–33). Routledge.

Daniels, N. (2000). Normal Functioning and the Treatment-Enhancement Distinction. *Cambridge Quarterly of Healthcare Ethics*, *9*(3), 309–322. doi:10.1017/S0963180100903037 PMID:10858880

EPFL. (n.d.). *Blue Brain Project*. https://www.epfl.ch/research/domains/bluebrain/

Featherstone, M., & Burrows, R. (Eds.). (1995). *Cyberspace/cyberbodies/cyberpunk: Cultures of technological embodiment.* Sage.

Fildes, J. (2009). *Artificial brain '10 years away'.* BBC News. http://news.bbc.co.uk/2/hi/8164060.stm

Fildes, J. (2009). *Bionic hand wins top tech prize*. BBC News. http://news.bbc.co.uk/1/hi/sci/tech/7443866.stm

Fischman, J. (2010). Bionics. *National Geographic*. https://www.nationalgeographic.com/magazine/article/bionics-robotics-medical-technology

Flanagan, M., & Booth, A. (Eds.). (2002). *Reload: Rethinking women + cyberculture*. MIT Press.

Fukuyama, F. (2002). *Our posthuman future: Consequences of the biotechnology revolution* (1st ed.). Farrar, Straus and Giroux.

Galić, D. (2009). *Kibernetski cvrčci spasioci*. Bug Online. http://www.bug.hr/vijesti/kibernetski-cvrcci-spasioci/97523.aspx

Gray, C. H., Mentor, S., & Figueroa-Sarriera, H. (Eds.). (1995). The cyborg handbook. Routledge.

Greguric, I., & Čatić, I. (2010). *Kiborgoetika – presjecište ili poveznica bioetike i tehnoetike*. Projekt MZT-a Primjena sustavnosne teorije u raščlambi opće tehnike (120-0000000- 1805). Fakultet strojarstva i brodogradnje.

Groothuis, D. R. (1999). *The soul in cyberspace* (Repr). Wipf ans Stock.

Grubin, D. (2007). *PBS series: The mysterious human heart*. The Replacement Heart. http://www.pbs.org/wnet/heart/episode1/replacement2.html

Gržinić, M. (1988). U redu za virtualni kruh. *Meander*.

Guger, C., Schlogl, A., Neuper, C., Walterspacher, D., Strein, T., & Pfurtscheller, G. (2001). Rapid prototyping of an EEG-based brain-computer interface (BCI). *IEEE Transactions on Neural Systems and Rehabilitation Engineering*, *9*(1), 49–58. doi:10.1109/7333.918276 PMID:11482363

Guirguis, M. (2013). AbioCor Artificial Heart. *BME, 181*.

Hanlon, M. (2007). *The Bionic Eye approaches: the next generation of Retinal Implants*. New Atlas. https://newatlas.com/the-bionic-eye-approaches-the-next-generation-of-retinal-implants/6855/

Hanlon, M. (2013). *Honda begins leasing Walking Assist Exoskeleton*. https://newatlas.com/honda-leasing-walking-assist-device-exoskeleton/27681/

Haraway, D. J. (2016). A Cyborg Manifesto: Science, Technology, and Socialist-Feminism in the Late Twentieth Century. In D. J. Haraway & C. Wolfe (Eds.), *Manifestly Haraway* (pp. 3–90). University of Minnesota Press. doi:10.5749/minnesota/9780816650477.003.0001

Hayles, N. K. (1999). *How We Became Posthuman: Virtual Bodies in Cybernetics, literature, and Informatics*. The Univ. of Chicago Press. doi:10.7208/chicago/9780226321394.001.0001

Hegel, G. W. F. (1991). *The encyclopaedia logic, with the Zusätze: Part I of the Encyclopaedia of philosophical sciences with the Zusätze* (T. F. Geraets, W. A. Suchting, & H. S. Harris, Trans.). Hackett.

Heidegger, M. (1977). *The question concerning technology, and other essays*. Garland Pub.

Hevesi, D. (2012). David Lederman, Pioneer of Artificial Heart, Dies at 68. *The New York Times*. https://www.nytimes.com/2012/08/29/health/research/david-lederman-pioneer-of-artificial-heart-dies-at-68.html#:~:text=David%20M.,He%20was%2068.

Ikeuchi, Y., Ashihara, J., Hiki, Y., Kudoh, H., & Noda, T. (2009). Walking assist device with bodyweight support system. *2009 IEEE/RSJ International Conference on Intelligent Robots and Systems*, 4073–4079. 10.1109/IROS.2009.5354543

Infonomia. (2011). *CYBORG LIFE - Talking to Kevin Warwick*. http://www.infonomia.com/articulo/videos/110

Jonas, H. (1987). Warum die moderne Technik ein Gegenstand fur die Philosophie ist. In Technik, Medizin und Ethik, Zur Praxis des Prinzips Verantwortung. Surhkamp.

Journal, M. (2013). *Global control through the RFID chip*. https://www.michaeljournal.org/subscription-boutique/leaflets/product/global-control-through-the-rfid-chip

Jutarnji list. (2016). *Amerikanci razvijaju kibernetičku telepatiju*. Jutarnji list. https://www.jutarnji.hr/life/znanost/amerikanci-razvijaju-kiberneticku-telepatiju-uz-pomoc-elektroda-u-mozgu-pricaju-racunalima-uskoro-ce-racunala-pricati-i-njima-98158

Kaku, M. (2011). *Physics of the future: How science will shape human destiny and our daily lives by the year 2100*. Doubleday.

Kennedy, P. R., & Bakay, R. A. E. (1998). Restoration of neural output from a paralyzed patient by a direct brain connection. *Neuroreport*, *9*(8), 1707–1711. doi:10.1097/00001756-199806010-00007 PMID:9665587

Krmpotić, Ž. (2007). *Belgijskom ovčaru namjestili bioničku šapu*. 24 sata. https://www.24sata.hr/news/belgijskom-ovcaru-namjestili-bionicku-sapu-24147

Kuikken, T. (2014). *A prosthetic arm that "feels"*. TedGlobal. https://www.ted.com/talks/todd_kuiken_a_prosthetic_arm_that_feels?language=en

Kunzru, H. (1997). You Are Cyborg. *Wired*. https://www.wired.com/1997/02/ffharaway/

La Mettrie, J. O. D. (1748). *Man a Machine* (C. G. Bussey, Trans.). Open Court.

Laborit, H. (1971). Kibernetika i ljudski stroj. In Zbornik radova. Izazov kibernetici. Centar za društvene djelatnosti omladine. RK SOH.

MacRae, M. (2014). *30 Years of Bionic Ears*. ASME.org. https://www.asme.org/engineering-topics/articles/bioengineering/30-years-of-bionic-ears

Mann, S. (1997). Smart clothing: The wearable computer and wearcam. *Personal Technologies*, *1*(1), 21–27. doi:10.1007/BF01317885

Marković, S. (2007). *Prvi službeni čovjek kiborg čuje trideset i šest boja*. 24 sata. https://www.24sata.hr/news/prvi-sluzbeni-covjek-kiborg-cuje-trideset-i-sest-boja-41616

McKeegan, N. (2008). *Second Sight announces milestone in groundbreaking retinal implant trial*. New Atlas. https://newatlas.com/retinal-implant-treats-blindness/8841/

Moles, A. A. (1971). Objekt, metoda i aksiomatika kibernetike. In Zbornik radova. Izazov kibernetici. Centar za društvene djelatnosti omladine RK SOH.

Mucić, M. (2008). *Prva bionička žena osjeća ruku koju nema godinama*. 24 sata. https://www.24sata.hr/news/prva-bionicka-zena-osjeca-ruku-koju-nema-godinama-79109

MYLE Team. (2016). *Japanese scienists invented artificial tongue to enable speech for oral cancer victims*. https://oralcancernews.org

N.J.D. (2009). *Kiborg kukac za prisluškivanje razgovora*. Moj Portal. http://www.mojportal.ba/novost/17424/kiborg-kukac-za-prisluskivanjerazgovora

Page, L. (2007). *DARPA to create brain-chipped cyborg moths*. The Register. https://www.theregister.com/2007/05/31/cyborg_chipped_terminator_moths_aiee/

Pandika, M. (2013). *The Promise of Brain-Machine Interfaces*. Ozy.com. https://www.ozy.com/the-new-and-the-next/the-promise-of-brain-machine-interfaces/3163/

Paravić, A. (2010). *Eksperiment stoljeća: Umjetni mozak bit će moguć 2018.* godine. Jutarnji list. http://www.jutarnji.hr/umjetni-mozak-bit-cemoguc-2018--godine/455125/

Perkowitz, S. (2004). *Digital People: From Bionic Humans to Androids*. Joseph Henry Press.

Portal, M. (2010). *Kontrola mislima: Izumljen telepatski čip*. Metro Portal. http://metro-portal.hr/izumljen-telepatski-cip/21393

Quick, D. (2008). *Touch Bionics launches world's first bionic finger*. The New Atlas. https://newatlas.com/touch-bionics-prodigits-bionic-finger/13554/

Rudež, T. (2006). *Bionički ljudi: Baterije pokreću motore koji otvaraju šaku i pomiču lakat.* Jutarnji list. http://www.jutarnji.hr/bionicki-ljudi--baterije-pokrecu-motore-koji-otvaraju-saku-i-pomicu-lakat/155158/

Rudež, T. (2008). *Svaki dio tijela sada se može umjetno stvoriti.* Jutarnji list. https://www.jutarnji.hr/naslovnica/svaki-dio-tijela-sada-se-moze-umjetno-stvoriti-3951129

Russell, K. (2014). See What Google Glass Apps Will Actually Look Like. *Business Insider.* https://www.businessinsider.com/what-google-glass-apps-will-look-like-2014-1

Savulescu, J. (2007). Genetic Interventions and the Ethics of Enhancement of Human Beings. In B. Steinbock (Ed.), *The Oxford Handbook of Bioethics* (pp. 517–528). Oxford University Press.

Scott, B. P. (1987). *Robotička revolucija*. August Cesarec.

Scudellari, M. (2012). *Missing Touch.* The Scientist. https://www.the-scientist.com/features/missing-touch-40531

Servant, H. (2016). *Bionic cat Oscar survived: Supervet, Noel Fitzpatrick.* Johnny Times. http://www.johnnytimes.com

Shelley, M. W. (1988). *Frankenstein, or, The modern Prometheus: With supplementary essays and poems from the twentieth century*. Orchises.

Singer, E. (2009). *The Army's Remote-Controlled Beetle.* Technologyreview. https://www.technologyreview.com/2009/01/29/95502/the-armys-remote-controlled-beetle/#:~:text=The%20insect%27s%20flight%20path%20can%20be%20wirelessly%20controlled%20via%20a%20neural%20implant.&text=A%20giant%20flower%20beetle%20with,to%20research%20presented%20this%20week.

Sisley, D. (2016). *Why this artist got an antenna implanted in his skull.* Dazed. https://www.dazeddigital.com/artsandculture/article/31102/1/why-this-artist-got-an-antenna-implanted-in-his-skull#:~:text=Neil%20Harbisson%2C%20who%20can%20now,the%20world%27s%20first%20cyborg%20artist&text=Harbisson%2C%20who%20was%20born%20completely,help%20him%20"hear"%20colour.

Someya, T. (2013). Bionic Skin for a Cyborg You. *IEEE Spectrum.* https://spectrum.ieee.org/bionic-skin-for-a-cyborg-you

Spence, R., & Jaworski, M. (n.d.). *Man With Eye Camera Partners with World's First 3D Printed Eye Prosthesis Company.* Eyebork. https://eyeborgproject.tv

Spohrer, J. (2002). Nano-, Bio-, Info-, Cogno-, Socio-Convergence to Improve Human Performance: Opportunities and Challenges. In Converging Tehnologies for Improving of Humkan Performance: Nanotehnology, Biotehnology, Information Technology and Cognitive Science.

Stableford, B. M. (2006). *Science fact and science fiction: An encyclopedia.* Routledge. doi:10.4324/9780203943588

Stelarc. (2021). *Third hand.* http://stelarc.org/

SynTouch. (n.d.). *Sensor Technology – BioTac Technologies.* https://www.syntouchinc.com/sensor-technology/

Telegraph. (2009). Goose gets a bionic leg in world-first operation. *The Telegraph.* https://www.telegraph.co.uk/news/newstopics/howaboutthat/5782747/Goose-gets-a-bionic-leg-in-world-first-operation.html#:~:text=Here%20is%20the%20world%27s%20first,heart%20to%20put%20it%20down.&text=Instead%2C%20they%20decided%20to%20operate,give%20her%20a%20bionic%20leg.

Telegraph. (2009). Telepathic microchip could help paraplegics control computers. *The Telegraph.* https://www.telegraph.co.uk/technology/news/6129185/Telepathic-microchip-could-help-paraplegics-control-computers.html

Time. (2009). *The 50 Best Inventions of 2009. The Eyeborg.* http://content.time.com/time/specials/packages/article/0,28804,1934027_1934003_1933989,00.html

Trifunović, J. (2009). *Bioničko oko: Čak 70% slijepih moći će progledati.* 24 sata. https://www.24sata.hr/news/bionicko-oko-cak-ce-70-slijepih-moci-progledati-139212

Tutton, M. (2009). *Paralyzed man 'turns thoughts into sounds.* CNN. http://edition.cnn.com/2009/HEALTH/12/16/brain.computer.speech/index.html#:~:text=London%2C%20England%20(CNN)%20-- ,time%20speech%2C%20according%20to%20researchers.&text=Guenther%20said%20his%20system%20converts,the%20condition%20hold%20normal%20conversations.

Vaughan, T. M. (2003). Guest editorial brain-computer interface technology: A review of the second international meeting. *IEEE Transactions on Neural Systems and Rehabilitation Engineering, 11*(2), 94–109. doi:10.1109/TNSRE.2003.814799 PMID:12899247

Večernji list. (2009). *Epson predstavio najlakše transparentne pametne naočale na OLED tehnologiji.* Večernji list. https://www.vecernji.hr/techsci/epson-predstavio-najlakse-transparentne-pametne-naocale-na-oled-tehnologiji-1062354

Vermeersch, E. (1971). Projekt umjetnog čovjeka. In Zbornik radova. Izazov kibernetici. Centar za društvene djelatnosti omladine RK SOH.

von Goethe, J. W. (2008). Faust: Part Two (Oxford World's Classics). Oxford University Press.

Warwick, K. (1999). Mind Machines. In S. Griffiths & J. Weiner (Eds.), *Predictions: 30 great minds on the future* (pp. 281–290). Oxford Univ. Press.

Warwick, K. (2003). *A study in cyborgs* (Vol. 16). Ingenia.

Warwick, K. (2004). I, cyborg. University of Illinois Press.

Wenker, O. (2002). Wearable Computing in Medicine. *International Journal of Medical Informatics, 1*(1), 1–3. PMID:11904243

Wiener, N. (1990). *The human use of human beings: Cybernetics and society*. Free Association Books.

Wolpaw, J. R., & Wolpaw, E. W. (Eds.). (2012). *Brain-computer interfaces: Principles and practice*. Oxford University Press. doi:10.1093/acprof:oso/9780195388855.001.0001

Yang, S. (2004). *UC Berkeley researchers developing robotic exoskeleton that can enhance human strength and endurance*. UC Berkeley News. https://www.berkeley.edu/news/media/releases/2004/03/03_exo.shtml

Yasenchak, D. (2013). *Filling the Colorless Void: The Cybernetic Synaesthesia of Neil Harbisson*. Valley Humanities Review.

Zehr, E. P. (2011). *Inventing Iron Man: The Possibility of a Human Machine*. Johns Hopkins University Press.

Zylinska, J. (Ed.). (2002). *The Cyborg experiments: The extensions of the body in the media age*. Continuum.

Zylinska, J. (2009). *Bioethics in the age of new media*. MIT Press. doi:10.7551/mitpress/9780262240567.001.0001

ENDNOTES

1 Simulacrum denotes a copy with no original, i.e. it has no basis in the real world but has its own ontological basis. This term is used by Baudrillard and Deleuze in their criticism of the postmodern culture which abolishes the boundaries between the real world and the illusory world – a simulacrum.

2 More information available at: https://movies2.nytimes.com/library/cyber/surf/022697surf-cyborg.html [5/3/2020]

3 This typology of cyborgs was created and presented in the scope of a research project Cyborgoethics. More details available in: Greguric, I., Čatić I. (2010). Kiborgoetika – presjecište ili poveznica bioetike i tehnoetike. Projekt MZT-a Primjena sustavnosne teorije u raščlambi opće tehnike (120-0000000-1805). Zagreb: fakultet strojarstva i brodogradnje., str. 25–74.

4 A cyborgized mouse with an implantet osmotic pump (Clynes and Kline, 1960, p. 27).

5 The photograph of a radio-controlled cyborg insect Mecynorrhina: a) a platform of a living bug (Mecynorrhina, 4-10g, 4-8 cm), b) an RF transmitter controlled by users via USB port on a personal computer, c) an RF receiver, d) a medium wave dipole antenna, e) a stimulating electrode on both optic lobes, f) basal flight muscles, g) electrode counter (Sato et al. 2009).

6 Alternative terms for bionic are biomimetics, biognosis, biomimicry, bionical creativity engineering. The first part of the word consists of a part of the Greek word βίον, which denotes a unit of life, and the adjective-forming suffix *–ic*, resulting in the meaning 'like life'.

7 Fischman (2010), photo by Mark Thiessen.

8 Claudia Mitchell and Jesse Sullivan had a talk during the press conference in September 2006 at the National Press Club. They underwent a revolutionary surgery which allowed them to regain the sense of touch in their hands through transplantation of sensory nerves (Scudelari, 2006).

9 More details about the operation and bionic arm rehabilitation at: Targeted Reinervartion in the Transhumeral Amputee [online]. Rehabilitation Institute of Chicago (Agnew et al., 2012).

10 I-Limb hand: Get a Grip on functionality (2009). Touch Bionics Inc.

11 Eric Jones showing his bionic fingers ProDigits (Quick, 2009).

12 More details about Oscar Pistorius will be presented in chapter 8: *Cyborgization in sports.*

13 Greg Gadson regained his full mobility with the help of electric limbs (Fischman, 2010)

14 The implant is wire-connected to the microphone/processor behind the ear (McRae, 2014).

15 The first hear transplant was performed by a South African surgeon Christian Barnard in 1967. More information: https://www.ncbi.nlm.nih.gov/pmc/articles/PMC6062759/

16 AbioCor Implantable. Texas Heart Institute at St. Luke's Episcopal Hospital website.

17 More on this in §4.4.3.1 *The brain-computer interface.*

18 The Blue Brain Project is the name of the first virtual brain in the world. Photo: www.aiyellowpage.com.

19 More details about the Biotech Campus research will be given in the interview at the end of the book.

20 Source: Bioethics archive. https://bioethicsarchive.georgetown.edu

21 The Codex of Medical Ethics and Deontology §6.4, Official Gazette 47/2004 [25/10/2004].

22 More: Patient Rights and Protection Act (ZZPP, NN 169/04) §6-18).

23 Kevin Warwick only uses the term *cyborg* as he regards treatment of humans to be identical to enhancement. His experiments are strongly imbued with the spirit of transhumanism..

24 Image from: https://wallpaperscraft.com/download/robot_nod_32_cyborg_5632/3840x2160 [22/6/2016].

Chapter 4
Hacking Human Beings

ABSTRACT

This chapter discusses the various areas of human enhancement application where such concerns arise, including physical (chip self-identification and payment, improving hearing with magnetic headphones, improving vision in the infrared spectrum, and use of chips for monitoring bodily functions), cognitive (deep brain stimulation, memory chips, manipulation of neurons and neural dust), and emotional level (mood-fixing, neuroprosthetics) and life-extending technologies (cyborg digital mind, cyborg digital body, cryonic preservation of the body).

INTRODUCTION

Hacking human beings can denote all acts of conducting scientific and technological experiments with the boundaries of the biological body and mind and changing the existing capabilities using digital implants in order to gain new functions or improve hidden ones. The fundamental standpoint of these experiments is the cybernetic understanding of the human body as a programmable system which can be upgraded with programmed digital implants. An entity which consists of biologically and digitally programmed implants is called a *cyber* cyborg. The aim of these experiments is to enhance human features at the physical and the cognitive level. The long-term aim is to expand the lifespan of the mortal human body and immortalize the mind in the computer interface.

The theoretical explanation and ethical justification of hacking human beings stems from the transhumanist philosophy as the metaphysics of the *scientific work* and the interested thought of the neoliberal right of every individual to independently (of everyone else) decide on the aspects of enhancements of his/her being. Hacking human beings and the overall historical reality leads us towards new anthropology and ontology of man. Life itself, older than both man and technology, is being pushed into a fragment of historical time or will be completely replaced by digital forms, marked as the communication of the machine with its environment. This communication will imply communication between an intelligent machine and an intelligent environment. A program will communicate with another program. Within the scope of *scientific work* as the substance subject of human history, there is no historical time as a 'space for human development' (Marx) as consciousness and self-consciousness implanted into a human being

DOI: 10.4018/978-1-7998-9231-1.ch004

will be replaced by the programs which will determine their behaviour and will control the rest of the life within them. The content of the program and the entity controlling its goals remain the crucial issues.

According to Savulescu, Sandberg and Kahane, there are several types of enhancement of man: physical enhancement, cognitive enhancement, expanding the lifespan, mood enhancement and moral enhancement (Selak, 2013, p. 76).

Physical enhancement implies "use of various biomedical means for overcoming the limitations of the body" (Selak, 2013, p. 78), such as doping. Cognitive enhancement refers to "increasing human cognitive capacities: intelligence, memory, and attention, using genetic intervention or neopharmacology" (Selak, 2013, p. 76), e.g. nootropics, neuron and cybernetic implants. Mood enhancement refers to the "widespread use of Prozac and other anti-depressants and reactive drugs" (Selak, 2013, p. 77) and mood enhancement chips. The aim of lifespan expansion is postponing aging and prevention of death. In the rest of this chapter, we shall describe several examples of hacking of human body and mind.

5.1 PHYSICAL ENHANCEMENTS

5.1.1 Playing with Dangerous Things[1]

The first near-field communication (NFC[2]) chip was self-implanted by Amal Graafstra, the founder of *Dangerous Things*, the company which sells sets of implantation chips (Fig. 1) with their price ranging from 50 to 110 USD. The chips enable their users to unlock doors, start their automobiles, store their IDs etc.

Graafstra (personal communication) claims that his sense of self has changed after implanting the chip. This technology is well-designed and transparent, which means it does not require control (rechargeable batteries) and is permanently built into us and becomes "a part of our expected ability…just like you expect to use your arm to reach for a pencil or use your legs to walk" (Graafstra, p.c.). In the future, he expects to be able to talk to the machines with the help of a chip. Graafstra believes that ethical approval is not necessary for chip implantation, much like you do not need to get approval for modifying your body with jewellery. In addition, he does not regard chip implantation to be a medical treatment, which is why it does not necessitate approval. People should decide on their own whether or not to use chips.

5.1.2 Self-Identification via Chip

The NFC chip can be programmed for just about anything (opening doors, unlocking computers, etc.) and can be purchased and ordered online and implanted at home. Andreas Sjöström uses the NFC microchip implanted into his arm to pass the security gates at the Stockholm airport. The chip contains the personal information of the passenger.

Through this experiment, the Scandinavian airline company is trying the generate new experiences for its customers through innovation. Sjöström states:

The biggest surprise was the feeling of being able to identify myself without anything other than my body. I didn't have to pull out anything. It gave me a new sensation, sort of a prenotion of what it will be like in the future when we don't have to reach out with physical objects to accomplish things. (Edwards 2016)

Figure 1. Dangerous things[3]

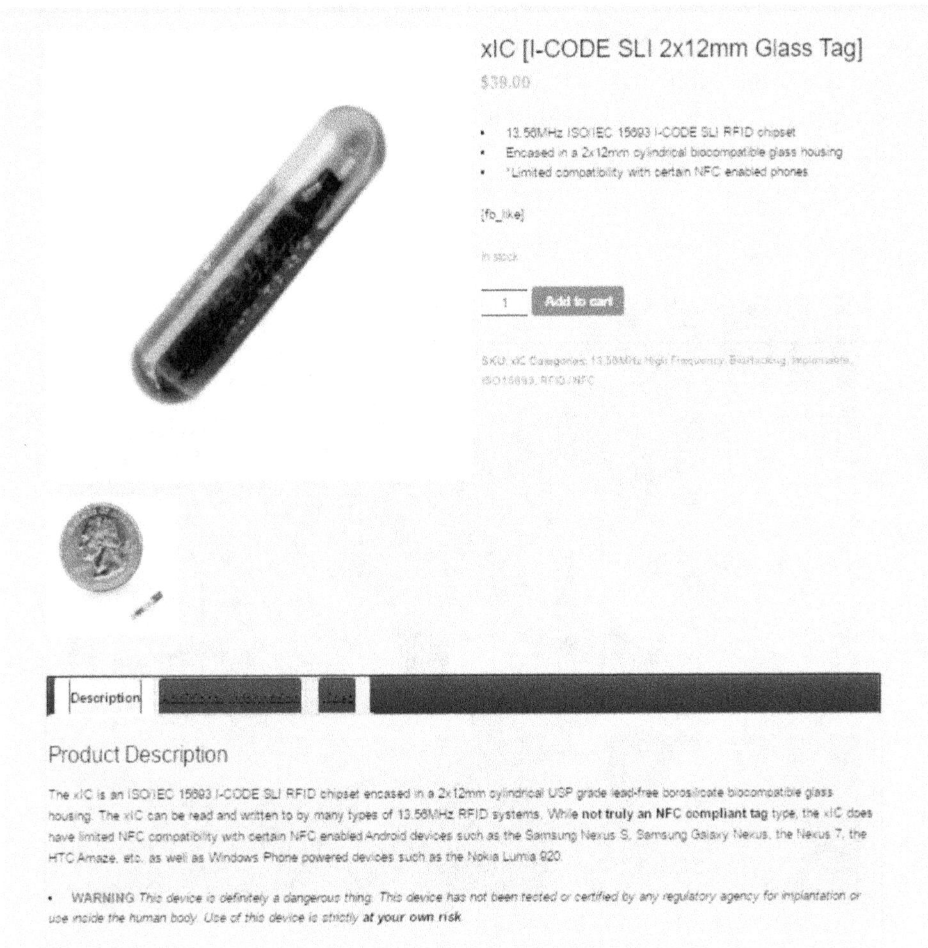

Figure 2. Use of an NFC chip in an airport (Liberatore, 2016)

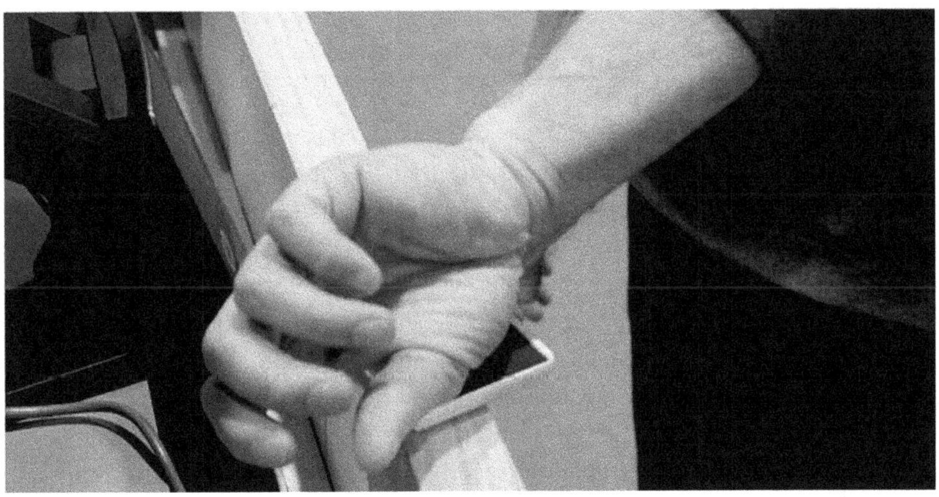

5.1.3 Payment via Chip

In 2011, Meow-Ludo implanted an NFC chip into his thumb (Redman, 2016). The chip enabled contactless payment and riding the public transportation. Two of the biggest areas which the chips can enable is authentication and activation of various things. In 2004, the VIP guests at the Baja Beach Club in Barcelona paid for their drinks using a chip.

Figure 3. Paying with an NFC chip (Jussi, 2005)

The Swedish railway companies have already introduced the possibility of paying for the fare using a microchip built into the passengers' fists. The conductors have a microchip for scanning, which allows them to check whether the passenger paid for their fare (Coffey, n.d.).

5.1.4 Magnetic-Field Perception

For hackers, achieving magnetic sensation means increasing sensory ability. This experiment achieves the magnetic-field perception through implantation of small but powerful magnets into the tip of the finger. The implant enables the collection of small metal objects (Fig. 4) and the sense of a magnetic field (Berg, 2012).

This experiment can be beneficial for individuals who deal with electronics. The thin magnetic sensors can offer progress in the field of robotics, security, and health tracking. Mark Fenske, a neuroscientist at the University of Guelph, claims: "A strong enough magnet will respond to changes in electromagnetic fields, which occur anywhere you have fluctuations in electrical current or movement of another magnet. Placing a magnet in the finger takes a form of information that normally isn't readily sensed by touch and converts it to movement or vibration that is readily sensed" (Neporent, 2013).

Figure 4. Magnetic implant for improved perception (Robertson 2014)

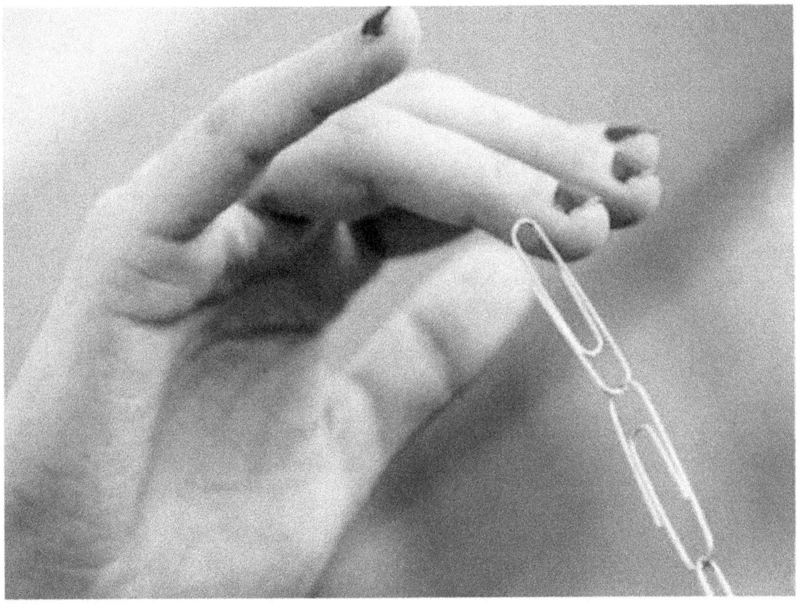

5.1.5 Hearing Enhancement via Magnetic Headphones

Hearing can also be enhanced using magnetic headphones. They are implanted into the auricle and cause vibrations of the ear bone. The magnets are stimulated by a magnetic coil worn like a necklace by the user (Fig. 5).

Rich Lee was implanted with magnetic headphones by the body modification artist Steve Maworth, whose surgical implantation pushed the boundaries of human capabilities. Lee explains: "This new synaesthesia of sorts is an exciting way to explore the world and develop new instincts about the way the world works around you" (Carmichael, 2013).

Lee plans to connect the magnetic headphones with his phone's GPS and microphone to be able to get directions about the person's location and hear their conversations (Arthur, 2013). For Lee, this is more than just an unusually intimate form of self-expression, it is a matter of survival.

5.1.6 Infra-Red Spectre Vision Enhancement

Humanity has so far been able to produce advanced technologies which allow us to observe only 2.3% of our reality, but what advancements would we achieve if we could perceive the remaining 97.7% of our reality (Lee, 2017)? What would we be able to see if we were born with this ability? It would be possible to see through severely tinted car windows or sunglasses, fog, dust etc.

During WW II, the US Navy fed the sailors with nutrients which substitute vitamin A with vitamin A2 in an attempt to increase their night vision. It is believed that that the human body which consumes increased amounts of only vitamin A2 will be forced to generate porphyropsin and gain the ability to see the infrared part of the spectrum (Patel, 2015).

Figure 5. Magnetic headphones for enhancing hearing (Lee, 2013)

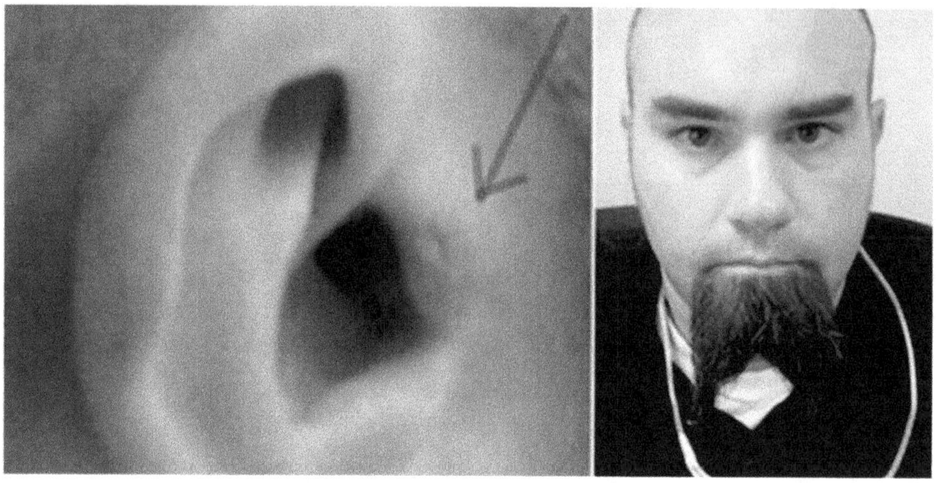

Rowlands, Tibbetts, Liccina and Galvin (Fig. 6) have shown with their research that it is possible to enhance your vision with the help of the protein complex porphyropsin. Using vitamin A2, human body is able to form its own pigments that enhance vision and expand the human perception of the world (Rowlands et al., 2016).

Figure 6. Human Near-infrared (NIR) visual perception (Rowlands et al., 2017)

5.1.7 Orientation Improvement

With the help of the ankle-worn external compass *North Paw*, it is possible to improve spatial orientation. The compass sends the information which the human brain accepts at a subconscious level, granting the wearer a real feeling of moving towards the desired goal. The compass constantly vibrates when the user is facing north.

The most important aspect of *North Paw* (Fig. 7) is creating a consistent sense of north-facing direction, which enables the brain to adapt to persistent device stimuli and feel the direction of north. The inventors are working on a chip that would be implanted into the body (Williams, 2013).

Figure 7. North Paw[4]

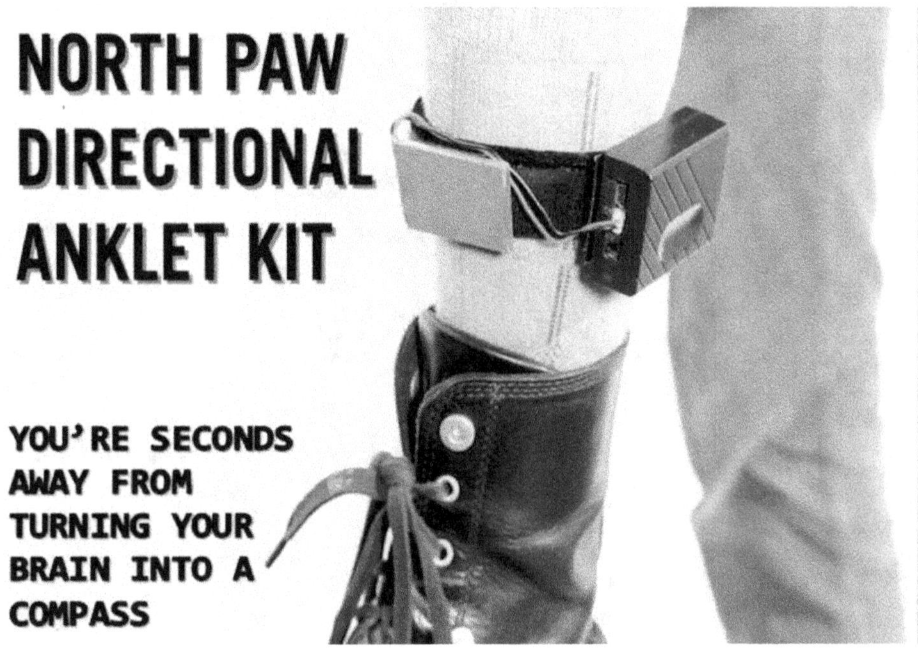

Sensbridge (sensebridge.net) are the manufacturers of electronic sets for personal hacking. The sets contain devices that enable people to sense their direction and see how their heart works.

5.1.8 Chip for Monitoring Bodily Functions

The company *Grindhouse Wetware* from Pittsburgh are researching chips (Fig. 8) for tracking human health. The inventors believe that implanted chips would allow us to measure body temperature, blood pressure, heart rate, chemical balance and other things in real-time. This would be a significant contribution to preservation of individual's health (Thompson, 2015).

Circadia (Fig. 9) is a computer, i.e. a biometric sensor implanted under the skin of the user's arm. It sends the biometric data from the body to the phone or tablet using wireless network. The device supervises, collects, and sends the data on pulse, temperature, and other biometric information. In the future, it will be used for transferring data onto a neural network, i.e. a neural interface (Rothman, 2013).

Biohacker Rich Lee from Utah designed the device *Lovetron 9000*, which he implanted under his pubic bone. This device allows his penis to start vibrating, which creates a completely new type of arousal. Lee has six implants that allow him to generate new sensations. In each ear, he has a hearing implants, he has a magnet at the top of two of his fingers, an NFC chip in his arm for controlling different devices, and a biothermal chip in his arm for measuring his temperature (Magazin.hr, 2016).

Figure 8. Grindhouse Wetware[5]

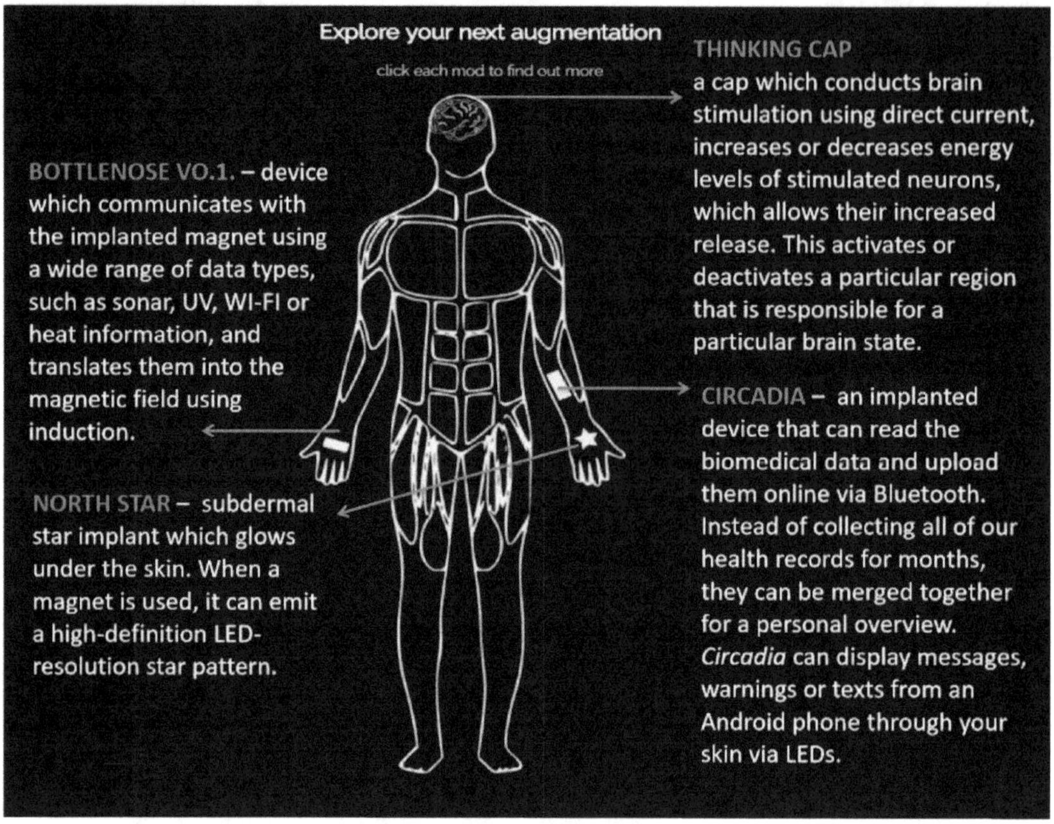

Figure 9. Circadia (Plenke, 2015)

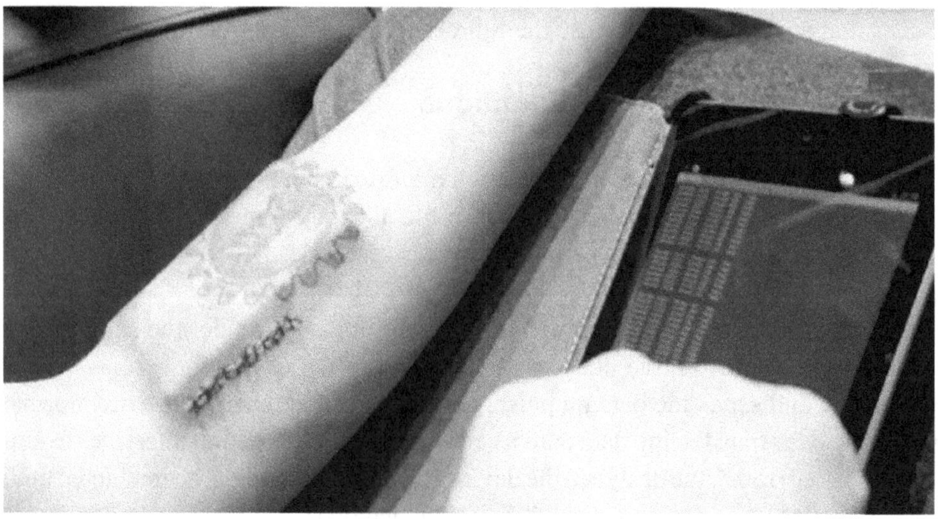

5.2 COGNITIVE ENHANCEMENT: HACKING THE BRAIN (TREATMENT EXAMPLES)

5.2.1 Deep Brain Stimulation

The researchers at the Cleveland Clinic have managed to treat the patients with OCD by inserting a stimulator which transmits electric impulses into the brain. The treatment has shown some success in treatment of patients (Orenstein, 2011).

Figure 10. Hacking the brain (Kurzweil, 2014)

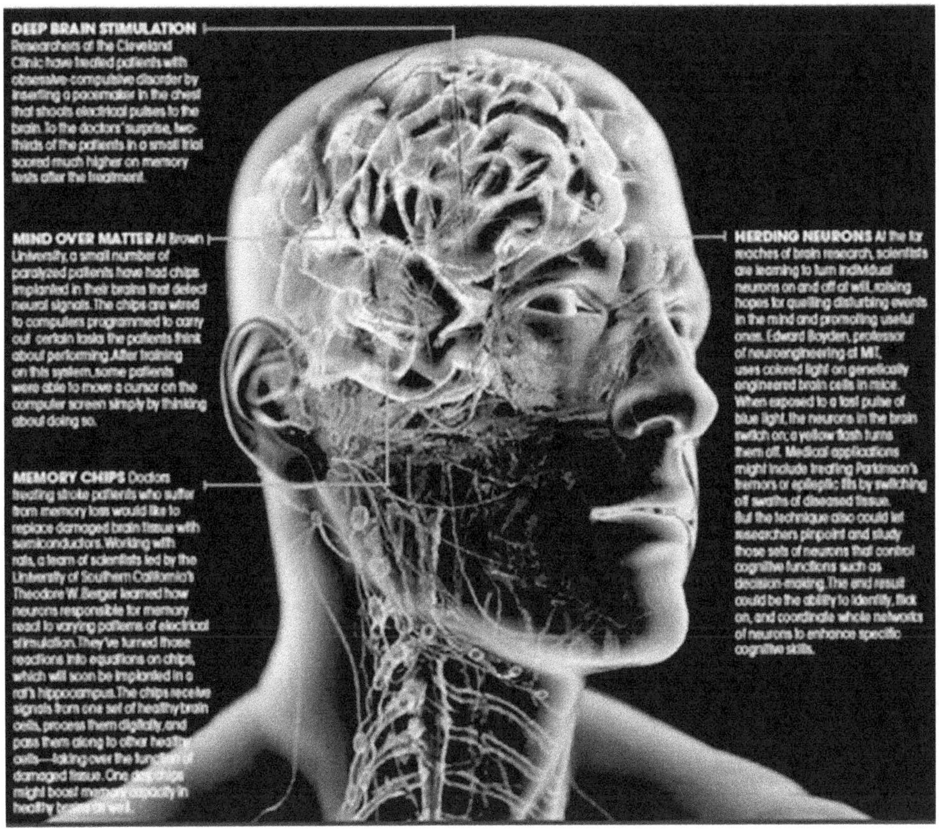

5.2.2 Memory Chip

Researchers at the Southern California University have discovered how the neurons responsible for disorders react to different patterns of electric stimulation. The chip receives the signals from healthy brain cells, processes them digitally and sends them to other healthy cells. Thus, the chips take over the function of the damaged tissue. Instances of stroke could be treated by replacing the damaged brain tissue with semi-conductors. In the future, the chips could even improve the healthy human brains by increasing its memory (Acey, 2013).

5.2.3 Neuron Manipulation

The scientists are exploring the possibility of activating and deactivating particular neurons at one's desire. At the MIT, the researchers are conducting this research by focusing coloured light onto genetically designed brain cells of mice. Faster pulse of the blue light activates neurones while the yellow flash deactivates them. This method could be used to treat Parkinson's disease or epilepsy by deactivating the diseased tissue. However, these experiments could also serve as a basis for manipulation of neurons which enable particular cognitive processes, decision-making in particular. This would mean that human beings could be directed and/or controlled through manipulation of particular neurons (Trafton, 2010).

5.2.4 Neural Dust

Neural dust is a wireless sensor that is implemented into the body next to muscles and organs. These sensors collect data on electric activity of nerves in real-time. So far, they have been proven successful in stimulation of electric nerve signals in rats. The researchers are developing tiny dust which could be implanted into the brain to stimulate nerves, control the appetite, treat epilepsy, control the function of different organs, etc. (Choi, 2016).

5.3 MOOD ENHANCEMENTS: HACKING THE BRAIN (ENHANCEMENT EXAMPLES)

5.3.1 Brain Wearables

Numerous portable devices could track heart rate, steps, body temperature or sleep, but *Mindflex* (Fig. 11) tracks the activity of the human mind. The creators of the brain wearables claim that the device can allow its users to increase focus, discover stress, or activate a video game directly with the mind. This is an external gaming console which uses the EEG technology to read the brain wave intensity over the EEG headset. The sensors are located on user's ears and forehead. Every act of increased concentration generates brain activity which can be used to move smaller objects with the power of one's mind (Rettner, 2016).

5.3.2 Thync – Mood Improvement

Thync (Fig. 12) is an external application which uses electroencephalography for measuring brain activity, changing the activity of neurons in the process. Thync operates by sending electronic signals through biocompatible materials into the brain nerves. The user controls the intensity and the duration of the signal via mobile phone app.

If a user decides to relax, he/she can use the *calm mode* option (post-meditation state), while the *energy mode* causes stimulation and clarity of mind, which makes the user more active and sociable (CBS News, 2015).

Figure 11. Mindflex (Mindflex, n.d.)

Figure 12. Thync (Thync, n.d.)

5.3.3 Neuroprosthesis

US company *Kernel* (Fig. 13) is working on a chip which would enhance the functions of the human mind (increasing the memory capacity) with the help of artificial intelligence. This chip imitates the electric brain signals, thus stimulating the neurons to communicate faster. The scientists argue that cognitive abilities could be increased this way.

Experiments have so far shown positive results with monkeys and human experiments are next in line. The chips are expected to help in treatment of stroke victims, but would also enable enhancements of healthy individuals (Vitas, 2016).

Figure 13. Neuroprosthesis (Radiant, n.d.)

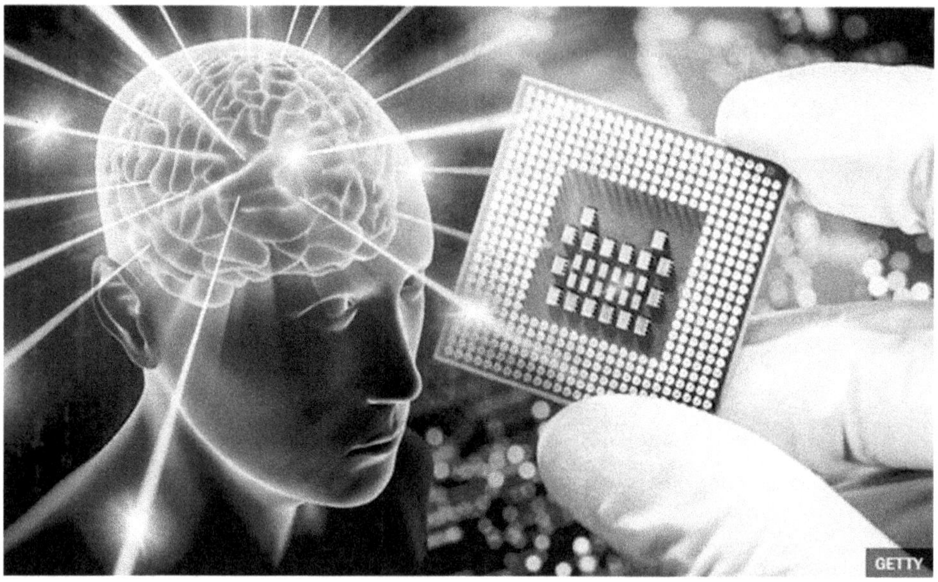

5.4 EXTENDING THE LIFE EXPECTANCY

5.4.1 Cyborg Digital Mind

The digital mind uses geo-mapping, biographic images, videos and documents uploaded onto a digital archive to be preserved for future generations. This will allow the creation of a computer avatar of a concrete individual.

To create a *Mindfile*, one needs to go through four stages: 1. upload biographic photographs, videos and documents into a digital archive which will be preserved for future generations; 2. organize them through geo-mapping and tagging to create your portrait (places you visited, people you met); 3. create a computer-based avatar which would fit your attitude; 4. connect with other people who want to increase the quality of their life (LifeNaut Project, n.d.). The avatar created this way will be a portrait of a computer-generated person – a (*cyber*) cyborg.

5.4.2 Cyborg Digital Body

The *Biofile* project offers completely free storage of the DNA material into a cybernetic file. The aim is to create a long-term storage of the genetic material and create the basis for generating a new body in the future which would be integrated with the information from the digital mind (*Mindfile*). After a person is declared dead, the technology of the future will merge the *Biofile* with the *Mindfile* into an entity that will be able to live forever that way.

To create the *Biofile*, you need to go through four steps: 1. Sign up for the LifeNaut BioFile project (Fig. 14); 2. The interested party is sent a bottle of mouthwash and a collection tube; 3. Live cells are collected from the sample and are stored at liquid nitrogen temperature (-190° C) for an indefinite period of time; 4. After a person is declared legally dead, their body will be grown via ectogenesis and their mind will be downloaded into it, which will allow the person to live forever (LifeNaut Project, n.d.).

Figure 14. The LifeNaut projects (LifeNaut Project, n.d.)

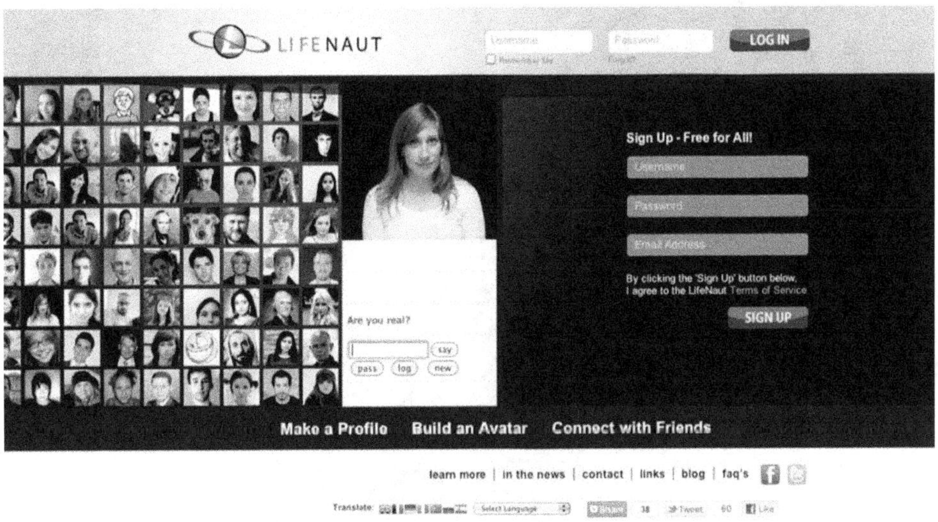

5.4.3 Cryonic Body Preservation

Medvedev and Pride are the leading scientists at the Russian Transhumanist Movement, organization for extending the natural lifespan. *KrioRus* is the only company for cryopreservation outside of USA that has been holding bodies of the deceased in liquid nitrogen since 2003 in order to be able to bring them back to life in the future. The price of immortality, i.e. freezing the entire body, ranges from 100,000 to 300,000 USD.

The best method for cryopreservation is replacing the water inside the body with a chemical which does not destroy information once it gets frozen. The bodies of patients are stored in tanks with double-sided chambers at a facility outside of Moscow, as shown in fig. 15 (LifeNaut Project, n.d.).

The cryopreservation procedure consists of several phases: 1. The cryogenics team provides cardio-pulmonary care immediately after the death and during the initial cooling. If a person dies before the cryosurgeon arrives, it is important to keep the body on ice; 2. Anticoagulant is injected to prevent the blood from thickening; 3. The cryosurgeon replaces the blood with the cryoprotectant (perfusion); 4. The head or the body are cooled in dry ice during several days until the temperature drops down to 120° C; 5. The cells are vitrified – turned into glass. This prevents any potential damage that might occur if the body got frozen while filled with water; 6. When the body or head reach the liquid hydrogen temperature (–195° C), it is moved into a cryogenic container that is filled with liquid nitrogen; 7. Liquid nitrogen is slowly added and the patient can be taken care of even without electricity; 8. Wait for singularity to happen (T.L.O., 2016).

Figure 15. Cryopreservation of the body (Reynolds, 2015)

REFERENCES

Acey, M. (2013). *Brain implants: Restoring memory with a microchip*. CNN. https://edition.cnn.com/2013/05/07/tech/brain-memory-implants-humans/index.html

Arthur, C. (2013). Music to your ears? Try a headphones implant like Rich Lee. *The Guardian*. https://www.theguardian.com/technology/2013/jul/04/headphones-implanted-ear-grinder-rich-lee

Berg, D. (2012). I Have a Magnet Implant In My Finger. *Gizmodo*. https://gizmodo.com/i-have-a-magnet-implant-in-my-finger-5895555

Carmichael, J. (2013). Why this guy implanted headphones in his ears. *Popular Science*. https://www.popsci.com/technology/article/2013-06/biohacking-implanted-headphones/

CBS News. (2015). Wearable tech aims to give you a jolt - or help you relax. *CBS News*. https://www.cbsnews.com/news/wearable-device-thync-helps-change-mood-says-ceo/#:~:text=If%20you%20turn%20to%20coffee,and%20it%20costs%20nearly%20%24300.

Choi, C. (2016). Wireless 'Neural Dust' Could Monitor Your Brain. *Popular Science*. https://www.popsci.com/tiny-wireless-implants-could-monitor-your-brain/#:~:text="Neural%20dust"%20wireless%2C%20implantable,only% 203%20millimeters%20in%20length.&text=Wireless%20powered%20implants%2C%20each%20smaller,scan%20and%20stimulate%20brain%20cells.

Coffey, H. (n.d.). Swedish commuters can use futuristic hand implant microchip as train tickets. *The Independent*. https://www.independent.co.uk/travel/news-and-advice/sj-rail-train-tickets-hand-implant-microchip-biometric-sweden-a7793641.html

Edwards, J. (2016). Now you can board a plane just by waving your hand. *News.com.au*. https://www.news.com.au/travel/travel-advice/flights/now-you-can-board-a-plane-just-by-waving-your-hand/news-story/12bc677a48c629e5742ebebae848d5f4

Francis, H. (2015). Chip implants beneath the skin bring a new meaning to 'pay wave'. *The Sydney Morning Herald*. https://www.smh.com.au/technology/chip-implants-beneath-the-skin-bring-a-new-meaning-to-pay-wave-20150528-ghbq71.html

Jussi. (2005). *Baja Beach Club*. Urban Eyes. http://project-urbaneyes.blogspot.com

Kurzweil, R. (2014). *Medical Nanobots Will Connect Brain to Cloud Computing*. Nano Brain Implant. https://nanobrainimplant.com/2014/02/08/medical-nanobots-will-connect-brain-to-cloud-computing-ray-kurzweil/

Lee, R. (2013). Geek surgically implants magnets inside ears, turns them into headphones. *TechBlog*. https://www.techeblog.com/geek-surgically-implants-magnets-inside-ears-turns-them-into-headphones/

Lee, R. (2017). Biohacking the eye to see infrared. *Omni Media*. https://vocal.media/futurism/biohacking-the-eye-to-see-infrared

Liberatore, S. (2016). Would you IMPLANT a chip so you can breeze through airport security? Dutch traveler first to use wireless tech. *Daily Mail*. https://www.dailymail.co.uk/sciencetech/article-3402048/Would-IMPLANT-chip-breeze-airport-security-Dutch-traveler-use-wireless-tech.html

LifeNaut Project. (n.d.). https://www.lifenaut.com

Magazin.hr. (2016). Pravi kiborg: Biohaker želi vlastiti penis pretvoriti u vibrator. *Net.hr*. https://net.hr/magazin/zdravlje/pravi-kiborg-biohaker-zeli-vlastiti-penis-pretvoriti-u-vibrator-32191e02-b1cd-11eb-99a9-0242ac140022

Mindflex. (n.d.). *NeuroSky*. https://store.neurosky.com/

Neporent, L. (2013). Magnet Implants: Sixth Sense or Nonsense? *ABC News*. https://abcnews.go.com/blogs/health/2013/05/27/magnet-implants-sixth-sense-or-nonsense

Orenstein, D. (2011). Deep brain stimulation helps severe OCD, but pioneer advises caution, compassion. *Brown University*. https://news.brown.edu/articles/2011/02/dbs#:~:text=Deep%20brain%20stimulation%20helps%20severe%20OCD%2C%20but%20pioneer%20advises%20caution%2C%20compassion,-February%2016%2C%202011&text=For%20patients%20most%20severely%20afflicted,when%20it%20never%20worked%20before.

Patel, N. (2015). No, these biohachers can't give themselves infrared vision. *Wired*. https://www.wired.com/2015/04/no-biohackers-cant-give-infrared-vision/#:~:text=Though%20the%20results%20were%20apparently,electronic%20snooperscope%20to%20see%20infrared.

Plenke, M. (2015). *These Young Cyborgs Are Building the Future of Modern Medicine*. Mic. https://www.mic.com/articles/125205/body-modification-is-the-future-of-modern-medicine

Radiant, D. (n.d.). *Cinemagraph*. https://express-uk.co/news/science/877457/brain-function-dementia-super-power-human-kernel

Redman, J. (2016). Bio-Implantable Bitcoin Wallets Becoming a Popular Storage Solution. *Bitcoin.com*. https://news.bitcoin.com/bio-implantable-bitcoin-growing-popular/

Rettner, R. (2016). Mind-tracking devices: Do brain wearables really work? *Live Science*. https://www.livescience.com/53840-do-brain-wearable-devices-really-work.html#:~:text=Independent%20experts%20say%20that%2C%20in,different%20parts%20of%20the%20body.

Reynolds, E. (2015). Brain freeze: The people putting themselves storage. *News.com.au*. https://www.news.com.au/technology/innovation/brain-freeze-the-people-putting-themselves-in-frozen-storage/news-story/3e1f493fc6370190760bb45542c3bdc6

Robertson, A. (2014). Cyborg conversion incomplete: my life with finger implants. *The Verge*. https://www.theverge.com/2014/7/31/5952647/cyborg-conversion-incomplete-my-life-with-finger-implants

Rothman, P. (2013). Biohacking/Grinder Update: Tim Cannon Implants Circadia 1.0. *Tech Insider*. http://www.businessinsider.com/grindhouse-wetware-launches-new-implantable-northstar-device-2015-11

Rowlands, Licina, & Galvin. (2016). Can we biologically extend the range of human vision into the near infrared? *Experiment*. https://experiment.com/projects/can-we-biologically-extend-the-range-of-human-vision-into-the-near- infrared#:~:text=We%20have%20developed%20a%20protocol,infrared%20vision%20to%20freshwater%20fish.&text=The%20cone%20cells %20are%20granted%20sharp%20color%20vision%20by%20the%20complex%20photopsin.

Selak, M. (2013). *Ljudska priroda i nova epoha*. Naklada Breza.

Thompson, C. (2015). A new cyborg implant may give users the power to control devices with their gestures. *Tech Insider*. https://www.businessinsider.in/a-new-cyborg-implant-may-give-users-the-power-to-control-devices-with-their- gestures/articleshow/49730562.cms#:~:text=The%20device%2C%20called%20the%20Northstar,to%20control%20devices%20using%20gestures.

Thync. (n.d.). *Roop*. https://roopbd.com/projects/thync.html

T.L.O. (2016). *Odrubite si glavu i zamrznite je sad – vratite se među žive kada čovječanstvo otkrije besmrtnost.* Index.hr. https://www.index.hr/vijesti/clanak/odrubite-si-glavu-i-zamrznite-je-sad-vratite-se-medju-zive-kada-covjecanstvo-otkrije-besmrtnost/932181.aspx

Trafton, A. (2010). Silencing the brain with light. *MIT News*. https://news.mit.edu/2010/brain-control-0107

Vitas, Z. (2016). Hoćemo li u mozgu imati čipove poput filmskih junaka? *Večernji list*. https://www.vecernji.hr/lifestyle/hocemo-li-u-mozgu-imati-cipove-poput-filmskih-junaka-1126659

Williams, A. (2013). North Paw: A vibrating, ankle-worn compass. *Gizmag*. https://newatlas.com/north-paw-vibrating-ankle-compass-kit/26459/

ENDNOTES

1 More information available in Francis (2015)

2 NFC (Near field communication) is a short-length wireless communication technology, i.e. a near-field communication technology which allows two devices to exchange data. Wireless payments are conducted via NFC chips.

3 Screenshot of the Dangerous Things website. Source: https://dangerousthings.com/?s=nfc&post_type=product, last accessed: 22/6/2016.

4 Source: North Paw Directional Anklet Kit from ThinkGeek (https://www.youtube.com/watch?v=ptWGVuRA0u8, last accessed 2/9/2016.

5 Grindhouse Wetware. Source: http://www.grindhousewetware.com/ [18/8/2016]

Chapter 5
Avatars as Bodiless Characters

ABSTRACT

In this chapter, the author introduces the reader to the importance of virtual reality in human life, avatars, and communicating with digital characters and demonstrates the pervasiveness of technology's penetration into our lives, not only physically, cognitively, and emotionally, but also environmentally. As the created interpreters and representatives of scientific work as the substance subject of scientific history, avatars participate, along with robots, cyborgs, and artificial intelligence, in the desubjectivization, biological denaturalization, and despiritualization of man and death of biological life. The 'cyborgization' of humans in virtual space extends the landscape of the discussion on cyborgoethics.

6.1 BODILESS MAN IN A VIRTUAL REALITY

Leibniz, that mathematical spirit, saw in the mystic elegance of the binary system that counts only the zero and the one, the very image of creation. The unity of the supreme being, operating by binary function in nothingness, would have sufficed to bring out of it all the beings.

McLuhan

For centuries, a significant part of human attention was focused on resolving the limitations of our mortal bodies. Many ancient religious traditions tell us that the final salvation of the soul is only possible in a permanently bodiless state. Immortality of the soul and its superiority over the material body was also the main topic of discussion within the philosophical schools of ancient Greece. Today, these discussions are primarily held in scientific laboratories that attempt to bring this human desire to life. Virtual cyberspace represents one of the possible pathways to the state of bodilessness. What makes it so special? A part of it certainly lies in the promise that virtual space will free man from the limitations of his mortal body and ensure the bodiless immortality of the soul. In this new digital culture, "we will socialize in digital neighbourhoods in which physical space will be irrelevant and time will play a different role" (Groothuis, 1999, pp. 9-10). This will represent the final triumph of technology over finiteness as "The very material world that so hampers the insatiable desires of the soul provides the wherewithal

DOI: 10.4018/978-1-7998-9231-1.ch005

for the technologies that aim to release the soul from its captivity" (Groothuis, 1999, p. 39), but with necessary adjustments:

[...] people believe technology works, that they rely on it, that it makes promises, that they are bereft when denied access to it, that they are delighted when they are in its presence [...] that they stand in awe of it, and that, in the born-again mode, they will alter their lifestyles, their schedules, their habits, and their relationships to accommodate it. (Postman, 1995, p. 19)

Thus, our technologies influence the way we perceive our world and its future as "Science explores: Technology executes: Man conforms." (Groothuis, 1999, p. 15).

Cyberspace[1] or virtual space was introduced for the first time in *Neuromancer* written by Willian Gibson. Gibson defines cyberspace as the space "in which computer-mediated communication occurs; that is, to the interface between digital bits and human consciousness – or between silicon and the soul" (Groothuis, 1999, pp. 13-14). According to Gibson, virtual space is a world without boundaries between the material and the virtual in which every identity is deconstructed:

Taking on various identities in varying circumstances is sanctioned by this new movement, for it exemplifies the death of belief in the unitary self, the hard ego, the irreducible center of personal identity. Identity is not fixed, but fluid; not singular but multiple; not prescribed, but protean; not defined but diffused. (Groothuis, 1999, pp. 26)

Instead of the natural-artificial, central-peripheral dichotomies, there is only gradience, the in-between steps that dissolve the memory of once clear and unambiguous identities. The actors of the virtual world "swept up by experiences that enable them to explore previously unexamined aspects of their sexuality or that challenge their ideas about a unitary self" (Groothuis, 1999, pp. 28). In place of such dividedness and disorientedness of the human personality, "the phrase swept up is the one to highlight. When we become engrossed in the capacities of a powerful new technology, our critical faculties may be overwhelmed by the pragmatics of making the thing work and by the sheer sensory delight of exploring new experiences (Groothuis, 1999, pp. 28). In the analysis of the virtual space, Turkle notes that this type of activity leads to a disturbance in personality devoid of its core.

Jaron Lanier argues that virtual reality is a simulation or a perceptive scenario that the human organism is exposed to in order to achieve the psycho-receptive reactions that can programmed and reprogrammed (Gržinić, 1988, pp. 91). What is new about the virtual space is its seemingly interactive nature, and this interactivity is what makes it psychologically real as a form social reality. Today, the term virtual reality has a two-fold meaning – on the one hand, it denotes the technology of the future that will represent pure information space in which the users will move across a 3D environment with a virtual body, and on the other hand, it refers to the full immersion technologies of today. French philosopher Paul Virilio notes that "we no longer even need to depart because 'everything arrives without our having to leave.' But what 'arrives' is already no longer a stopover or the end of the trip; it is merely information, information-world, no, information universe!" (Groothuis, 1999, p. 74). Michael Heim argues that

[...] the computer network simply brackets the physical presence of the participants, by either omitting or simulating corporeal immediacy. The stand-in body reveals only as much of ourselves as we mentally wish reveal. The avatar self lacks the vulnerability and fragility of our primary identities. The stand-in

self can never fully represent us. The more we mistake the cyberbodies for ourselves, the more machine twists ourselves into the prostheses we are wearing. (Groothuis, 1999, p. 101)

The inhabitants of the virtual world are, thus, virtual themselves and without a body as a unifying anchor and as digital alchemists have tools at their disposal that their predecessors could only dream of. Computer screens are their magic mirrors which depict alternate realities diverging from one another in myriad ways. The possibility of having multiple electronic personalities in the bodiless world of virtual communities results in blurred identities, lost in the technology which augments human capabilities, but on a global level threatens the domination of man by fighting for autonomy:

Given that the respective identities of humans and machines can be so easily confused in cyberspace[2]*, one wonders just what kind of community this might be. Simulated personality may be fascinating but how can interaction with a bot help civilize us, help us work through challenges and difficulties with real people? Moreover, if the nature of much cyberspace communication is such that it can be duplicated by a bot, how deeply personal and truly communicative can it be – even when it is between humans? (Groothuis, 1999, p. 136)*

Computer expert Winn Schwartau also expresses his concerns about this: "The essence of our very being is distributed across thousands of computers and data bases over which we have little or no control" (Groothuis, 1999, p. 140). In the digital world, bits appear as living beings and man is just a fleeting moment.

In this respect, McLuhan argued long ago that the interfaces of modern communication are simply added to the body: "To say that any technology or extension of man creates a new environment is a much better way of saying that the medium is the message. Moreover, this environment is always invisible" (Groothuis, 1999, p. 23). In the same vein, Biocca regards the development of appropriate interfaces as progressive embodiment. The process of progressive embodiment occurs when there is increased social integration of the interface, which means that the interface is integrated into everyday activities at work, at home or on the street. The end goal is a perfectly natural interface, as opposed to 'unnatural environment, which causes the so-called cyborg's dilemma:

The more natural the interface the more "human" it is, the more it adapts to the human body and mind. The more the interface adapts to the human body and mind, the more the body and mind adapts to the non-human interface. Therefore, the more natural the interface, the more we become "unnatural," the more we become cyborgs. (Biocca, 1997)

William Gibson "calls the body 'wetware' and the embodied world 'meatspace'" (Groothuis, 1999, p. 42), and argues that the body is just 'meat' that should abandoned in the bodiless happiness of virtual reality. In his novel, Gibson "describes a computer addict who laments losing his ability to load his consciousness into the computer matrix. For Case, who'd lived for the bodiless exultation of cyberspace, it was the Fall. In the bars he'd frequented as a cowboy hotshot, the elite stance involved a certain relaxed contempt for the flesh. The body was meant. Case fell into the prison of his own flesh" (Groothuis, 1999, p. 42). If we reject the human body and willingly accept the participation in the transformation of "flesh into silicon and consciousness into bits – cyberworld without end" (Groothuis, 1999, p. 43), we could achieve digital immortality. However,

The quintessential error of those hoping for a digital immortality is their confusion of the material brain with the immaterial mind. They reduce all mental activity – intending, hoping, willing, rejoicing, lamenting, loving – to nothing but the arrangement of the meat. But, as philosophers have long argued, this confuses the subjective consciousness with the objective organism. Brain states accompany mental states in our physical bodies. (Groothuis, 1999, p. 44)

Michael Heim also adds:

We cannot locate the anchor for our reality check outside this fluctuating, changing world. No universal divinity ensures an invariant stability for things. But we need some sense of metaphysical anchoring, I think, to enhance virtual worlds. A virtual world can be virtual only as long as we can contrast it with the real (anchored) world. (Heim, 1993, p. 133)

Baudrillard describes the exchange of direct experience of human beings in the natural community with machines. This, according to Baudrillard, leads to estrangement in which every desire and knowledge are provided to another user. The multimedia and news reports allow us to see what others have already seen so much so that there is nothing that someone else had not seen before. We gave the machines a task to watch instead of us, and we will soon give computers the task to make all the decisions for us (Groothuis, 1999). We can say that virtual reality reflects the Western philosophical tradition, the Platonic worldview in which the physical tactile world is nothing else but a reflection of a purer, more ideal domain of the perfect form. For some, the emergence of new information technologies represents the realization of this model. The artificial language of the binary code, separated from the everyday conventions of material communication can strive towards the higher perfection of the absolute form: "At the computer interface, the spirit migrates from the body to a world of total representation. Information and images float through the Platonic mind without a grounding in bodily experience" (Heim, 1993, p. 101). Physical information can be transformed into pure forms of bodiless intelligence and data: human ontology is digitalized.

Heim believes that

Today's computer communication cuts the physical face out of the communication process. Computers stick the windows of the soul behind monitors, headsets and datasuits [...] The living non-representable face is the primal source of responsibility, the direct, warm link between private bodies. Without directly meeting others physically, our ethics languishes. (Groothuis, 1999, p. 48)

Due to lack of physical connection, we can wonder "In this age of disembodied communication in cyberspace, how should we honor and respect the physical presence of persons" (Groothuis, 1999, p. 47). The artificial is offered as the only alternative for the postmodern soul. There is no real self nor a valid suggestion of a moral and spiritual progress (Groothuis, 1999). Despite all of that, "[...] some are heralding cyberspace as a means to release the soul from the prison of the body. They hail the digital age as ushering in a new era of disembodied intelligence that is free to explore and inhabit an electronic realm of almost infinite possibilities. Are they right?" (Groothuis, 1999, p. 35). Seneca said: "To be everywhere is to be nowhere. To adapt this to cyberspace, we might say: 'To be virtually everywhere is to be literally nowhere'" (Groothuis, 1999, p. 74). If "cyberspace is kept closely fastened to the real world, and if we refuse its temptations to exchange the virtual for the literal, it can be our servant. Otherwise, it will become a demanding and all-consuming media master" [which consumes our body and soul, I.G.] (Groothuis, 1999, p. 143). As we speak, the speed and the amount of data flowing through the internet network make it far superior than the outdated human being.

Our analysis of cyborgization of the entire historical reality (both nature and society) leads to the conclusion that the virtual world rests on its own ontology of the digital code which aims to become its "metaphysical principle" (Baudrillard, 1993, p. 57) which will erase the differences between the virtual and the real world.

6.2 ESTABLISHING AN AVATAR IN ACTUAL AND VIRTUAL REALITY

The Sanskrit word *avatar* originally denoted the personification of a god, which in Hinduist philosophy represents a manifestation or an incarnation of a divine being in multiple forms on Earth. In the 1856 novel *Avatar*, Théophile Gautier describes a disabled young man treated by a mysterious doctor who, having recently returned from India, treats the young man by replacing the body with a brand new one using electromagnetic equipment. Avatar was represented as an alien being in Poul Anderson's novel *The Avatar* (1978). Neal Stephenson's 1992 novel *Snow Crash* was the first novel to use the term *Metaverse* for virtual reality (Casilli, 2005, pp. 183-209).

In today's cybernetic reality, avatar denotes the presence of a human agent in a technological environment, a technologically surrogate of the user's body. Depending on the characteristics used in the process of interaction, avatars can be divided into material and digital avatars (Fig. 1). We will show different examples and manners of their use in the virtual and the real world but the emphasis will be on digital avatars which should, according to the promises of our futurologists, take us into digital immortality.

Material avatars are traditionally used to describe the human form of God in its interaction with mortal beings. They can be two-dimensional material avatars, like the images of the god Vishnu in the form of a fish (Mastya), or three-dimensional material avatars, like the Mohini statue (Vishnu's avatar). In the Hindu mythology, a divinity or the supreme being does not come down to earth in its real form but is manifested as a man or an avatar. The avatar of the god Vishnu has had ten manifestations on Earth – ten avatars or incarnations appear in particular time periods in order to preserve the established order (Javaram, n.d.).

Digital avatars are also categorized into several groups (Fig. 2): therapeutic avatars, disease-treating avatars (therapeutic avatars for anxiety and schizophrenia) and interactive avatars which imitate human communication and have a human-like appearance (hologram avatar, emotional avatar, real-time avatar)

A *therapeutic avatar* is a computer-generated avatar which helps in interaction with patients to treat various disorders, such as anxiety, schizophrenia, etc. The *Anxiety Disorders Clinic* in Florida uses the virtual simulation program to help children with anxiety disorders. The children participating in the interaction with avatars have to respond to social scenarios of varying difficulty and use interactive exercises to develop certain social values (Warnke, 2012). *Center for Disease Control and Prevention* wants to prepare the participants (using images, sounds or smells of disasters) for emergencies in which they are exposed to simulated outbreaks of diseases or accidents (Wilcox, 2007). Psychiatrist Julian Leff has developed a schizophrenia treatment therapy in which patients are faced with a computer-generated avatar that has their own voice. This allows the patients to confront themselves in order to control the avatar (Weston, 2013). Therapeutic avatars have had some success in the domain of mental health and

Figure 1. Typology of avatars[3]

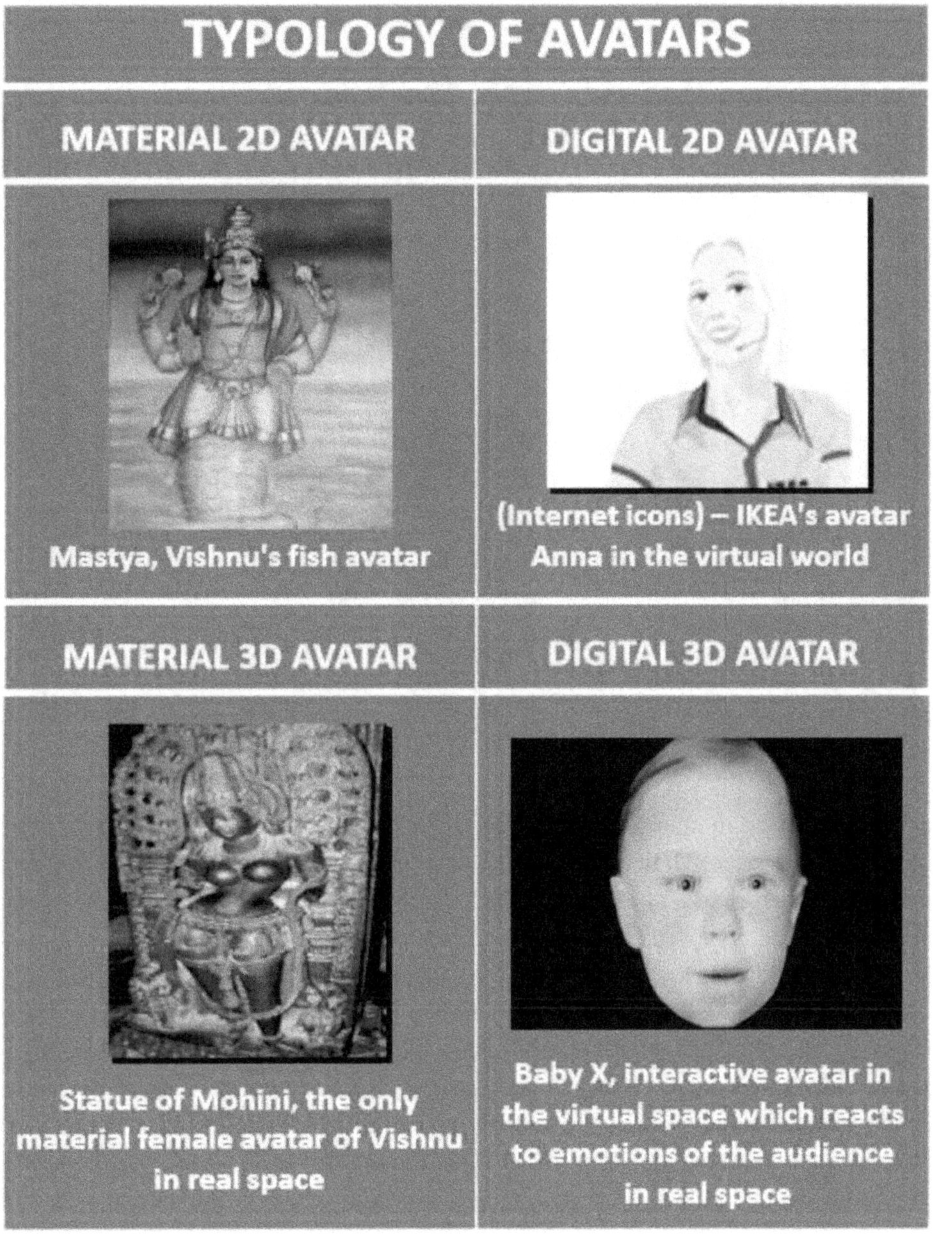

can be used for physical wellbeing as well, e.g. for showing the morbidly obese persons what they might look like after losing weight.

A *hologram*[5] *avatar* is an interactive avatar projected as a holographic image on a plastic PMMA screen in the form of a 3D high-resolution projection in colour. The aim of holographic avatars is, for instance, to help the passengers in airports with flight information. Some avatars have proximity sensors which activate messages when somebody is in their vicinity, while others continuously repeat their pre-programmed speech patterns, such as security regulation, boarding information, etc. (Schweber, 2012).

Figure 2. Typology of digital avatars[4]

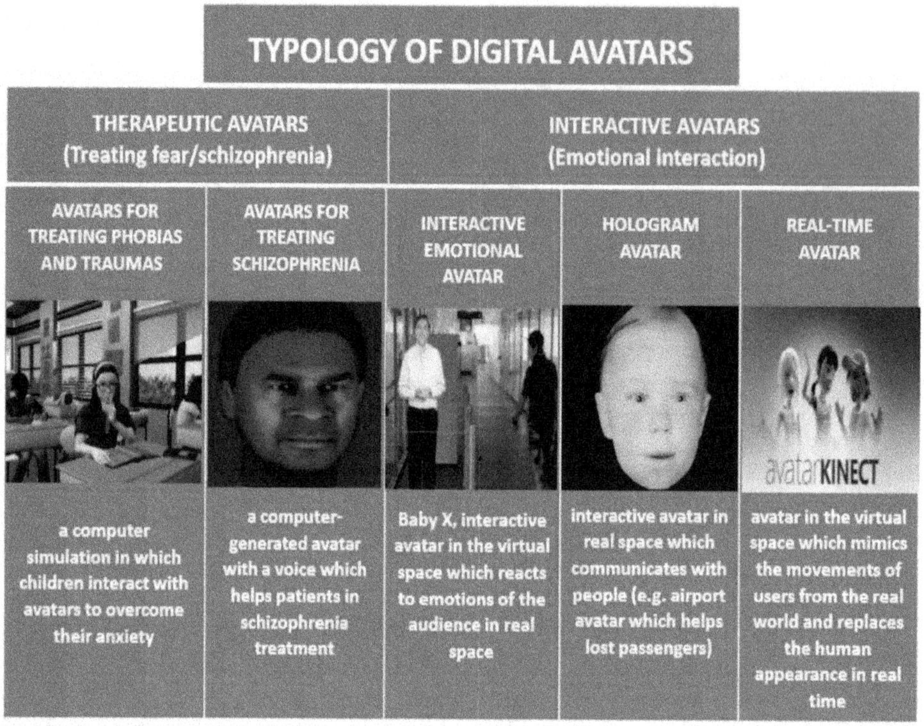

An *emotional avatar* is an interactive avatar in virtual reality which reacts to emotional expressions of the audience in a real space. Mark Sagar and the laboratory of animation technology at the University of Auckland present the research of emotional interaction through an interactive avatar *Baby X*. *Baby X* has an artificial brain and is moved through the artificial nervous system via input. If a person stands in front of the screen, the face of the baby reacts to its facial expression. It can control the lights and the sound in the room. The future goal of this interactive avatar is to connect it with other artificial intelligence (AI) systems that can provide an additional level (O'Neill, 2011). As an example of this, we should mention the avatar *Zoe* which was created at the *Toshiba Research Laboratory* and the Department of Engineering at the *Cambridge University*. *Zoe* is an emotional avatar in the form of a talking head, which serves as a personal digital assistant. The aim of the avatar is to repeat realistic expressions of human emotions, which allows the user to create an avatar which will change its emotions (happy, angry, scared) and voice (accent, speed, pitch) to match the feeling that the user wants to simulate. By writing personalized messages and choosing the emotions, the face will recite the suggest text and show the selected emotion. This technology could be used as an aid for the hearing impaired, in video games, audio-visual books, online lectures and other user interfaces. In the future, the users will be able to upload their own face and voice, thus personalizing their own emotionally realistic avatars, and make more natural-like gestures (Cellan, 2013).

A *real-time avatar* (Fig. 46) mimics the movement of the user in the real world and can replace the human form in real time. Avatar *Kinect* is an online video chat application for Xbox that uses the Kinect sensor to track face and lip movement and transforms them into an on-screen user avatar (Johnston,

2013). Apart from Avatar *Kinect*, there is also the Japanese avatar software which replaces the human appearance in real time, records a video about a person and automatically gives it a new hairstyle and clothes in real time (Frucci, 2007).

An *online user* avatar represents a computer-generated user avatar in the virtual world, like video games, forums, blogs and social networks. Computer avatars for blogs/forums represent forum users (personality, ambitions, online social status, etc.). Computer avatars are used by players of video games, such as *Ultima IV* (1981), a series of games in which the avatar represents a user and can be modified at will. *Kavena* (2004) was the first game to allow combining the social network and the 3D virtual world with the ability of playing; *Second Life* (2003) is an online 3D virtual world in which the users generate content etc.

Figure 3. Typology of real-time avatars[6]

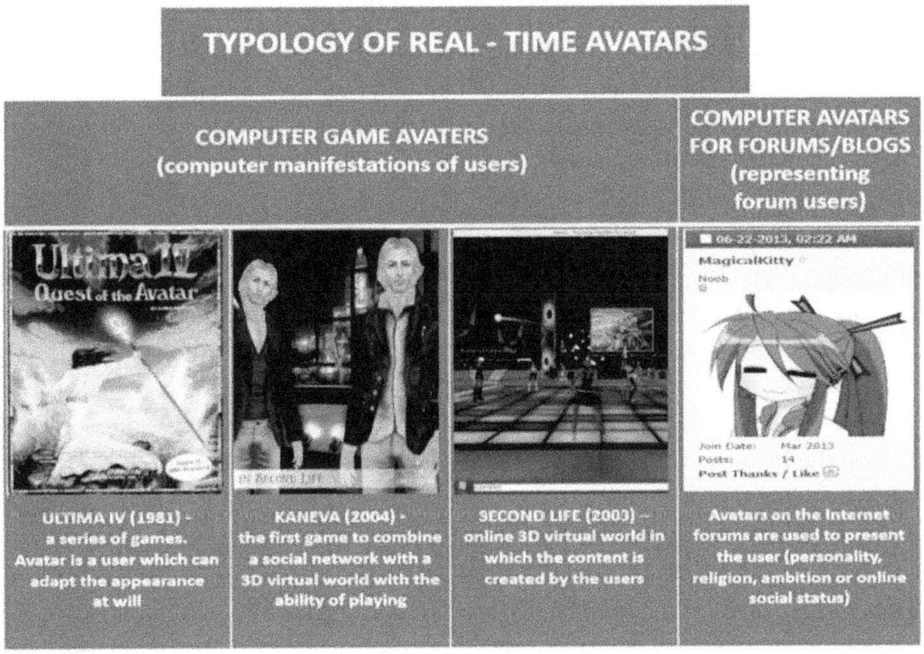

In the future, general physicians will be human-like avatars. They will appear on private home screens when necessary and diagnose the disorders. Since their program will have the general genetic picture of its patient host, it will be able to diagnose 95% of the most common ailments. The health state of every patient will be tracked using chips implanted into human environment that will be available to healthcare professionals (Kaku, 2011, p. 25). Apart from its application in treatment, everyday life, and games, use of avatars brings about a negative aspect as well: a change in person's attitudes and beliefs due to their emergence in the virtual reality. Cases of sexual harassment by avatars are well-known, as well as cases of identity thefts and negative reactions of the brain in virtual reality. The brain often perceives the immersion into the virtual reality as a physical presence in a virtual environment: "That takes care of two of our senses—visual and auditory. Another high-resolution sense organ is our skin, and "haptic" interfaces to provide a virtual tactile interface are also evolving" (Kurzweil, 1999, p. 104).

Philosophy is facing a challenge of rethinking the relationship of identity and physical body in virtual reality. We have shown the subjectivity of the real life can be overcome with a simulated avatar in virtual reality. The difficulties of readjustment of the brain to the perception of the reality outside of the simulated reality leads to the so-called *Alternate World Syndrome* (AWS), "an acute form of body amnesia which can become chronic *Alternate World Disorder* (AWD)" (Heim, 1995, p. 67). According to Heim, AWS occurs because

AWS mixes images and expectations from an alternate world so as to distort our perceptions of the current world, making us prone to errors in mismatched contexts. The virtual world obtrudes upon our activities in the primary world, and vice versa. The responses ingrained in the one world get out of synch with the other. AWS shows the human being merging, yet still out of phase, with the machine. (Heim, 1995, p. 67)

In the future, it will be possible to use the Kinect sensor to copy a real person, which will enable hackers to record and steal personal identities, but also to create our doppelganger avatars. Artificial bodies could also be misused in various forms of criminal activities, which will increase the risk of sexual harassment or even rape of virtual entities, i.e. avatars. One of the examples of taking over someone else's online personality that leads to virtual rape is the situation

[...] when a MUD player devises a way to possess another character textually, thus becoming its online ventriloquist, as it were. Such a perpetrator is then the only typing out messages for both his character and the one normally played by another person. The real world player at the other end sits at the screen in amazement and then disgust when she finds her onscreen character submitting to sexual acts she neither instigated nor consented to have. (Groothuis, 1999, p. 94)

If an avatar performs a sexually explicit gesture or sexually harasses a user, the question is how the court could prove the rape by an avatar whose real-life user can live in another country with different legislature under which electronic evidence are not enough for prosecution. According to research conducted at the Stanford University, women who play video games using sexy female avatars are prone to sexual objectivization of themselves in real life and are more inclined to accept the myth that they are partially responsible for their own rape. This research is based on the phenomenon known as the *Proteus effect* which describes to what extent human behaviour is adapted to the digital identity. By embodying characters in the virtual world, personal attitudes and beliefs are increasingly adapted when an avatar resembles the person playing the game (Fox et al., 2013).

We shall describe a few more avatars which are being put to different use. The company *Medical Learning* is developing haptic interfaces with interactive virtual patients to enable their physicians to get a feel for different diseases (Kurzweil, 1999, p. 104). The *motion capture* technology, as used in the Microsoft Kinect games, is used to create real experiences with special haptic suits and physical sensors, "Such a suit could also provide a thermal response and thereby allow the simulation of feeling a moist surface" (Kurzweil, 1999, p. 104). *Beam me up Scotty* is the international project financed by the European Commission which aims to explore how humans can visit remote locations via the internet and feel immersed into the new environment (The Telegraph, 2014).

The *Sweetie* avatar was created by the *Terre des Hommes* organization with the aim of raising awareness of exploitation of children via web cameras. During the course of ten weeks, the avatar revealed 1000 individuals who were willing to pay children for committing sexual acts (Tech Times, 2013). Using this method, face and online behaviour of the avatar could help identify people and check their digital identity, which is crucial for legal actions.

Various research projects about human behaviour are related to the emergence of avatars. Blascovich examines the reaction of the brain in the virtual environment. It was discovered that the brain often fails to distinguish the virtual from the real experience. For instance, in the experiment with a virtual hole, none of the participants wanted to jump into the hole, despite being informed that the hole is only virtual (Blascovich & Bailenson, 2011). Time spent in the virtual world can have numerous implications on brain development.

Later in the twenty-first century, as neural implant technologies become ubiquitous, we will be able to create and interact with virtual environments without having to enter a virtual reality booth. Your neural implants will provide the simulated sensory inputs of the virtual environment—and your virtual body—directly in your brain [...] This will be the essence of the Web [...] A typical "web site" will be a perceived virtual environment, with no external hardware required. You "go there" by mentally selecting the site and then entering that world. (Kurzweil, 1999, p. 105).

An advocate of scientific evolution of man and the co-founder of transhumanism Nick Bostrom lists three arguments for the simulation argument:

1. The chances that a species at our current level of development can avoid going extinct before becoming technologically mature is negligibly small. 2. Almost no technologically mature civilisations are interested in running computer simulations of minds like ours. 3. You are almost certainly in a simulation. (Bostrom, 2009, p. 40)

Realization of similar proposals has already taken place through the quest for information immortality by transferring the human consciousness into holographic avatars. This is the scope of the Russian project *2045 Initiative*, which should be completed by 2045. The main goal is "to create technologies enabling the transfer of an individual's personality to a more advanced non-biological carrier, and extending life, including to the point of immortality" (2045 Strategic Social Initiative, n.d.). The project founder Dmitry Itskov wants to achieve immortality of cybernetic beings and deems it possible and crucial for stopping aging, death, and overcoming the basic boundaries of physical and mental capabilities that currently limit our physical bodies. The *Avatar* project is divided into four stages. The *Avatar A* stage (2015-2020) includes controlling the human replicas through the brain-computer interface which will be able to send orders directly from the mind and back into it. A similar project is currently being conducted by the US *Defense Advanced Research Projects Agency*, which invested 7 million dollars to develop an interface that would allow the US soldiers to control the avatar with their minds using "interfaces and algorithms to enable a soldier to effectively partner with a semi-autonomous bi-pedal machine and allow it to act as the soldier's surrogate" (Rahman, 2012). The *Avatar B* stage (2020-2025) aims to transplant the human brain into an artificial body *post mortem*. The immovable patients will healthy brains will be able to fully regain their mobility by moving into a virtual body using their new synthetic bodies and various bioelectronic devices. The *Avatar C* stage (2025-2035) predicts the ability to transfer knowledge, memory, personality and individuality into an artificial body. Itskov is hoping to find the means to download the consciousness into synthetic versions, which will grant humans cybernetic immortality and offer the possibility for ordinary human beings to enhance their brain and manipulate their memories. In the final stage *Avatar D* (2035-2045), human life will be reduced to holographic bodies – avatars, matter of independent computer minds, bodies of different forms. A holographic body will be able to walk through walls and move at the speed of light, while the nanorobot bodies will be able to take on several different forms (Segal, 2020).

Research shows that mankind is increasingly losing control over computer avatars. Mark Meadows explains:

Part of the danger lies in how we control our avatar and how our avatar controls us. As people become more involved in the roles and rules of their avatar, they can also lose control of their alternative personality they have invented for that system. The alternative personality can become predominant and begin to take over the primary, daily one. People sometimes prefer their avatar personas to their 'real' ones. Your role as an avatar can take control of your life as a person. (Meadows, 2008, pp. 82-83)

We see a similar scenario in the British TV show *Black Mirror,* in which the wife replaces the husband's dead body with a robot – a digital simulation based on his Twitter conversations. After logging into his former Twitter profile, the machine talks like the deceased husband. Achieving cybernetic immortality is scrutinized by V. Bush in the article *As We May Think* published in the *Atlantic Monthly*:

[...] the physicists should develop a "memex." This would be "a device in which an individual stores all his books, records, and communications, and which is mechanized so that it may be consulted with exceeding speed and flexibility." It would be, Bush wrote, an "intimate supplement to his memory." Bush dreamed of scientists wearing glasses that could automatically record those things "worthy of the record." (Turkle, 2010, p. 299)

A similar ground-breaking feat was accomplished by Steve Mann in the late 1970s. Using his digital goggles, he made a video record and put it onto a network to be equally available to everyone. Today, the analogue multi-purpose *Google Glass* is already being sold world-wide. During the 1990s, Gordon Bell and Jim Gemmell from Microsoft started the project *MyLifeBits* which, in their view, allows the creation of a life collection which would allow our descendants to get to know our life:

Bell wore voice-recording equipment and a camera programmed to take a new photograph when it sensed (by a change of ambient light) that Bell was with a new person or in a new setting. MyLifeBits recorded Bell's telephone calls, the songs he listened to, and the programs he watched on radio and television. When Bell was at the computer, it recorded the Web pages he visited, the files he opened, the messages he sent and received. It even monitored which windows were in the foreground of his screen at any time and how much mouse and keyboard activity was going on. (Turkle, 2010, p. 300)

Using the online service called *Lifenaut,* it is possible to digitally clone and archive human personality. The goal of *Lifenaut* is to create an online version of the user so that he/she could live forever as an avatar with whom the future generations will be able to communicate after the user's death (Finley, 2013). Turkle (2010, p. 302) mentions several issues that we should pay attention to when making the digital life archive – what happens with memory in a completely archived world (do we have to memorize less things if technology is doing that for us), whether life in the digital culture is transforming into an archive creation strategy (how will we lead the life we are recording – are we leading a life that we would love to have recorded? Bell mentions a positive example of a digital life archive that enabled him to have cleansing of sorts when getting rid of all memories by storing them into a computer (Turkle, 2010, p. 302). This could pave the way towards digital immortality in which humans will become similar to robots, devoid of all memories which they will be able to upload when necessary. Will this lead to situations in which a man does not exist if he/she is not digitalized? "If it's not in your database, it doesn't exist? That's the sort of eerie philosophical proposition Bell's project raises" (Turkle, 2010, p. 302).

Avatars have been portrayed like the means of man's purpose in the form of virtual doppelgangers which can, like every other means have both positive or negative impacts on the life of man. If we go back to the original meaning of the word *avatar*, denoting the embodiment of God, we come closer to the true essence of avatars of today. In the *age of scientific humanism as naturalism*, they most certainly do not embody a religious divinity which in our age does not have the place or meaning it used to have in recent past. The avatars of our age embody *scientific work* as the supreme deity in the age of *scientific humanism as naturalism*. In this respect, avatars are not the doubles or simulations of man in virtual or real space, but ontologically established bodiless beings, new mediums and interpreters of *scientific work*. As the created interpreters and the representatives of *scientific work* as the substance subject of scientific history, avatars participate, along with robots, cyborgs and artificial intelligence, in the de-subjectivization, biological denaturalization, and despiritualization of man and death of biological life. We have to emphasize that the possibility of choosing *humanism as naturalism* or *scientific humanism as naturalism* still remains within the scope of our life responsibility.

6.3 COMMUNICATING WITH DIGITAL CHARACTERS

Augmented reality connects the real world with computer apparitions using special goggles. Scientists at the *University of Washington* in Seattle are working on a prototype of internet contact lenses which would enable their users to have a direct communication with the internet and download all wanted information and content, as well as manage our computer files. Glasses are being implanted with software for recognition of objects and living beings, which will allow the user to have all the information about the person they had just met. A tiny camera records the surroundings and sends the images to the internet via wireless connection, according to one's desires (Turkle, 2010, p. 299). In the age of *scientific humanism as naturalism*, both material and digital avatars are established to represent persons, animate characters of users of a computer or other persons in the virtual space. As an example, 'revival' of the deceased via computer interface, Russian programmer Eugenia Kuyda managed to 'revive' her tragically deceased best friend Roman Mozurenko using a computer. Eugenia gathered all the digital memory about her late friend – his photographs, newspaper articles, text messages, Facebook messages, and put them into a special program she created to generate an app to which she can talk. She can talk with her friend about his life or chat with him as you would normally do, and he will answer in his usual tone, according to Eugenia. In this respect, digital avatars bridge the gap between illusion and reality.

REFERENCES

Baudrillard, J. (1993). *Symbolic exchange and death.* Sage Publications.

Biocca, F. (2006). The Cyborg's Dilemma: Progressive Embodiment in Virtual Environments. *Journal of Computer-Mediated Communication, 3*(2). doi:10.1111/j.1083-6101.1997.tb00070.x

Blascovich, J., & Bailenson, J. (2011). *Infinite reality: Avatars, eternal life, new worlds, and the dawn of the virtual revolution* (1st ed.). William Morrow.

Bostrom, N. (2009). Are you in a computer simulation? In S. Schneider (Ed.), *Science fiction and philosophy: From time travel to superintelligence* (pp. 38–42). Wiley-Blackwell.

Casilli, A. (2005). Les avatars bleus. *Communications, 77*(1), 183–209. https://doi.org/10.3406/comm.2005.2269

Čatić, I., Greguric, I., & Bahtijari, D. (2013). *Kiborzi*. Roboti i Avatari.

Cellan, R. (2013). Zoe – Cambridge's emotional talking head. *BBC News*. https://www.bbc.com/news/technology-21827924

Finley, K. (2013). If You Plug Twitter Into a Digital Avatar, Can You Live Forever? *Wired*. https://www.wired.com/2013/09/lifenaut/

Fox, J., Bailenson, J. N., & Tricase, L. (2013). The embodiment of sexualized virtual selves: The Proteus effect and experiences of self-objectification via avatars. *Computers in Human Behavior, 29*(3), 930–938. https://doi.org/10.1016/j.chb.2012.12.027

Frucci, A. (2007). Japanese Avatar Software Replaces Your Hair, Gives You Facepaint in Real Time. *Gizmodo*. https://gizmodo.com/japanese-avatar-software-replaces-your-hair-gives-you-337734

Groothuis, D. R. (1999). *The soul in cyberspace*. Wipf ans Stock.

Gržinić, M. (1988). U redu za virtualni kruh. *Meander*.

Heim, M. (1993). *The metaphysics of virtual reality*. Oxford University Press.

Heim, M. (1995). The Design of Virtual Reality. In M. Featherstone & R. Burrows (Eds.), *Cyberspace/cyberbodies/cyberpunk: Cultures of technological embodiment* (pp. 65–78). Sage.

Jayaram, V. (n.d.). *The Concept of Avatar or Incarnation in Hinduism*. https://www.hinduwebsite.com/incarnation.asp

Johnston, C. (2011). The loneliest talk show host: Ars tests Avatar Kinect on the Xbox 360. *Ars Technica*. https://arstechnica.com/gaming/2011/07/the-loneliest-talk-show-host-ars-tests-out-avatar-kinect-on-the-xbox-360/

Kaku, M. (2011). *Physics of the future: How science will shape human destiny and our daily lives by the year 2100*. Doubleday.

Kurzweil, R. (2000). *The age of spiritual machines: When computers exceed human intelligence*. Penguin.

Meadows, M. S. (2008). *I, Avatar: The Culture and Consequences of Having a Second Life*. New Riders.

Negroponte, N. (1995). *Being Digital*. Hodder & Stoughton.

O'Neill, R. (2013). Baby X: is she the future of AI? *ZDNet*. https://www.zdnet.com/article/baby-x-is-she-the-future-of-ai/

Postman, N. (1993). Technopoly: The surrender of culture to technology. Vintage Books.

Rahman, A. (2012). DARPA Investing $7 Million In Real life 'Avatar' Project for U.S. Soldiers. *The Tech Journal*. https://thetechjournal.com/science/darpa-investing-7-million-in-real-life-avatar-project-for-u-s-soldiers.xhtml

Schweber, N. (2012). At la Guardia a Smiling Helper Materializes, Digitally. *The New York Times*. https://www.nytimes.com/2012/08/09/nyregion/la-guardias-digital-avatar-gives-passengers-airport-information.html

Segal, D. (2013). Dmitry Itskov and the Avatar Quest. *New York Times*. https://www.nytimes.com/2013/06/02/business/dmitry-itskov-and-the-avatar-quest.html

Strategic Social Initiative. (n.d.). http://2045.com

Tech Times. (2013). Computer avatar Sweetie helps nab over 1000 online child predators. *Tech Times*. https://www.techtimes.com/articles/1295/20131107/computer-avatar-sweetie-helps-nab-over-1000-online-child-predators.htm

The Telegraph. (2014). Beam me up, Scotty: teleportation could become reality. *The Telegraph*. https://www.telegraph.co.uk/news/science/science-news/10863929/Beam-me-up-Scotty-teleportation-could-become-reality.html

Turkle, S. (2010). *Alone together: Why we expect more from technology and less from each other*. Basic Books.

Warnke, K. (2012). New software will help children cope with anxiety disorders, social interaction. *Central Florida Future*. http://www.centralfloridafuture.com/news/new-software-will-help-children-cope-with-anxiety-disorders-social-interaction-1.2744590#.U4s75ElBuM8

Weston, D. (2013). Avatar therapy helps silence voices in schizophrenia. *UCL News*. https://www.ucl.ac.uk/news/2013/may/avatar-therapy-helps-silence-voices-schizophrenia

Wilcox, L. S. (2007). Will avatars offer answers? *Preventing Chronic Disease*, *3*(4). http://www.cdc.gov/pcd/issues/2007/jul/07_0060.htm

ENDNOTES

1 The prefix *cyber-* comes from the word *cybernetics*, which denotes the study of automatic systems.

2 The term *cyberspace* used here is equivalent to our use of the term *virtual reality*.

3 This image was created for the purpose of the roundtable discussion on "Cyborgs, Robots and Avatars" (Čatić et al., 2013), held at the Europe House Zagreb.

4 This image was created for the purpose of the roundtable discussion on "Cyborgs, Robots and Avatars" (Čatić et al., 2013), held at the Europe House Zagreb.

5 Holography was invented by Hungarian scientists Dennis Gabor in 1948. A hologram is a collection of all possible views on a scene within a unique plane of light-modelled samples. When the light passes or is reflected from the plane, the scene is optically reconstructed in the environment (Negroponte, 1996).

6 This image was created for the purpose of the roundtable discussion on "Cyborgs, Robots and Avatars" (Čatić et al., 2013), held at the Europe House Zagreb.

Chapter 6
Cyborg Anthropology

ABSTRACT

In this chapter, the body is re-envisioned as a cybernetic organism, and new types of exchange of information, matter, and energy are anaylsed. Through the cybernetic understanding of a human being as a machine system, it is possible for man to modify and replace lost organs or functions and create a partially artificial being. When the body becomes a scientific and technological creation and real relationships become a cyborg prosthesis of the non-human, the body and the social relations become a problem of cybernetic functioning.

BACKGROUND

Man is an invention of recent date. And one perhaps nearing its end… one can certainly wager that man would be erased, like a face drawn in sand at the edge of the sea.

Michel Foucault

In anticipation of Foucault's fears, scientists are on the verge of creating an artificial being that will imitate man's appearance, engagement, behaviour, and actions. Two possibilities of enhancement are being developed today. One was already described – integrating man with the machine, making him almost artificial through identification of his behaviour and computer program. On the other hand, we have artificial beings which are almost human and which mimic the features and the appearance of man, but which are not human but non-human. This second possibility will be described in the following chapters.

George Grant expresses his attitude towards man's dependence on technology. He adopts Nietzsche's ideas from the *Will to Power*, according to which the new world originated from a pure will for technology, which is in reality an abstract and demonic will for accumulation and disintegration, as a dynamic essence of the modern society. According to Nietzsche, through development of will, the very *technique* as the core of the modern age, "the horizon had finally been wiped clean" (Kroker, 1985, p. 27). Nietzsche knew that "a new and intensified version of technical domination was being prepared for the inhabitants of the empire of technology was being prepared for the inhabitants of the empire of technology. He also

DOI: 10.4018/978-1-7998-9231-1.ch006

recognized, tragically so, that with the coming-to-be of the society of the pure drive to technique, there could be no turning back to the recovery, however urgent, of the "inner restraints" of western culture" (Kroker, 1985, p. 27). In this world, human beings "were reduced to the "commandments" of the will to technique" (Kroker, 1985, p. 28). The tragedy of the new world lies in the fact that we will be forced to live within the purview of the technological society as one of its means. With respect to this, Grant argues: "We can hold in our minds the enormous benefits of technological society, but we cannot so easily hold the ways it may have deprived us, because technique is ourselves" (Kroker, 1985, p. 28). His analysis reveals the meaningless vision of life within the modern technological society: "All descriptions or definitions of technique which place it outside ourselves hide from us what it is" (Kroker, 1985, p. 28).

According to McLuhan, "Media, by altering the environment, evoke in us unique ratios of sense perceptions. The extension of any one sense alters the way we think and act - the way we perceive the world. When these ratios change, MEN CHANGE" (McLuhan 1969 qtd. in Kroker, 1985, p. 56) In contrast to "Grant's haunting image of trapped beings, 'half-flesh/half-metal', McLuhan speaks, of a 'cosmic man' caught up in a more cosmopolitan and Darwinian struggle among competing media of communication which gain primacy as extensions of the 'social and psychic faculties'" (Kroker, 1985, p. 53). In the age of *scientific humanism as naturalism*, we become human beings living inside the mediated environment of the technostructure whose content is "a red herring distracting our attention from the essential secret of technology as the medium, or environment, within which human experience is programmed" (Kroker, 1985, p. 56). In this view, technology is "an 'extension' of biology: the expansion of the electronic media as the 'metaphor' or 'environment' of twentieth-century experience implies that, for the first time, the central nervous system itself has been exteriorized" (Kroker, 1985, p. 57). This way,

when an organ goes out (ablation) it goes numb. The central nervous system has gone numb (for survival). We enter the age of the unconscious with electronics, and consciousness shifts to the physical organs, even in the, body politic. There is a great stepping up of physical awareness and a big drop in mental awareness when the central nervous system goes outward. (Kroker, 1985, p. 74)

All of McLuhan's writings represent an attempt to show that the "technostructure is an extension or 'repetition' of ourselves. In his essay, 'The Gadget Lover', McLuhan noted precisely why the Greek myth of Narcissus is of such profound relevance to understanding the technological experience" (Kroker, 1985, p. 57). Young Narcissus[1]

mistook his own reflection in the water for another person. This extension of himself by mirror numbed his perceptions until he became the servomechanism of his own extended or repeated image. The nymph Echo tried to win his love with fragments of his own speech, but in vain. He was numb. He had adapted to his extension of himself and had become a closed system. Now the point of this myth is the fact that men at once become fascinated by any extension of themselves in any material other than themselves. (Kroker, 1985, p. 57)

McLuhan believed that only an educated perspective can help us resist the syndrome of the upcoming technostructure. If the human destiny is to live inside its own central nervous system in the form of electronic simulation of consciousness, then the human challenge is to respond to entropy as creators.

Oswald Spengler argues that "[...] every triumph over Nature incites to yet others. The soul of this beast of prey is ever hungry, his will never satisfied — that is the curse that lies upon this kind of life, but also the grandeur inherent in its destiny" (Spengler, 1932, p. 70). The beasts of pray first thought "their desire was to 'know God,' and yet it was the forces of inorganic Nature — the invisible energy manifested in all that happens — that they strove to isolate, to seize, and to turn to account." (Spengler, 1932, p. 83). But this prey did not satisfy them. They created a small world which moves by its own power just like the large world, and obeys only the command of man. *Deus sive natura*, the final triumph over God or nature. "To build a world oneself, to be oneself God — that is the Faustian inventor's dream, and from it has sprung all our designing and re-designing of machines to approximate as nearly as possible to the unattainable limit of perpetual motion" (Spengler, 1932, p. 85).

Spengler believes that man does not face nature as a being with imperfections, as claimed by Arnold Gehlen, but as a rebel and challenger. The emergence of culture is not a result of strategy of unburdening man threatened by nature, but an expression of an unconditional will to rule. The belief in technology takes over a religious character. It is "eternal and immortal like God the Father, it delivers mankind like God the Son, and it illumines us like God the Holy Ghost" (Spengler, 1932, p. 86). At the same time, this disposition leads man into increasing alienation from nature and an endless struggle against nature and its enslavement. Spengler views the progress of technological emancipation as an act of stepping out of nature and sees world history as a "fateful rift between man's world and the universe" (Spengler, 1932, p. 44). This is how man's tragedy begins as "The fight against Nature is hopeless and yet — it will be fought out to the bitter end." (Spengler, 1932, p. 55). Man has lost the control over development long ago: "The lord of the World is becoming the slave of the Machine, which is forcing him — forcing us all, whether we are aware of it or not — to follow its course. The victor, crashed, is dragged to death by the team." (Spengler, 1932, p. 101). We are born "into this time and must bravely follow the path to the destined end [...] The honourable end is the one thing that can not be taken from a man." (Spengler, 1932, p. 114).

The technologization of the human body and spirit is depicted rather vividly in Jean Dominique Bauby's work *The Diving Bell and the Butterfly* (Bauby, 1998). The author discusses his experience of existing on the verge of death. Bauby was supposed to spend the fateful evening, which he discusses, in the company of his son. Instead, he suffered a heart attack which put him in a coma and left him completely paralyzed. All of his life functions were reduced to blinking and turning his head (Čović, 2004, pp. 69-70). It caused conflicting emotions within him, a mixture of happiness over his existence, but also anxiety over physical immobility. It is this existential rift between the spirit and the body – the butterfly and the diving bell[2] – that serves as a warning sign for the connection of spirit with the body and which should be overcome by moving our spiritual wings immediately before we face the end of our natural life (Čović, 2004, pp. 84-85).

In the age of cyborgization, the diving bell as a metaphor for the dependence of the body is being transformed into a metal diving bell[3] (Čović, 2004, p. 83), which is trying to protect us with its impenetrable shield of technology from all misfortunes of life, making our bodies a protective armour of the spirit. The metal diving bell is a result of the scientific and technological paradigm which has emerged and changed our bodily existence. Through cyborgization of the body and implementation of the technology into our originally biological bodies, the metal diving bell tries to protect the created being, beings which owe its existence and extension of existence to the diving bell. Through technological enhancements of our organs and functions and by reshaping and enhancing our diving bell, we question the very lifelines of the butterfly, i.e. the spirit that resides in the metal diving bell. The diving bell and its orientation on the technical replacement and enhancing represents a huge danger and causes dilemmas over the value and the meaning of life.

However, *scientific work* and its procedures interfere with the physical body of man by building the diving bell around it, but also by interfering into his nervous system, thus bringing into question the very human spirit and turning it into a metal butterfly which cannot fly and fulfil its task because of all the metal it contains. In addition to creating a metal diving bell around our body and giving it functions it does not inherently possess, the procedures of the *scientific work* aim to enhance the mental capacities, which are imbued with the program of the scientific and technological mind. By implanting chips into our bodies, we endanger our original being and turn our lively spirit into a *metal butterfly* whose ideas, thoughts, and decisions are under control.

Man with implanted technological parts thus becomes an increasingly automating, self-mobile contraption that will be programmable according to one's desires. Human mental capacities will also be the target of mechanization in the future, so that the machines could think in man's stead, giving them subjective and legal responsibility. This vision leads towards the end of man as a subject, the end of biological and historical man, as prophesized by Francis Fukuyama (Fukuyama, 1992), and even Paul Virilio, who claims that "One day the day will come when the day won't come" (Virilio, 2008). Another almost Frankensteinian prediction comes from Čatić (2010): "The industry of inorganic and organic polymers has allowed the cyborgs, through mediation of chip, plastic and rubber industry, to decide on the ban on all natural human beings. The deadline is 31st December, 2049".

The concept of enhancing human evolution is used by the

[…] techno-utopians […] as a metaphor. 'In the same way' that fish came out of the sea and learned to walk, we are going to learn to augment our intelligence or beef up our muscles. For the fish, it happened slowly and 'automatically'. For us it will happen quickly. And we will choose it, but through a nearly automatic process that these prophets sometimes call cultural evolution. (McKibben, 2004, p. 218).

Ray Kurzweil claims that "For tens of thousands of years, humans had created tools by sharpening one side of a stone. […] It took our species tens of thousands of years to figure out that by sharpening both sides, the resultant sharp edge provided a far more useful tool" (McKibben, 2004, pp. 218-219). For Hans Moravec, the cultural revolution is "a self-accelerating cycle that is reaching escape velocity" (McKibben, 2004, p. 219). Thus, the cultural evolution enables us to decide and choose an end goal for ourselves (McKibben, 2004, p. 219). Mark Alan Walker suggests that we apply the technology "to create beings who are as far removed from us in intelligence as we are from apes" (McKibben, 2004, p. 225). But, have we already been changed? "Some techno-zealots argue that it's too late to draw the line: we're already engineered. People with hearing aids or artificial joints are protorobots, the insist; if you work at a computer linked to a network you've already embarked on an 'unimistakably doubled artifuclation' that 'signals the end of traditional identity.'" (McKibben, 2004, p. 163). According to David Ross, "we ought to aspire to discard the body and upload the mind into 'the world-wide Cyberspace Web.'" (Groothuis, 1999, p. 43).

The transhumanists advocate a leap into an enhanced future but, according to McKibben, they do not know why they want to make the leap. Some of their arguments are:

[…] it will allow us a deeper understanding of what truly we are, says Rodney Brooks. Because, says Gregory Stock, as our new biology allows us to 'pierce the veneer of inside things, we may reach the naked soul of man.' Because, says J. Hughes, reengineered minds will 'permit us to think more profound and intense thoughts'. (McKibben, 2004, p. 225).

In their fourth national convention of the Extropian movement in 1999, the techno-utopians discussed their ideas and their "board of directors included Marvin Minsky and Ray Kurzweil, and the keynote speakers at its very first conference meeting were Hans Moravec and Eric Drexler, [...] Gregory Stock, [...] and Roy Walford" (McKibben, 2004, p. 200). Max Moore came up with the term "Extropy to the opposition of entropy [...] picked his new name as a sign of commitment 'to what my goal is: always to improve, never to be static" (McKibben, 2004, p. 200). He added: "I was going to get better at everything, become smarter, fitter and healthier. It would be a constant reminder to keep moving forward" (McKibben, 2004, pp. 200-201).

The futurologists imagine "uploading themselves into computers and then, in computer environments, becoming 'anything you like' [...] They even foresee a kind of ethical Singularity, a moment when unhappiness is banished" (McKibben, 2004, p. 102). In his book *The Hedonist Imperative*, futurologist David Pearce predicts the abolishment of suffering:

[...] nanotechnology and genetic engineering will eliminate aversive experience from the living world. Over the next thousand years or so, the biological substrates of suffering will be eradicated completely as we strive for the neuro-chemical precision engineering of happiness for every sentient organism on the planet. (McKibben, 2004, pp. 102-103)

Instead of a "genetically transmitted predisposition for sadness, anxiety and malaise, it will become purely an issue of (post)human decision whether unpleasant modes of consciousness are generated in any form or texture whatsoever" (McKibben, 2004, p. 103). In their book *Beyond Humanity*, Gregory Paul and Earl Cox wonder: "Why do we only have one heart, why not two in case one fails? Or two livers?" (McKibben, 2004, p. 111).

Rodney Brooks believes that

Twenty years from now [...] when we have neural connection from our brain to the Internet, we will be able to think the lights off downstairs instead of having to stumble down in the dark to switch them off, and in the morning, we will be able to think the coffee machine on, down in the kitchen. (McKibben, 2004, p. 120)

In addition, humans will implant themselves with smart identification chips that will grant them a multitude of possibilities:

your body temperature might give your stereo system cues as to your mood and it would select appropriate music; the chip could also compute how much of your body weight is fat, and offer suggestions for diet recipes to the refrigerator (McKibben, 2004, p. 121).

We have already expressed our attitude about the boundary between treatment and enhancement of man using scientific and technological procedures. Some scholars find the line between them to be

[...] so blurry as to be no line at all. If you say you're going to use it for disabilities, but you are not going to use it for character traits, then is schizophrenia a disability but depression a character trait? [...] When is dwarfism a disability, and when is being short a social problem? Once it's accepted that parents have a right to use germline intervention to change a predisposition to shortness into a predisposition to

average height [...] could you argue that they didn't have a right to predispose their child towards above-average height? Is the federal government going to start setting height limits? (McKibben, 2004, p. 126)

Once we go down the path of enhancing transformation

[...] there will be no way, and no reason, to turn back. We'll do what our programming indicates, never knowing how much choice we really have. We'll be like obsessive-compulsives, for whom some accident of wiring or chemistry has overridden the ability to choose. (McKibben, 2004, p. 210)

Thus enhanced, "We'll do what we're supposed to do: we'll be brainy, or brawny or pious. We may not feel sad – we won't necessarily want to be liberated from it – but we'll live in a world where our humanity really has vanished" (McKibben, 2004, pp. 210-211). For instance,

If you didn't know how you felt, or if you felt how you did because your rejiggered cells were pumping out designer proteins – in that kind of world, what would democracy mean? (McKibben, 2004, p. 199)

In his talk *The Ultrahuman Revolution: Amendments to the Human Constitution*, Max Moore highlighted the shortcoming of the human constitution – from unfairness of dying, limited senses, poor memory, etc. – and suggested seven amendments:

The elimination of death in favour of eternal life topped his list, followed by increased perception through novel and improved senses, enhanced memory and intelligence, improved emotional responses and reshaped motivational patterns. Such gains would come initially through biotechnology and later through integrating other advanced technologies into our bodies [...] his menu of enhancements will move us from a human to an ultrahuman condition. (McKibben, 2004, p. 201)

But the question is whether enhancements would stop at that particular level or whether "From the human, we jump to the ultrahuman and someday, doubtless, to the double superultrahuman" (McKibben, 2004, p. 202). According to the extropists, the reason "we have minds: to push forever ahead, transforming ourselves ever and again into something new" (McKibben, 2004, p. 202) represents the desire for constant progress, without which we would not be able to live better, as predicted by Bostrom:

[...] orgasms and aesthetic-contemplative pleasures whose blissfulness vastly exceeds what any human has yet experienced and love that is stronger, purer, and more secure than any human has yet harboured and values that will strike us as being of far higher order than those we can realize as unenhanced biological humans. (McKibben, 2004, p. 202)

As enhanced beings, we would be immortal and have superintelligence. According to Bostrom "Brains can be enhanced. And, in principle, machines can be made to process information as efficiently – or more efficiently than – biological nervous systems" (Bostrom, 2010, p. 111). Under the term intelligence, Bostrom refers to "the panoply of cognitive capacities, including not just book-smarts but also creativity, social intuition, wisdom, and so on (Bostrom, 2010, pp. 111-112). He also lists several other methods that could be used to enhance the human brain:

There are of course the traditional means: education and training, and development of better methodologies and conceptual frameworks. Also, neurological development can be improved through better infant nutrition, reduced pollution, adequate sleep and exercises, and prevention of diseases that affect the brain. We can use biotech to enhance cognitive capacity by developing pharmaceuticals that improve memory, concentration, and mental energy; or we could achieve these ends with genetic selection and genetic engineering. (Bostrom, 2010, p. 112)

In this manner, we could one day create "superintelligence, a general intelligence that vastly outperforms the best human brains in every significant cognitive domain." (Bostrom, 2010, p. 113). According to Bostrom, the machines would have the advantage in terms of speed of data processing and memory capacity:

[…] an artificial neuron can operate a million times faster than its biological counterpart. […] artificial intellects can be easily copied and each new copy can – unlike humans – start life fully fledged and endowed with all the knowledge accumulated by its predecessors. (Bostrom, 2010, pp. 112-113)

To move from the level of human intelligence to the level of superintelligence, Bostrom mentions the scenario based on

[…] the Singularity hypothesis, some sufficiently advanced and easily modifiable machine intelligence (a 'seed' AI) applies its wits to create a smarter version of itself. This smarter version uses its greater intelligence to improve itself even further. The process is iterative, and each cycle is faster than its predecessor. The result is an intelligence explosion. Within some very short period of time – weeks, hours – radical superintelligence is attained. (Bostrom, 2010, p. 114)

Such a breakthrough would be of critical importance:

Superintelligence would be the last invention biological man would ever need to make, since, by definition, it would be much better at inventing than we are. All sorts of theoretically possible technologies could be developed quickly by superintelligence – advanced molecular manufacturing, medical nanotechnology, human enhancement technologies, uploading, weapons of all kinds, lifelike virtual realities, self-replicating space-colonizing robotic probes, and more. It would also be supereffective at creating plans and strategies, working out philosophical problems, persuading and manipulating and much else besides. (Bostrom, 2010, p. 114)

McKibben believes that the enhanced man will no longer have the consciousness that would make him different from the machines:

The tensions, in other words, between our unlimited desires and our capacity for self-restraint would simply disappear. We'd be on the more track. If you're designed to be athletic, you will always choose more speed and power, and you'll choose them for your children. The tension between that athletic part of you and the other parts will simply disappear; you won't question yourself at mile 23 of the marathon. If you're designed for piety, the temptations of the world may barely arise […] They keep us from being all one way, one thing. From specializing emotionally. (McKibben, 2004, p. 211)

The human body, however, will be stronger than ever: "We'll press 840 pounds on the leg machine. We'll see in six dimensions. We'll have eyes all over our heads. We'll have a box that cranks out anything we want. We'll live forever" (McKibben, 2004, pp. 211-212). But the question is whether man really wants that. Immortality might lead to overpopulation: "elimination or slowing of human mortality would aggravate the problem" (McKibben, 2004, p. 161) but also create infinite time: "Without mortality, no time. All moments would be equal." (McKibben, 2004, p. 159). Ethicist Leon Kass believes that "The new immortals [...] would not be like us at all. A new species, *Homo permanens*. It will be by far the biggest change we've ever faced. But will it be a change for the better?" (McKibben, 2004, p. 157)

Once man accepts the possibility of reshaping the body to his desires, he'll see that "there's nothing beyond tinkering. The list expands exponentially, till there's not a feature of the human body that can't be enhanced in some way or another" (McKibben, 2004, p. 24). Improving visual and auditory perception could be used to "improve artistic potential [...] the ability to see the ultraviolet range or the infrared range – which would greatly enhance a person's night vision [...] Some plump for eyes in the back of the head on the theory that it would make driving safer" (McKibben, 2004, p. 24). In the same vein, the former director of Emerging Technology Threats bureau at the US *Department of Defense* claimed that "were he given the permission by a medical ethics board, he would try to engineer a person to have wings" (McKibben, 2004, p. 25).

Ian Pearson predicts: "Homo Cyberneticus will soon emerge with a full duplex link between man and machine. This creature will in turn merge with 'Homo Optimus' the genetically engineered elite race of people who are smart, agile, and disease-resistant. Together they will form 'Homo Hybridus' which will have no trouble displacing 'Homo Ludditus'" (McKibben, 2004, pp. 87-88). The same prediction of the dawning age is made by Kurzweil in his book *The Age of Spiritual Machines*

By 2019, most interaction with computers is through gestures and two-way natural language spoken communication... people are beginning to have relationships with automated personalities as companions, teachers, caretakers, and lovers. By 2029, there is growing discussion about the legal rights of computers and what constitutes being human. Machines claim to be conscious and these claims are largely accepted. And by 2099? Well, by 2099 there is no longer any clear distinction between humans and computers. (McKibben, 2004, p. 76)

In his attempt to convince us that continuing our evolution is the right step, Hans Moravec argues that it might not be wise to remain attached to the human body in the post-biological world due to the evolutionary pace of the machines (Moravec, 1988, p. 100). This process of externalization incudes the transmission of one's mind and consciousness into the machine system, such as robots or computer disks. The result of mental transmigration is an immortal and potentially multiplied bodiless human existence (Moravec, 1988, p. 102). In the trans- and post-humanist perspective, medicine will not be necessary as everything that is will become a technical matter.

From the perspective of immortality, the issue for the transhumanists is not whether enhancing cyborgization is good and justified, but what should the artificial man look like. The absurdity of the transhumanist endeavour is all the more serious as they try to use technological 'gymnastics' as answer to the question of overcoming the metaphysical character of the activity in the world. Implanting objects and physical software into the enhanced man would erase all Western metaphysics from the thought of the cyborgized man, which would bring him back to his pre-metaphysical, ancient source of purposeful existence. In this concept, man does not exist as subject, as a collector of entity from being. Against him stands the established, artificially manufactured being with an artificially created prosthesis of thought, which operates in a program controlled by – whom? How did this happen? How did the question of power of the master of the scientific and technological development disappear from the world stage of transhumanism in the age of the largest appetite of the *will to power*?

The advocates of enhancements believe that "nearly everyone will see the Golden Age shimmering enchantingly in the distance and will not dream of relinquishing his ticket…Before long, the objectors will include only a handful of eccentrics. […] our agonizing about playing God will give way to a new chorus: When can I get a pill?" (McKibben, 2004, p. 223). George Gilder believes that only transhumanist enhancement can save man:

If we don't establish a transhuman future, then Planet Earth is doomed to a rather dismal Malthusian extermination from disease, hunger, and lack of resources. If we don0t build a transhuman future, billions of people WILL DIE… What do you intend to do about it? (McKibben, 2004, p. 112)

As opposition to the transhumanist perspective, Fukuyama believes that extending the lifespan would cause major changes in the internal structure of the society and lead to the formation of new hierarchies. The war between generations would join the class and ethnic struggle to become the most important dividing line in the society. This would also change the people's attitude towards death:

Death may come to be seen not as a natural and inevitable aspect of life, but a preventable evil like polio or the measles. If so, then accepting death will appear to be a foolish choice, not something to be faced with dignity or nobility […] Will they cling desperately to the life that biotechnology offers? Or might the prospect of an unendingly empty life appear simply unbearable? (Fukuyama, 2002, p. 71)

His fears about the misuse of potentials of cybernetics were also highlighted by Wiener in *The Human Use of Human Beings – Cybernetics and Society*, but since the first edition of his book was publish Wiener is

[…] delighted to see that awareness on the part of a great many of those present of the social dangers of our new technology and the social obligations of those responsible for management to see that the new modalities are used for the benefit of man, for increasing his leisure and enriching his spiritual life, rather than merely for profits and the worship of the machine as a new brazen calf. (Wiener, 1990, p. 162)

The essence of technological future was described by Heidegger (qtd. in Žižek, 2007, p. 36): "what if we are actually cast into this world, if we never feel at home in it, but displaced and dislocated, and what if this dislocation is our constitutional, original state, the very horizon of our being" (Žižek, 2007, p. 36) and what if

The body must become a cyborg to retain its presence in the world, resituated in technological space and refigured in technological terms. Whether this represents a continuation, a sacrifice, a transcendence, or a surrender of "the subject" is not certain. (Bukatman, 1993, p. 247).

Maybe "consciousness has another function. Maybe it is the gift – the specialness – that allows us to eventually put a brake on this kind of evolution. To slow ourselves down, to keep ourselves from driving down certain roads [...] our gut revulsion at the coming 'enhanced' world is consciousness trying to save itself" (McKibben, 2004, pp. 221-222) because if we

Turn into engineered automatons, then consciousness as we know it – including the ability to make our own decisions, to say no – will eventually disappear. We will have reduced it to meaninglessness [...] But if, on the other hand, our nay-saying ability proves strong enough to help limit our desires, and hence our technologies, consciousness may survive into the deep future. That's the immortality that should interest us, the 'evolutionary goal' we should target. (McKibben, 2004, p. 222)

In that case, we

[...] would need to choose to stop; it won't happen automatically [...] We would need to pass laws, form the committees, enforce the regulations" (McKibben, 2004, p. 220) because what makes us unique human beings is "that we can restrain ourselves. We can decide not to do something that we are able to do. We can set limits on our desires. We can say, 'Enough'.(McKibben, 2004, p. 205)

7.1 CYBERNETIC LIFE

In §3.3 *Scientific humanism as naturalism*, we have portrayed the manifestations of the procedures of *scientific work* in relation to man as the mediator of its scientific and technological establishment of everything that is. This realized ontological work belongs to the highest controller and helmsman – cybernetics, which as the *techno-logic scientific work* defines its historical world of scientific truth and freedom. The crucial question is whether today's man in his scientific, technological and cybernetic constitution will find the place of his realization or whether he will forever remain the means and the material of *scientific work* as the self-producing, self-establishing, self-realizing and self-defining supreme being into *Nothing*.

We believe that rethinking the essence of cyborgization, especially the enhancement of man is of the vital interest for modern thought as it unites the discussions on ontology, anthropology, life, and ethics as inescapable guidelines on our quest to answer the question what man is and why he in his essence is still not fighting to get out of the firm grasp of the metaphysical fate of thought and life. By presenting and analysing the real forms of cyborgization, we have shown how scientific and technological enhancements lead towards further realization of the world, now undergoing the final stage of reshaping man into a non-human being – a thing, an artificial being, a being without human essence – a non-being which in the timelessness of its duration takes over the spatial domination over the world of biological life.

The second line of self-movement of *scientific work* in reality is manifested through the procedures of cybernetic enhancement and reshaping of man's body and mind, in the simulation of timeless space, whose void is filled with binary signs without human beings. Based on the principles of cybernetic understanding of the human body as a system, we will present the gradual transformation of the human in to non-human and technological regulation of immortality of new life.

7.1.1 Body as a Cybernetic System

When the body becomes a scientific and technological creation and real relationships become a cyborg prosthesis of the non-human, the body and the social relations become a problem of cybernetic functioning. When a part of technical appliance is broken down or worn out, it can be replaced by an identical new part and the system will work as new. Does the same happen with man during implementation of artificial technology to replace a diseased (worn out) part or during augmentation with an artificial organ or limb?

Through systematic analysis of the human organism in the spirit of cybernetic understanding, replacing organs or limbs in the human body belongs to the domain of maintenance, or in cases of failure (disease), return to the original state (health). Čatić (1997) analyses the human body as a system and compares it to the *ideal state*. Figure 1 (from Čatić, 1999) represents a coarse outline of every system, even that of the human body.

The analysis distinguishes between two types of replacement organs – those that constitute the performance system and are responsible for the exchange of matter and energy, and those that constitute the information system, primarily the brain. All the phenomena of this world can be portrayed as the flow of mass, energy and information (Table 1). The examples of the comparison come from Plato and Petrić.

Table 1 (Čatić et al., 1991) exemplifies the basic laws of the system theory. The attributes of the body and state come from Plato, and those of the soul and virtues from Petrić. The guardians imply the military, the police, the administration and the church. The features of the system will be used for systematization of organs intended for replacing worn out organs or limbs or when implanting completely new ones that the human beings do not inherently possess.

The concept of human culturology is based on the idea that everything that is created, i.e. not immediately available in the nature is culture. There are three cultures: phytoculture, animal culture and human culture. Human culture is divided into material and spiritual culture. (Čatić, 1996). Medicine and technology are connected by three fields – functionality, maintenance and the ethics of general technology. Medicine has thus far generally involved maintaining the state of health of man exposed to injuries and diseases. However, research focused on man's production (functionality) have become increasingly frequent. One of the areas of production of man is artificial insemination (Čatić et al., 2007 & 2008). This matter, however, is beyond the scope of this book.

A human being can get injured or sick and parts of its organs or limbs can get worn out. Disease can lead to complete failure of a particular organ or a particular limb, while several types of injuries can also occur. For the purpose of this analysis, we will take into account those injuries which result in separation of limbs. For instance, if a finger(s) is/are separated from hand, this injury can be fixed by attaching thus separated extremity and practically regain the functionality of hand (return to the original state). Some diseases may cause the affected limb to be removed, e.g. to sever a leg or replace a diseased kidney. They organs and limbs removed in this process often have to be replaced. A severed leg or a fist can be replaced by an artificial product. Implanting manufactured polymer-based parts of the body is becoming increasingly common. This can mean either physical replacement of limbs and organs that are missing with complete or partial restoration of their functions, or augmentation of particular functions of the organism which restore the state regarded as normal for human beings. But this is not the final stop of attempts to extend the existing functions of the organism through artificial parts, that is, of adding organs or functions not characteristic of the human species, which allow human individuals to

Figure 1. An example of a coarse outline of the system of man's body[4]

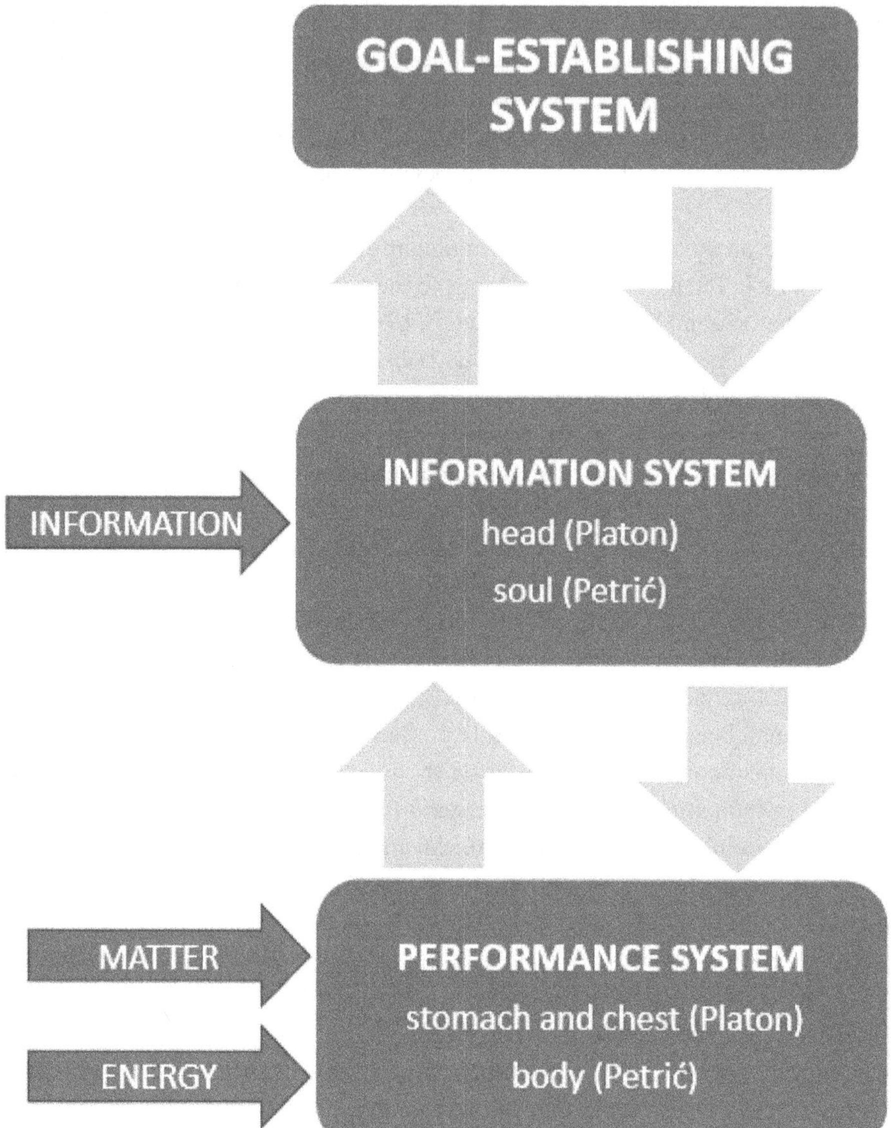

Table 1. Features of the system with examples[5]

System features	Body	Soul	Virtue	State
information	head	reason	wisdom	rulers
mass	chest	will	bravery	guardians
energy	stomach	lust	moderation	producers

thrive in otherwise inhospitable environments (Čatić, 2001). Implanting any artificial parts constitutes an act of cyborgization of a living being.

7.1.1.1 Replacing Existing and Implanting new Organs and Limbs

One of the crucial areas of modern medical science is the development of artificial organs, limbs or parts thereof. Artificial organs are created for the purpose of being implanted into the human body to replace its affected biological counterparts, while artificial limbs are used to replace the biological ones which were removed in any way, which falls under the domain of reconstructive procedures. There are two types of artificial organs: artificial organs with human origin and those with non-human origin. The aim of cyborgization, i.e. implantation of artificial parts into the human body is two-fold: reconstruction or creation of an enhanced human being. New insights about the way in which the human body functions have enabled us to create completely new human beings which ultimately become immortal (Jutarnji list, 2009). In this section, we discuss the *homo cyborg* (bionic cyborg), a human being whose lost or damaged biological organs and functions are replaced and restored by bionic implants. We present it like a system, as assumed and understood by cybernetics.

Through reconstructive procedures (cyborgization), *homo cyborg* is brought into the state of health and normal functioning that had been taken away from him. Today, there is almost not a single organ or limb that cannot be replaced by its artificial equivalent, except for the brain. In previous chapters, we have listed numerous positive aspects of nanotechnology and surgery whose implantation into a biological organism allowed the affected persons to return their organism into its normal state, but have also opened up numerous questions regarding the third and the fourth category of cyborgs. In Table 2, we list the potential future implantation of artificial organs, limbs and implants.

7.1.1.1.1 Organs for Exchange of Information

For the sake of consistency in describing the whole human body in the cybernetic perspective, we will repeat some of the artificial organs and functions that were described in the chapter on types of cyborgs.

a) Brain Prosthesis

Replacing a part of the brain is not as easy as replacing limbs, but even this should be possible in the future. Theodore Berger (Berger and Glanzman, 2005) created a computer chip that could replace the hippocampus, a part of the brain in charge of short-term memory and spatial memory. It is often affected by conditions such as the Alzheimer's disease and stroke, which is why the hippocampus implant could be of vital help in maintaining the normal function for those severely affected. In the future, Berger wants to enable the creation of long-term memory chips which would be implanted into the brain. The scientists at the Ohio State University and the Feinstein Institute in USA have implanted a patient with an implant which stimulates the part of the brain in charge of movement. The patient stimulates and moves the muscles with his mind. To increase the functionality, an electronic sleeve was made to help the stimulation of arm muscles (Vitas, 2016).

Table 2. Possible and future implantation of artificial organs, limbs and implants (cyborgization)[6]

Cyborgization	Organs for exchange of information	Organs for exchange of matter and energy	Organs of movement	Other
Possible organ transplantation	● brain prosthesis S5, 1, K2 ● neurostimulator S5, 3, K2 ● RFID chips S5, 5, K3 ● artificial eye S5, 22, K1 ● artificial ear S5, 22, K1 ● deep brain stimulation S4, 1, K 2 ● spinal neurostimulation S4, 7, K2 ● vagus nerve stimulation S4, 12, K2 ● hearing aid S4, 13, K1 ● brain-computer interface S4, 14, K3 ● silicone voice prosthesis S4, 2, K1 ● Google contact lenses	● artificial teeth S5, 2, K1 ● artificial heart S5, 4, K1 ● artificial lungs S5, 16, K1 ● artificial liver S5, 7, K1 ● artificial pancreas S5, 8, K1 ● artificial bladder S5,10, K1 ● pace-maker S4, 10, K1 ● ventricular assist device S4, 11, K2 ● transportable artificial kidney S4, 8, K1 ● artificial trachea S4, 6, K2 ● artificial jaw	● artificial hip S5, 12, K1 ● artificial suit S4, 3, K2 ● artificial (smart) leg S5, 14, K1 ● prosthetic (robot) leg S5, 13, K2 ● artificial hand S5, 19, K1 ● artificial fist S6,16, K1 ● artificial fingers S5, 15, K1 ● artificial spinal discs S5, 6, K1 ● artificial muscles S5, C, K3	● artificial stomach S4, 9, K1 ● implanted defibrillator S4, 4, K2 ● drug dosage system S4, 6, K2 ● artificial ovary S5, 9, K1 ● artificial lips S5, 20, K1 ● artificial nose S5, 21, K3 ● artificial breasts S5,18, K1 ● artificial bone marrow S5, D, K1 ● endoscopic capsule S4, 5
Future organ transplantation	● artificial tongue S5, G, K1 ● artificial brain S5, f, K1	● artificial cells S5, 11, K1 ● artificial blood S5, A, K1 ● artificial kidney S5, B, K1 ● artificial skin S5, E, K1 ● artificial blood vessels S5, E, K1	● artificial bone	● artificial sperm

* SX – number of figure, XX –number of section, KY – cyborgization category (1–4)

b) Neurostimulator[8]

A neurostimulator is a device used for treating neurological disorders. It is implanted into the person's body below the collar bone and it sends the signals to electrodes implanted near the spinal cord or peripheral nerves, thus enabling stimulation of the spinal cord via tiny electric nerves used to treat the headache. Scientists in Boston are developing implants and neurostimulation devices that include new treatments for deafness and chronic pain. The neurostimulator has been approved by the US *Food and Drugs Administration* for treating chronic pain. Today, this device is an integral part of medical systems implemented in deep brain stimulation and the vagus nerve stimulation treatments.

Figure 2. Examples of human body implants[7]

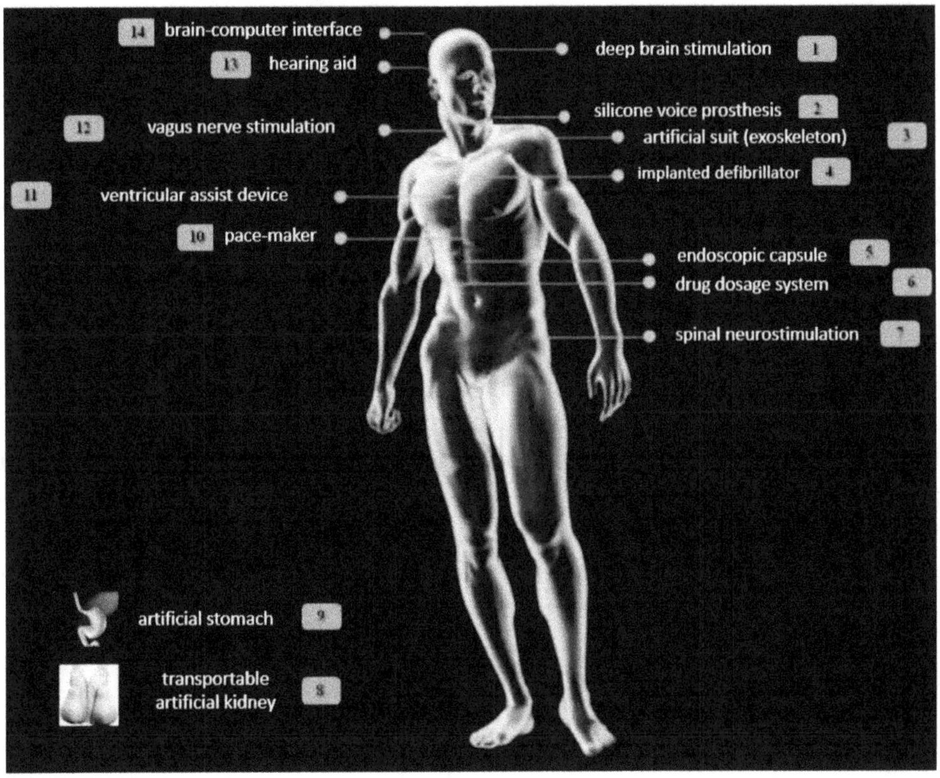

c) Radio-Frequency Identification Chips[9]

The application of the radio-frequency identification (RFID) chips was described on the example of Kevin Warwick as a cyborg. According some estimations, the application of RFID chips will surge in near future, allowing every user a reliable identification during different transactions, enabling patients to acquire data about the medical state of patient and locate cancerous cells in the human body. In addition, they would be used for tracking movement, recording verbal conversations, and changing the physiological processes of the body, etc. The RFID chip will enable the mental, emotional, and even physical programming of the human consciousness and perception of reality, and it will be possible to terminate particular physiological processes in the human body from a remote location, which could lead to complete control over life and death. These chips are typically 0.05mm x 0.05 mm in size and have the shape of a grain of rice. The chip consists of a transponder, storage system for reading information, and a rechargeable lithium battery that has an unlimited lifespan. The chips are implanted onto neural endings in the forearm, fist or head; the implantation leaves no scars and the chip is imperceptible from the outside. It can only be detected using an appropriate scanner for reading the data it contains. If it were to be surgically removed, the tiny capsule would break and the person would be contaminated with lithium and other chemicals in the microbattery, the GPS would detect the removal and alert the police. An available option of the device will be to use remote control to release the lithium from the microchip capsule, which would lead to poisoning and eliminating unwanted members of the population.

d) Artificial eye

As mentioned earlier, the bionic eye is a result of cooperation of medicine, science and technology with the aim of enabling the blind to regain vision. The artificial eye is connected to the visual nerve of the afflicted eye and it consists of a silicon chip inserted into the eye, spectacles, and a tiny belt-worn computer. The most famous one is the *Argus II* – a chip with 60 electrodes connected to the retina, which receives the signals from the external camera. The miniature chip is surgically inserted behind the retina and connected with a video camera and a transmitter on the spectacles. The electrodes in the chip translate the image into electric signals which are transferred into the brain through the optic nerve, after which the person can recognize light or contours. The implant does not require any replacements as it has a life-long duration, while the battery in the mini-computer can be recharged like a cell phone (Ko, 2010).

Thanks to Argus II, patients suffering from macular degeneration or *retinitis pigmentosa* can partially regain their sight, recognizing white, black and grey colour. The researchers are trying to produce a camera the size of a pea that could be implanted into the eyeball, replacing the natural tissue with an artificial product. An artificial biosynthetic cornea was made and successfully integrated into the human eye and it can restore the damaged nervous tissue and restore clear sight. The cornea was made from a combination of human collagen and it fuses with the cells and the nerves of the natural cornea after a certain period of time.

e) Artificial Ear (Cochlear Implant)

The artificial ear represents one of the first implants of the bionic man which became available in 1969, but has undergone multiple improvements ever since. Today it is used by hundreds of thousands of people across the world. The artificial ear consists of a digital microprocessor, which sends the signal into the brain of a deaf or hearing-impaired person. The way it works is very simple: the microprocessor antenna sends the signal into the external receiver which is located under the skin. The receiver of the stimulator channels the signal from the antenna receiver to the electrodes in the cochlea. The electrode network ultimately stimulates the nerves to transmit the sound signals to the brain (Fischman, 2010).

f) Deep Brain Stimulation

Deep brain stimulation is a procedure in which the deep parts of the brain are stimulated. The procedure targets the parts which lead to neurological and psychological issues. Deep brain stimulation is made possible by a neurostimulator implemented deep into the brain and the generator, which resembles a pace-maker, implanted near the collar bone. The generator sends very weak electric impulses towards the neurostimulator which sends them forward to the neuron cells. This procedure aims to change the area of the brain that is not functioning properly. This suppresses the symptoms of the Parkinson's disease and enhances the motor functions of the patients. Physicians in Canada use this method to treat dystonia, epilepsy, depression and Alzheimer's disease. The results show that two out of six patients with a disorder in the hippocampus have reported improvements as the hippocampus stopped deteriorating and started growing. A hospital in Los Angeles was the first hospital in the world to transmit the whole operation in which the patient was implanted with a neurostimulator. The patient was awake throughout the whole operation and could have followed the entire procedure. As much as 70 thousand people all over the world have tried the deep brain stimulation therapy (Glas Amerike, 2007a).

g) Spinal Neurostimulation

In this procedure, electric impulses are applied to particular groups of nerve cells to achieve a therapeutic effect. The neurostimulator uses low-voltage direct current created by the subdermal generator. The therapy stimulates the nerve cells of particular parts of the central, peripheral or autonomous nervous system. The electric impulses are conducted to the point of stimulation with a system of thin subdermal electrodes. The expected results depend on the aims of therapy. It can stimulate, muffle or synchronize the functioning of nerve cells. It is used for treating neuropathological pains, dystonia, Parkinson's disease and other disorders (Surgery section, n.d.).

h) Vagus Nerve Stimulation

Vagus nerve stimulation (VNS) in the neck is intended for treating severe forms of epilepsy. It includes implantation of a device similar to a pace-maker which generates electric impulses to stimulate the vagus nerve, one of 12 cranial nerves which transfer the information from and towards the brain. The vagus nerve serves many organs and structures, including the larynx (voice box), ears, tongue, sinuses, oesophagus (food tube), lungs, heart and digestive system. In addition to epilepsy, vagus nerve stimulation helps in treatment of depression and obesity (Rielo, 2011).

i) Hearing Aid

The hearing aid is an electronic device which enhances the sound from the environment and facilitates hearing. It consists of a miniature microphone, amplifier, and a speaker. The most contemporary hearing aids are programming digital electronic devices, i.e. miniature computers placed in the ear canal which can automatically adapt to current changes in the sound of the environment and enable better hearing and speech understanding. The scientists are developing a hearing aid in the form of a capsule equipped with a sophisticated equipment and implanted subdermally (Mayo Clinic Stuff, 2011).

j) Brain-Computer Interface

The brain-computer interface (BCI) called *BrainGate* enables the disabled patients to control the virtual limbs via brain-implanted microchip and an electroencephalograph (EEG) device, thus communicating with their environment (NJD, 2011; Greguric, 2010).

Braingate behaves just like any other cybernetic device which receives information and transforms them into work orders. A brain chip reads the patients' thoughts and sends them via the EEG device to the computer interface which produces further orders for activities in the patient's vicinity, such as turning on the TV, opening the door, etc. The user has to wear the head-piece with electrodes which measure the electronic activity in the brain (EEG). For now, the BCI only allows one-way communication between the brain and a computer (NJD, 2011). Researchers are working on a new type of device which should enable two-way communication. The company *BAE Systems* has constructed a pilot helmet for the RAF which is currently not being used for treatment purposes but in aerial combat. This sophisticated helmet called *Striker Display Helmet* is equipped with an optical guidance system which allows the pilot to control the projectile at super-sonic speeds. When the target appears in the visual field, the pilot gives the voice command to fire and then guides the projectile with his sight as long as the target is within his view (NJD, 2011).

k) Silicone Voice Prostheses (Blom-Singer Method – Voice/Speech Restitution)

Numerous people who lost their speech for various reasons have used the electro-larynx as a successful method of voice rehabilitation. The second method of voice and speech rehabilitation was developed in 1978. This method is based on the use of external speech prostheses (silicone as an inorganic polymer). A small incision is made between the trachea and the oesophagus and a removable silicone voice prosthesis is inserted into it. It is actually a one-way valve which allows the redirection of air current from the lungs to the end points of the oesophagus and the mouth cavity where voice is formed (InHealth Technologies n.d.).

7.1.1.1.2 ORGANS FOR EXCHANGE OF MATTER AND ENERGY

a) Artificial Teeth

When teeth are worn out, dental implants can be inserted into the empty space left behind them. These are completely biocompatible titanium implants that are fused with the bone and they replace the roots of teeth. Researchers from the University of Tokyo claim they are at the brink of a revolutionary accomplishment – growing completely new teeth from stem cells. The rats' jaws were implanted with cells programmed to grow new teeth that can be grown to full functionality (Lauc, 2020).

b) Artificial Jaw

The doctors in the Netherlands have successfully implanted a lower jaw created by a 3D printer. The jaw was created by heating and fusing the titanium powder with laser, layer by layer. Once all layers were fused together, the jaw was given a bioceramic coating. The patient was able to speak and swallow the day after the surgery (BA/VLM, 2012).

c) Artificial Heart

In 1945, Willem Kolff conducted the first haemodialysis and started a new area of medical inventions, such as artificial kidneys, bladders, pancreas, while also inventing the first artificial heart. *AbioCor* is an artificial heart made of titanium and plastic in the *AbioMed* laboratory, which weighs only 900 grams. In 2001, *AbioCor* was implanted for the first time to patient R. Tools, who lived for only 151 day after the surgery, even though the device could last up to 18 months. *AbioMed* is working on its improved version *AbioCor II*, the bionic heart which should last up to five years. The artificial heart is actually a hydraulic pump which is implanted into the chest. A permanent artificial heart was given to a 15-year-old boy in Rome, which was the first such case in history. It is four centimetres long and is connected to the aorta through the left ventricle. It is powered by a battery which the patient wears around the waist (Grubin, 2007).

d) Artificial Lungs

The respirator or artificial lungs is the machine which breathes for the patient. When the patients' muscles undergo significant deterioration, they need a respirator in advanced stages of the disease. A device known as *NovaLungs* LA is located outside the body and stimulates the work of the lungs to take enough oxygen and remove carbon dioxide. The machine allows the patients to survive for up to 30 days, which gives the patients and their doctors precious time for further treatment. The scientists at Harvard have created primitive artificial lungs which allow test animals to breathe for several hours. They are made of human lung cells and have developed into different types of lung tissue. After the transplant, the lungs in rats functioned for about six additional hours. In the next stage of research, this procedure will be repeated with stem cells. The scientists at the University of Texas were the first to grow human lungs inside of a laboratory. The lungs were grown within 3 days using a more complex procedure which involves tissue harvested from the deceased persons (T.V., 2014).

e) Artificial Trachea

Swedish scientists have managed to grow the first biological trachea from stem cells. The trachea is modelled using 3D technology, so that the blocks of grown cells were laid over the mould until the entire organ was created. The trachea was successfully transplanted. This type of transplantation facilitates the treatment as the patient does not need a donor and the chance for body to accept the organ made from its own tissue is very high (I. S., 2011).

f) Artificial Liver

Researchers and physicians have so far been unable to create a device which would replace biological liver. For the past 50 years, research has been focussed on creating a laboratory-grown organ made out of biological cells. The US scientists have managed to use the cells from rat liver to grow a transplantable liver. The research team confirmed that the cells were successfully grown and transplanted, but have emphasized that the transplanting the liver is just a minor part of its normal function. The biggest issue in potential transplant is connecting the artificial liver with the cardiovascular system. Biologist Takanoria Takabea and his colleagues at the University of Yokohama City have successfully created a miniature biological liver from the stem cells from human skin, which were reprogrammed in the embryonic phase. The mini-biological liver was transplanted into a mouse and conducted many functions of the normal liver (Z. J., 2010).

g) Artificial Pancreas

The US *Food and Drugs Administration* has approved the usage of the first artificial pancreas. It will be of crucial use for diabetics as it automatically controls the insulin level. This system called Mini Med 530 G can dialyze 93% of hypoglycaemia and will sound the alarm if the sugar level is too low. The artificial pancreas consists of a part for continuous glucose tracking and an insulin pump. After the sensor sends the signal to the transmitter, it sends it to the control box which delivers the information to the pump about the amount of insulin to release. This invention helps the insulin-dependent diabetics lead a more normal life and allows them to avoid life-threatening side-effects, such as too low or too high blood sugar levels (University of Cambridge Research, 2011).

Scientists at Cambridge have shown that artificial pancreas can regulate the level of glucose in blood for children with Type 1 diabetes. The device which makes the so-called artificial pancreas significantly decreases the risk of blood glucose level dropping too low. In US alone, over 23 million people suffer from diabetes and over four million people need to continuously monitor their blood sugar level with injections. This machine could remedy this situation.

h) Artificial Bladder[10]

Scientists at the *Wake Forest University* lead by Anthony Atala have managed to completely reconstruct the bladder by using the laboratory-grown live tissue. After the transplant, it is adapted to the body and it starts to function. Research has involved five boys with damaged urinary tubes and the new organ worked successfully after transplantation in all five cases. This is a regenerative personalized procedure as the organ is created from the sample of patient's tissue. The patient actually gets his own new biological organ. It is a promising turn of events in the medical field which gives hope that someday we will be able to regenerate other organs as well.

i) Pace-Maker

The era of successful implantation of pace-makers started during the 1950s and 1960s and it allowed the patients who had heart rhythm disorders to return to their everyday activities. What a pace-maker does is it exclusively stimulates the heart using electric impulses, thus giving the heart its rhythm. The artificial heart electrostimulator consists of two parts:

1. The pace-maker body – pulse generator, which consists of a battery and an electronic components system.
2. Electrodes, which span from the pace-maker onto which they are attached to the top of the heart or one its parts, where it comes into contact with the heart tissue and sends its electric impulses (MedicineNet, n.d.).

j) Ventricular Assist Device

First produced in 1960, and implanted as a help for the heart cavities to pump blood towards the body, devices like *Heath Mate II* enable long-term ventricular assistance. Their aim is to provide a hemodynamic support for patients in the final stages of refractory insufficiency of the left ventricle. It is a rotating system with an axial pump that can generate the flow of up to ten litres per minute (Texas Heart Institute, n.d.).

k) Transportable Artificial Kidney

The automated wearable artificial kidney (AWAK) has been developed in the USA and its use could liberate the persons suffering from kidney failure from dialysis. The device is compact and light enough to be attached to a belt. It works better than the traditional dialysis as it can be used throughout the entire day. Medical teams are also working on growing biological kidneys but no reliable data on achieved results are available beyond the fact that preliminary testing is in progress (Oregon State University, 2004).

7.1.1.1.1 ORGANS OF MOVEMENT

a) Artificial Hip Joint

A hip destroyed by osteoarthritis can be replaced with an artificial hip joint. In a traditional procedure, a surgeon cuts off the top of the femur which forms the ball and joint capsule, after which an artificial hip capsule is implanted. It can be made of metal and plastic, while the ball and the screw are made of metal. Hip replacement is mostly performed with older patients whose hips are affected by osteoarthritis, while it is not recommended to younger patients as they are more active and the artificial hip is under more strain. The latest procedure in use is mixing bone marrow stem cells with purified bone grown from another patient. The dead tissue is then removed from the hip and replaced with new tissue (Mužić, 2009).

b) Artificial (Smart) Leg

There are two general concepts of producing smart legs. The first is the *Power Knee*, the artificial leg which represents the most modern bionic technology. It is adapted to the manner and speed of human walking and terrain type and it can prevent people from falling down. The second option is the C-leg or artificial, smart leg constructed by the company called Otto Bock. It is run by a smart battery – a program which does not allow it to completely discharge. The movement is assisted by sensors and a microprocessor which give the users stability and enable them to ride a bicycle and change their speed during sports activities (Fischman, 2010).

c) Prosthetic Leg

This is an electric limb which uses motorized springs to replace biological legs. The movement-tracking sensors and ankle-control software allow the patient's movement to be adjusted. The device allows to patient to sit down or get up naturally. *Cheetah* prosthetic legs are made of a carbon polymer and plastic matrix composite. The most famous man with prosthetic legs is the Olympic medal winner Oscar Pistorius (nicknamed *Blade Runner*), the athlete who went down in history in 2005 as the first disabled sportsman who competed with the able-bodied athletes (Weiner 2008).

d) Artificial Hand

The artificial hand is a prosthesis which amplifies the electric signals in the muscles to start the electric motors which move the joints on the hand and fingers, thus enabling the sense of touch to be transmitted and received. The artificial hand was developed by Todd Kuiken. To move this artificial limb, the nerves left after the amputation are used to transmit the signals from the brain. This method is known as targeted muscular reinnervation. The patient imagines a hand movement and the muscle activity in the rest of the hand is deciphered by the computer and causes actual movement. The researchers are working on perfecting the arm by using the surviving nerves which would transmit the sense of warmth, vibration, and pressure from the artificial arm to the brain (Fischman 2010).

e) Artificial Fist

The prosthetic fist *I-limb* is made from aluminium and is controlled via smartphone. All fingers are movable and the thumb can be rotated. The artificial fist is provided with a smart phone app with 24 different types of fist movements. The fist is powered by electric impulses which are gathered by the electrodes in the joint of the hand when the muscles are contracted and the app monitors the movements. Another version of the artificial fist is made out of light plastic. The fist is run by five motors located between the root and the joint of fingers and a sensor which receives signals from the patient's brain (Fildes 2009b).

f) Artificial Fingers

Bionic or artificial fingers are moved using the micro-electric sensors which register signals from the remaining muscles in the arm. They are covered in robot skin and enable the normal functioning of fingers. They are made out of strong but light plastic and are moved by miniature motors. In 2003, a pianist lost her fingers due to bacterial infection. After the *ProDigits* bionic fingers were implanted, she was able to play the piano once again (BBC News 2009).

g) Artificial Spinal Discs

Creation of artificial spinal disks represents a huge step forward in spinal surgery, especially in terms of treating the degenerative disease of the intervertebral spinal disc. By implanting an artificial disk, damage is reduced to neighbouring discs and joints. This is because the artificial replacement disc enables the movement and preservation of the spine and facilitates a more normal distribution of fatigue across the spine. The materials used are similar to those used for replacing hips or knees and they cause no negative reactions in the body. After the implantation of metal discs, separated by polyethylene spacers, the bones merge together above and around them. In many countries, disease-ridden spinal disks are replaced with artificial ones produced by a 3D printer. The operation is managed by sophisticated software, planning every detail of the procedure (Howard & Kristen, 2010).

h) Artificial Bones

A US company has come up with a way to produce bones via 3D printer from stem cells of the patients' fatty tissue, and the whole structure spends three weeks in the bioreactor during the period of bone growth. Clinical testing on humans should start in a few years, pending approval of the US *Food and Drug Administration* (Solar, 2016).

i) Artificial Muscles

MIT scientists Mehmet Kanik, Sirma Örgüç, Polina Anikeeva, Yoel Fink, Anantha Chandrakasan and C. Cem Taşan have created super durable and powerful artificial muscles that could one day be found in robots, exoskeletons and prosthetics. Unlike the usual approaches in the development of artificial muscles, for which hydraulic systems, servo motors, metals that remember shapes and polymers that respond to stimulants, MIT has developed a fibre-based system of extremely low weight that has a very rapid response. The scientists performed this with a special fibre extraction technique to combine two different polymers to obtain a single fibre thread. The key thing in the process was the merging of two materials that have very different thermal expansion coefficients, which means they stretch differently when heated. It is a similar principle used in many thermostats, where a tape of bimetal is used to measure temperature. Scientists say that the possible application of these materials has almost no limitations, so they could be found in humanoid robots, industrial machines, prosthetic limbs and exoskelets that would provide humans with superhuman power without the need for engine launchers and the large and heavy equipment used in such systems today (Chandler 2019).

j) Artificial Suit (Exoskeleton)

Bionic HAL (*Hybrid assistive limbs*) are bionic clothes and exoskeleton which help the disabled persons (Fig. 2). They increase their strength and replace injured limbs. Their creator is Yoshiyuki Sankai, the founder of the Japanese company *Cyberdyne* which developed the bionic suit. The exoskeleton weighs 23 kilograms and allows the user to move freely and lift products ten times heavier than those he could lift with his own muscles. His joints are powered by motors which increase the user's strength and they are charged with a batter with a duration of five hours. Skin sensors detect electric activity that the brain sends through the nervous system when ordering an action. Through the sensors of neural impulses, the exoskeleton receives the order to move and is activated a fraction of second before the muscles in the human organism. There is also the possibility of adapting the sensors which detect motor neural signals on skin surface to track vital signals and potentially protect the patient or call for help. HAL is already available to buyers in Japan, but its producer does not want to allow the military use of this products. Its primary users are persons who face difficulties moving, who suffered from heart attack or a stroke, or industry workers (Ricker 2006).

7.1.1.1.4 OTHER

a) Artificial Stomach

In 2006, a team of British researchers headed by Martin Wickham constructed the first artificial stomach which faithfully mimics food digestion in the organism. The device is constructed from sophisticated types of plastic and metals resistant to stomach acids and enzymes and can be fed using real food. Controlled by a computer, the stomach mimics the physical and chemical reactions occurring during digestion (even stomach contractions, vomiting), which should help the scientists understand how the human body dissolves food. Based on our knowledge on glucose absorption speed in the cardiovascular system, this could lead to new procedures for treating diabetes (Jutarnji list 2010).

b) Implanted Defibrillator

A defibrillator is a medical device used in cases of heart arrythmia, atrial fibrillation, or ventricular tachycardia. The system consists of a casing which contains the batter and the electric circuitry, and the catheter which contains the electrode connected to the heart. The casing is implanted below the left collar bone. It is implanted subdermally and produces an electroshock if the heart rhythm is disturbed, thus stimulating regular heartbeat. The device offers new hope to cardiac patients as its success rate is around 98% (University Of Rochester Medical Center, 2001).

c) Endoscopic Capsule

It is normal for examination of the bowels to be performed endoscopically. The *PillCam* capsule is now available, which has a tiny camera. After the patient swallows it, it transmits the images of the bowels to the external computer (Mone, 2008).

d) Drug Dosage System

To prevent chronic muscle pain and cramping, a patient is implanted with a pump which delivers the drugs directly into the spine. Google contact lenses help track sugar levels for diabetes sufferers. Using them, the sugar level is kept under control more easily. The Drexel University is working on miniature robots for delivering drugs inside the human body. The robots are supposed to move through the bodily fluids with the help of the magnetic field. They could be useful for delivering drugs to fight cancer (Večernji list, 2016).

e) Artificial Ovary

US scientists have used bio-engineering methods to manufacture an artificial ovary that has all the functions of a natural one. Since the artificial ovary in laboratory conditions secretes sexual hormones, scientists believe it will be the better choice for women than hormonal therapy which is being used to replace the lack of oestrogen and progesterone. Another team of scientists at the Wake Forest University is working on developing a biological ovary from rat tissue. The plan is to implement the ovary into the animal to test its function in a living organism (Brown University, 2010).

f) Artificial Lips

Lip augmentation falls under the purview of aesthetic surgery whose aim is to change the size and appearance of lips through use of silicone (inorganic polymer) implants. Various skin polymer implants are injected under the patient's skin, from collagen (polymer), plastic to crystal gels. Artificial lips belong to the domain of aesthetic enhancement (Plastic Surgery Advisor, n.d.).

g) Artificial Nose

UK scientists have managed to grow a biological nose from fatty tissue stem cells. So far, no nose has been transplanted using this method. Most of the growth has thus far been conducted on the patients' forearms or head. According to BVC, the doctors in India were able to replace a 12-year-old boy's damaged nose which was growing on his forehead. In the first stage, the boy was implanted with a piece of metal which freed up space for the growth of a new nose. After this step, a chemical for supporting growth was injected (Elsevier, 2009). In the second stage, a cartilage was taken from one of the fingers on his arm and it was used for growing the nose. After three months, the doctors removed the nose from the forehead and implanted it onto the boy's face. The doctors believe that the nose will be fully functional (HINA, 2016b). However, there also exists an artificial nose, a device which NASA developed for the purpose of air quality control on the space station. It can be used to detect cancer-affected brain cells. The device is able to recognize molecules diluted in the air and differentiate between brain cells affected by carcinoma from healthy ones. Researchers are hoping this will enable easier diagnostics and early detection methods for brain cancer. If implanted into human beings, it will enhance human capabilities (Kučković, 2014).

h) Artificial Breasts

Implanted breasts are prostheses which change the size and form of female breasts for cosmetic or reconstructive reasons (e.g. after breast removal due to malicious diseases). According to the data of the *American Society of Plastic Surgeons* (ASPS), breast enhancement surgery is the most common surgical procedure (Walsh 2010).

7.1.1.2 FUTURE IMPLANTATION OF ARTIFICIAL ORGANS AND LIMBS

7.1.1.2.1 Organs for Exchange of Matter and Energy

a) Artificial Bone Marrow

In 2014, German scientists reported that they managed to create the environment for the production of artificial bone marrow. While there exist numerous methods for growing stem cells, growing blood-forming cells is an extremely complex task as it is quite demanding to provide an environment in which they can thrive. The scientists created a spongy bone-like structure and they added a sequence of proteins similar to those in the biological bodies and the so-called mesenchymal stem cells from the umbilical cord. A significant problem for this procedure stems from the fact that they mature very quickly into adult blood cells which have a short lifespan and cannot self-replicate but have to be constantly reproduced by the bone marrow. The scientists believe that it will take them around ten more years before this finding can be put to use (All Headline News, n.d.).

b) Artificial Cells

Scientists at the *Institute of Molecules and Materials* at the Radboud University in the Netherlands have managed to create the first artificial plastic cells which have organelles for performing various chemical reactions. Just like the cells in our body, after the reaction in the organelles the chemicals went into cell plasma to be processed in another place in the cell. The researchers are currently trying to solve the problem of artificial cells producing their own energy. Several teams of chemists are working on artificial cells projects throughout the world. Some are trying to create cells that contain cytoplasm, while others are building them from polymers. The general aim of this research is to understand the circumstances in which evolution of chemical elements transitions into life (NASA, 2003).

Figure 3. Possible and future implantation of organs and limbs (Greguric and Čatić, 2010)

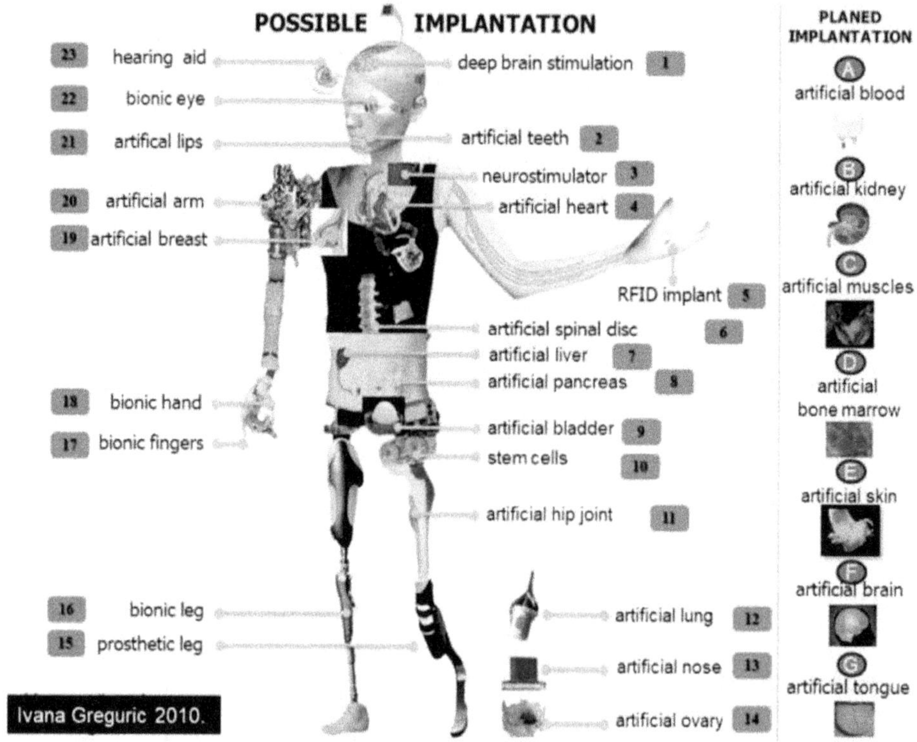

c) Artificial Sperm

Scientists at the *Kallistem* company in Lyon have reported successful production of functioning human sperm from parts of genetic material. This could be the solution for male infertility. The experts in the field are somewhat sceptical about this finding (Capar, 2015).

d) Artificial Blood Vessels

Scientists at the *Cambridge University* have developed three main types of cells which make up the walls of blood vessels, using patients' cells in the process. The artificially grown blood vessels represent a possible alternative to a bypass or a stent, but can also be used for fixing the arteries of patients facing potential limb amputation. Artificial blood vessels could be produced in a laboratory or the stem cells could be simply injected into the heart to form the walls of blood vessels (Horvat, 2012).

e) Artificial Blood

After six years of research, Romanian scientists report to have found the recipe for artificial blood, which consists of water, albumin salt, and hemerythrin – protein extracted from sea worm tissue. Scientists find the preliminary results on mice to be promising as the mice showed no signs of the disease after being injected with artificial blood. It is believed that the first testing will be conducted with humans in two years. British scientists plan to create unlimited quantities of artificial human blood from embryonic stem cells that will be used for emergency transfusions but without the risk of infection. Creation of erythrocytes from embryonic cells in laboratory conditions could eradicate the risk of diseases related to blood and blood components. Making sure that artificial blood is completely safe and free from pathogens, like the HIV and hepatitis virus, is made much easier with this method. The first results of research with artificial erythrocytes show that they carry oxygen just as well as the red blood cells from blood donors (Connor, 2009).

f) Artificial Kidney

Researchers at the *Massachusetts General Hospital* have taken a mouse kidney and washed out old cells from it. They kept the kidney for 12 days in a device which imitated the conditions inside the mouse body. In laboratory conditions, the kidney produced 23% of urine produced by a real kidney. After transplant into the body of a mouse, effectivity dropped down to 5%, but the scientists have reasons to believe even this is better than haemodialysis. The second team of researchers worked with a human kidney from which old cells were removed and replaced with new ones from the patient's body. A kidney processed like this is expected to have a high chance of being accepted by the body without the immunosuppressants even after being processed using this method (Allday, 2010).

g) Artificial Skin

In about twenty years, artificial limbs could be given skin sensitive to temperature and touch. Carbon nanotubes whose width is equal to about 1/10000 part of human hair are the most effective known conductors of heat and electricity. Carbon nanotubes would be distributed inside the skin made out of soft plastic. The senses are collected via active endings of non-dead nerves and sensors which differentiate temperature from pressure. The nanotechnology will serve to create the ingredients of the waterproof skin which will be laser-formed to look like real skin. Scientists at the MIT have developed transparent elastic skin used to cover wrinkles and bags under the eyes (Yang, 2010, Solar, 2016).

7.1.1.2.2 Organs for Exchange of Information

a) Artificial Tongue

The electronic tongue uses microspheres, tiny sensors which change colour when exposed to different targets, such as certain types of sugars. It cannot replace the taster to decide whether something tasted good or not, but it can ensure that the chemistry of good taste is consistently replicated. Scientists are also working on an artificial tongue for human use. It is made out of the larynx muscle, connected to the source which emits nerve signals much in the same way the pace-maker does and has so far been tested on pigs (Farrier, 2009).

b) Artificial Brain

The biggest challenge in creating bionic organs is definitely the human brain. In recent years, electronic brain implants were used with several patients. The most famous implant is definitely *BrainGate,* a four-square-millimetre chip which is inserted into the brain cortex. The scientists working on the *Brain-Gate* project are trying to wire-connect the motoric cortex of the completely immobile patients with a computer. Connecting a computer with their mind would enable the patients to manipulate the objects in their immediate surroundings. The researchers are planning to create an artificial hippocampus, the part of the brain where memories are stored, so that they could implant it to people who lost their memory. Henry Makram is the lead researcher at the Blue Brain project, which was conducted on mice in 2005. The aim of the project is to thoroughly explore the area of mental diseases and find methods for curing the diseases by understanding the way in which brain functions. The team of researchers working on this revolutionary experiment has spent the last 15 years studying the functions of the brain with the main focus on the neocortex. To describe the model of the brain and explain the rules that govern human communication and which parts of the brain are interconnected, the team members have created a three-dimensional model on a super computer. After understanding how it functions, artificial brain will be made of silicone (inorganic polymer), gold and copper (metals) and will, according to the researchers, have the ability to develop thoughts independently of external input and will have the ability to feel. The scientists fear that this brain might be more evil than good. This would open up numerous ethical questions related to this project, whose ultimate goal is to enhance and reshape human brain (Fildes, 2009a).

Scientists at the *Salk Institute* in California have established that human brain requires very little energy, only 20 watts, and at the same time it has 1 petabyte of memory capacity, which is roughly equal to all information found on the entire internet. This fact should be taken into consideration in all attempts to create the human brain (Šerić, 2016). Austrian scientist Jurgen A. Knoblish is researching the methods to grow human brain. The secret of his recent success in treating brain diseases lies in his growth method which uses the substance called the *matrigel*, obtained from the mouse cells affected by carcinoma, which is perfectly suitable for developing cells that would later become a part of the brain. The real revolutionary improvement in this research lies in cell reprogramming, i.e. returning the cells from their completely mature state into pluripotent stem cells, which is done by simple genetic manipulation. This method does not require use of stem cells from embryos (Vitas, 2017). The question remains how the person might react when he/she finds out that their cells might be used to create an artificial brain.

7.2 SOCIETY AS A CYBERNETIC ORGANISM

The age of *scientific humanism as naturalism* is presented as the entire reality of man cyborgized by the procedures of *scientific work*. Human and social relations are gradually established through chips, mass media, networks and other means of communication as mediated cyborgized social relations. Cybernetics as the science of general rules of control and communication processes, ways of receiving and using information in organized systems – machines, living organisms, and their mutual relationship towards one another assumes that the human nervous system processes reality as a set of information collected in its central processor made of flesh and blood – the human brain. For cybernetics, human brain is neither natural nor supernatural organ of thought which raises man from his organic nature and enables him to have a meaningful relation towards being, other beings and other humans; it is a binary processor which consists of input and output information. The body is seen as a cybernetic system which will be enhance with neurocognitive and physical capabilities in the new cyborgized environment.

Since the human brain is seen as a binary processor which operates based on the principle of input and output information, human behaviour can be observed and directed both at the individual and general level. Cybernetics has taken a task upon itself to take the human nervous system conceived in this manner (as an automaton) and expose it to deterministic, linear, and mathematical reshaping. The cybernetic science abolishes the metaphysical existence of universal principles which formed the template and the corrective method of behaviour of man and science, and every difference between the animate and the inanimate and the possibility of knowing. Understanding man's life, sense and activity is reduced to a binary system in terms of external stimuli which send information into the processor in accordance with its general means of control and functioning. The internal processes of brain's understanding are trivialized and the soul as the energy of vivacity of life and its emotions are excluded from the cybernetic understanding of man, despite being a constant point of discussion.

Cyborgization of human relations and their mediation is based on Wiener's idea that it is possible to cyborgize human thoughts and body. Merging man and cybernetic technology is manifested in the transition of individual human consciousness into virtual reality. This human consciousness can live forever without a body and be fused with the cybernetic consciousness of other people, become a part of the general mentality. Computers have thus far been unable to offer the processes of intelligent human thought (computers operate on the principles of logic and programs) but they could, as a result of human consciousness diving into the virtual space of computer influence, affect the human consciousness and direct it to function according to the established logical rules. There is an increasing number of studies showing that the virtual reality technology affects the human brain as it creates enough stimuli which show the existence in virtual world in which the user is submerged. In this virtual reality, a person can leave its body, personality, and identity; travel through timeless space and do as they feel without the need to obey any moral procedure or legal limitations which apply in the real world. The human brain opened and directed in this manner is susceptible to the influence of the entire communication structure of the mass information society. It is not just the brain but the human body as well that are transformed into a communication network based on reproduction and the exchange of signals in time and space. This virtual cyborg represents the synthesis of a living being with a lifeless cybernetic network in the virtual space. Uploaded body is dematerialized in the network as a sign among other signs which remain silent about the biological man. In the cyberspace as information space, data are formed so as to give the operator an illusion of control, mobility and access to information in which it is possible to get the feeling of feedback and simulated world in an interconnected simulation with a large number of users.

One of the existing forms of cyborgization of society described and advocated by Luciano Floridi is focused on the influence of the digital ICT in the forming of self, our relationship with others and our interaction with the world. Nanotechnology, Internet, smartphone apps, touch screen, augmented reality, drones, autonomous cars, etc. all form a part of our environment which affects how our intellectual and physical reality is created and shaped, changes our self-understanding and our way of life. In other words, it changes both our culture and our anthropology.

The fourth industrial revolution, which Floridi calls the *infosphere*, leads man into hyperhistory as ICT causes fundamental changes in the course of history, organisation, understanding and the way of life. There was no ICT in ancient history, but there have been ICTs throughout history – they were used to record and transmit information, but the society was still primarily dependent on other types of technology related to the primary resources and sources of energy. In hyperhistory, there are ICTs and they record, transmit, and above all, process information with increasing autonomy, while the nation states become vitally dependent on them and the information as the primary resource for their progress. One can conclude that the benefits and the added value of the ICT have transitioned into dependence and reliance on ICT.

The infosphere itself is based on bilateral communication, i.e. the machine-machine and man-machine interaction, also known as HCI (human-computer interaction). In the machine-machine communication, the ICT system controls all functions, which is the case, for instance, with autonomous cars. In the human-computer interactions, ICT is used to create, facilitate, and enhance the communication between human users and computer systems. The amount and diversity of generated information perceivable by individuals is trivial in comparison to what is happening in all bilateral interactions. According to Floridi:

A few years ago, researchers at Berkeley's School of Information estimated that humanity had accumulated approximately 12 exabytes of data in the course of its entire history until the commodification of computers, but that it had already reached 180 exabytes by 2006. According to a more recent study, the total grew to over 1600 exabytes between 2006 and 2011, thus passing the zettabyte (1.000 exabytes) barrier. This figure is now expected to grow fourfold approximately every three years, so that we shall have 8 zettabytes of data by 2015. Every day, enough new data are being generated to fill all US libraries eight times over. Of course, armies of ICT devices are constantly working to keep us afloat and navigate through such an ocean of data. These are all numbers that will keep growing quickly and steadily for the foreseeable future, especially because those very devices are among the greatest sources of further data, which in turn require, or simply make possible, more ICTs (2014, p. 13):.

This is why Floridi believes we have entered the age of the zettabytes, also known as the age of "big data", which causes "a mixed feeling of apprehension about the risks, excitement at the opportunities, and astonishment" (Floridi, 2014, p. 14). The "big data" are assets, resources for exploitation and no one is forcing us to digest every available byte. One must learn where to find "the new patterns with real added-value lie in their immense databases, and how they can best be exploited for the creation of wealth, the improvement of human lives, and the advancement of knowledge (Floridi, 2014, p. 15). The problem of "big data" is not just a matter of power of computers but also the power of the human brain. Human memory is not just the matter of storage but also the effectiveness of data management. As old versions of files are erased or replaced with new versions, the Web becomes a place with no memory of its own history (Floridi, 2014, p. 14) and when "most of our knowledge is in the hands of this forgetful memory, we may find ourselves imprisoned in a perpetual present" (Floridi, 2014, p. 18). Since the lifespan of data storage units is limited, we are facing the possibility of "big data" becoming dead despite being transferred onto new storage units in regular intervals. One might wonder which data will be transferred to the new storage and who will be the one making this decision. A large part of the hyperhistory may deliberately be forgotten or erased. It might not matter for the future hyperhistory of cybernetic beings, who should be able to function in a pre-installed program of the *scientific history.* The very program itself will probably predict which data might be relevant for the cyborgized society and which data should simply be deleted. Floridi (2014, pp. 40-41) further argues that:

ICTs make us think about the world informationally and make the world we experience informational. The result of these two tendencies is that ICTs are leading our culture to conceptualize the whole reality and our lives within it in ICT-friendly terms, that is, informationally... the most obvious way in which ICTs are transforming the world into an infosphere concerns the transition from analogue to digital and then the ever-increasing growth of the informational spaces within which we spend more and more of our time. (2014, pp. 40-41)

Life in the infosphere will be under the complete influence of the ICT which not only changes (or cyborgizes) our world, but also creates a new reality and promotes the information-based justification of every aspect of our world and our lives within. As the interfaces become less and less apparent, the boundary between the analogue and the digital world becomes increasingly blurry.

The digital online world is spilling over into the analogue-offline world and merging with it. This recent phenomenon is variously known as 'Ubiquitous Computing', 'Ambient Intelligence', 'The Internet of Things', or 'Web-augmented things'. I prefer to refer to it as the onlife experience. It is, or will soon be, the next stage in the development of the information age. We are increasingly living onlife. The gradual informatization of artefacts and of whole (social) environments means that it is becoming difficult to understand what life was like in pre-digital times. (Floridi, 2014, p. 43)

In the near future, an increasing number of objects and human beings will be IT entities. This is reflected in the upwards trajectory of RFID chips sales. According to the "report by Market Research Company InStat the worldwide production of RFID increased more than 25-fold between 2005 and 2010 to reach 33 billion tags" (Floridi, 2014, p. 45). In Floridi's view, creating new digitalized societies could generate "new forms of discrimination between those who can be denizens of the infosphere, and those who cannot, between insiders and outsiders, between information-rich and information-poor. It will redesign the map of worldwide society, generating or widening generational, geographic, socio-economic, and cultural divides, between Generation Z^+ and Generation Z^-" (2014, pp. 48-49). Multitudes spend their time online by transmitting themselves, communicating with one another, which leads us to conclude that for a cyborgized man to exist means to be interactive. Through their presence on Facebook, Google, LinkedIn, users shape and update their personal identity, find their kindred spirits according to their desires and life preferences.

Consciousness appears in the digital aspects of the social media and "nothing is too small, irrelevant, or indeed private to be left untold. Any data point can contribute to the description of one's own personal identity. And every bit of information may leave a momentary trace somewhere, including the embarrassing pictures posted by a schoolmate years ago, which will disappear, of course, like everything else on this planet, but just more slowly than our former selves will." (Floridi, 2014, p 63).

However, Floridi does not believe that the users immersed in the virtual worlds have lost touch with their naturality. The main reason is that the nature of man is on its own a product of human manipulation through history, and secondly, "because social media also represent an unprecedented opportunity to be more in charge of our social selves, to choose more flexibly who the other people are whose thoughts and interactions create our social personality, to paraphrase Proust, and hence, indirectly, to determine our personal identities. Recall how the construction of your social self (who people think you are) feeds back into the development of your self-conception (who you think you are), which then feeds back into the moulding of your personal identity (who you are). More freedom on the social side also means more freedom to shape oneself" (Floridi, 2014, p. 64). The matter at stake in this case is creating a personal information identity in the online environment. Floridi finds that the questions of identity and veracity in different times and circumstances are focused on the imposed social goals. "Who are we, who do we become, and who could we be, once we increasingly spend our time in the infosphere?" (Floridi, 2014, p. 65). He believes that "the right interface seems to be offered by an informational conception of the self" (Floridi, 2014, p. 68). The basis of identity is the unity of consciousness and continuity of memory for "as long as you are a thinking entity, you are the specific thinking entity that is going through such specific mental processes" (Floridi, 2014, p. 68). If we allow our consciousness or memory to be hacked we cease to be ourselves. Identity is a story created through social interactions – we give each other identities, especially through our socialization via ICT networks, which is why Floridi (2014, p. 69) claims: "The self is seen as a complex informational system, made of consciousness activities, memories, or narratives. From such a perspective, you are your own information" and as such are inevitably and fundamentally affected by the ICT patterns. The self, understood as an application of information concept of self, reflects the mind-body dualism. Floridi is aware of the body's meaning as "is no development of the self without the body, but once the latter has given rise to a consciousness, the life of the self may be entirely internal and independent of the specific body and faculties that made it possible" (Floridi, 2014, p. 70). In the explanation of the body, Floridi makes reference to the American physicist John Archibald and his for formulation "it from bit", which symbolizes the idea that every aspect of the physical (material) world is made up of information and not of matter.

Through cybernetic metaphysics (though he discusses modern philosophy as an equivalent of third-order technology), Floridi strives to overcome the Cartesian mind-body duality and justify the cyborgization of the biological body and mind by replacing them with information identity, that is, establishing the body in the virtual reality. As a "moderate optimist", he uncritically accepts the scientific and technological progress and find the infosphere as cybernetic life in the virtual reality to be the true essence of man.

A particularly notable aspect of cyborgization of historical reality is Max Weiser's notion of "ubiquitous computing" (Kaku, 2011, p. 17), which is based on the fact that chips will become increasingly affordable in the near future, thus allowing their implementation in the overall human reality – clothes, furniture, walls and everything else, including human bodies. Connected to the internet, chips will be able to fulfil all human wishes and desires. This way, the cyborgized reality has been given a new lifeform. A change of paradigm from a single chip in a computer to myriads of chips distributed across the human reality will create a new lifeform of reality to be imbued with artificial intelligence. To paraphrase Hegel, what is real will become mental and what is mental will become real. Hegel's metaphysic of the absolute spirit is realized in *scientific humanism as naturalism* in a new, cybernetic way, far beyond the realms of his philosophy.

According to Weiser's idea, every flat surface will be a potential screen of the ubiquitous computing. "The destiny of computers—like other mass technologies like electricity, paper, and running water—is to become invisible, that is, to disappear into the fabric of our lives, to be everywhere and nowhere, silently and seamlessly carrying out our wishes" (Kaku, 2011, p. 19). Through computers and chips implanted into the human reality, man will be able to converse with inanimate entities as the whole reality will be connected to artificial intelligence and established onto virtual reality via computers. It is expected that:

> *The rapid rise of computer power by the year 2100 will give us power like that of the gods of mythology we once worshipped, enabling us to control the world around us by sheer thought. Like the gods of mythology, who could move objects and reshape life with a simple wave of the hand or nod of the head, we too will be able to control the world around us with the power of our minds. We will be in constant mental contact with chips scattered in our environment that will then silently carry out our commands. (Kaku, 2011, p. 19)*

Even today we can live in the virtual world through the contact of our avatar with other avatars, we can buy virtual things, fall in love in virtual reality. The prediction is that future man of cyborgized reality of *scientific humanism as naturalism* will be able to visit unreal world, walk on other planets and feel computer-generated objects. Human relations will be replaced by delivery of information between connected cybernetic chips over which an individual has no control. The individuals will no longer be the representatives of social structures but actors of social networks and chip-based reality of *scientific humanism as naturalism*. Cybernetic medication of social relationships is the information communication project of connecting the participants in the network via universally comprehensible language, which conveys the messages in a completely pragmatic way. This mediation of artificial non-human communication via instrumentalized language which transmits knowledge as application of skills and not as understanding or knowledge about the marked objects. Baudrillard argues that this language testifies about the scientific and technological destruction of being.

The information communication society as a cyborgized relationship between humans constitutes reality with a computerized character of thought which transforms the human history into *scientific history* and establishes the binary genetic code on it. The social relations in the reality are replaced by coded information, which opens up a new anthropologic and ontological view of the world. Man is not, as Marx believed, a collection of social relations, *zoon politikon*, but a lonely individual, an apparition without society and kin, caught in the web of the corporative structure of a communication system. In the virtual reality, there is no space-time of human intersubjective freedom of relations but media mediation in which power and rule stem from the control of power provided by *scientific work*. Application of this science and this knowledge establishes the technologically predictable progress of global society of *scientific humanism as naturalism* in which biological life is an object of the scientific and technological experiment, with the ultimate goal of completing the human nature in the non-human form of life, cyborgs, robots and apparitions in real and virtual space.

In his *Physics of the Future*, Kaku discusses augmented reality which is at the doorstep of *scientific humanism as naturalism*. The use of a computer program will create the illusion of new world and the possibility of reshaping human body. Smart glasses (whose prototype already exists) will make a whole new world around us come to life. Augmented reality of our cyborgized world will be fused with the real world with computer apparitions. The glasses will create the illusion of disappearance of real objects so much so that invisible will be made visible. For instance, the *Institute of Technology in* Beijing has restored in virtual reality the beautiful garden of perfect light which was destroyed by the British and French forces in 1860. Kaku believes that by mid-century, people will live in a mixture of real and virtual world. Smart glasses will be used to dive into virtual world, keeping track of real events. *University of Keio* in Japan is working on a prototype of glasses that fuses imagination and reality. The glasses enable the illusion of disappearance of objects before the user's eyes. Nicholas Neecke has created the illusion of virtual walk across the city of Basel. Like a time-traveller, the user walks the town streets, sees the images of old building and people. The computer determines his position and shows him the medieval scenes in appropriate spots through the glasses. Smart glasses will enable see-through vision in 360° view, with various other information in offer. Memorizing things will no longer be necessary, people will not need their cell phones, watches, TVs, as all technological devices and their icons will be projected into the smart glasses' lenses. These are just some of the examples of the future cyborgized reality of man.

The character of space-time of *scientific humanism as naturalism* is the *eternal recurrence of the same* which follows from the essence of *scientific work*. The cyborgized or bio-cybernetically created body of the immortal scientific worker stops being a matter of science of life and drifts into the domain of scientific and technological solution for the simulation of life. *Scientific humanism as naturalism*, in this case the information society with its systems of digital networks, is from a philosophical standpoint manifested as a transcendental condition of existence of an individual in reality and virtuality of connected communication participants. In this society, the primal communication which depends on the immediate experience of participants in the communicative act is replaced with the network as a dehumanized means of mediation between virtual bodiless actors. Baudrillard argues that the replicas of the participant of a connected society wander the networks in which they will never face their original as the virtual world has no bearing to reality. This is not a matter of fighting entropy and survival in the form of virtual reality and artificial intelligence but a matter of meaning of life and man in general.

The digital code is a superimposed superstructure, the immanence of relations, and man is an object, a thing, a product of manipulation and experimentation. This is why Baudrillard rightfully wonders whether the virtual world is just a powerless illusion whose aim is to remove all uncertainties. *Scientific work* with its scientific and technological affiliations is no longer capable of giving its object a final meaning or sense. The object of revelation is revealed and hidden in the depths of the experiment, and the scientific revelation as a means of a means of *scientific work in scientific history* is increasingly losing ground in its fight with real-world objects. Humans united in the electronic space-time more and more believe in the reality of virtual and augmented reality and events, which are mediated not by human but virtual means of communication. In this manner, the virtual affects the real behaviour of man as an input stimulus – an information that is no longer questions but serves as a foundation for our reaction to events.

7. 3 HUMAN ENHANCEMENT TECHNOLOGIES

If you look at automata which have been built by men or which exist in nature, you very frequently notice that their structure is controlled to a much larger extent by the manner in which they might fail and by the (more or less effective) precautionary measures which have been taken against their failure. (Von Neümann, 1966)

In the previous chapters, we have described the procedures for cyborgization of man deemed medically and ethically justified for health reasons. The primary criterion in those procedures is restoration of organism's lost organs and functions so as to enable it to have normal life, equal to that of other healthy individuals. When a healthy human being is cyborgized to enhance or reshape its mental or physical capabilities, cyborgization no longer serves to preserve life but functions as a scientific and technological experiment and has its causes in the self-reproduction of *scientific work* which no longer needs biological man in its *scientific history* but a scientific worker more appropriate for scientific work – a cyborg or a robot.

The field of human enhancements is an emerging paradigm within the domains of medicine and bioengineering whose aim is to develop technologies for overcoming current human cognitive of physical capabilities (Naam, 2008). Technologies developed within the field of enhancement are called the human enhancement technologies (HET). They refer to every method or procedure whose aim is to increase the human individual realm of possibility beyond the current *status quo*. An important factor that affects the development of human enhancement technologies is the convergence of four fields of research: nanotechnology, biotechnology, IT, and cognitive sciences. This point of convergence is also known as NBIC and its aim is to enhance human bodies at an individual level. This refers to:

1. Enhancing capabilities of cognition and perception – achieved through implantation of neurochips and neurotechnology.

   a. enhancing cognitive capabilities – implies use of sensory perception, infrared spectre vision, enhancing memory, thoughts, imagination and decision-making.
   b. Enhancing capabilities of perception or affection – implies mood, social behaviour, trust, compassion enhancement, etc.
2. Enhancing physical capabilities – primarily refers to increasing strength, endurance, precision, temperature resistance, etc. These modifications are achieved through implantation of prosthetic and bionic technology, pharmacologic substances, and genetic engineering.
3. Enhancement whose aim is to gain superhuman capabilities – implies obtaining abilities that are not characteristic for human beings, including eternal life.

Procedures for enhancement of human capabilities aim to help the human body and mind in fighting the entropy of the world, which is why they create human bodies with augmented capabilities, enhanced mental and physical features. Human beings become objects of procedures of evolution of life programmed by *scientific work* in the *age of scientific humanism as naturalism*. Future research of procedures of human enhancement are nano(bio)-technology (e.g. nanobots, artificial immune systems), anti-aging, and life extension (tissue engineering); genetic engineering (e.g. artificial chromosomes); implants and prosthetics (cyborg technologies, brain-computer interface); super-soldiers (DARPA). We believe that analysing future technological development and approaches used in those procedures of human enhancement and reshaping necessarily involves:

1. questioning ethical, medical, social and philosophic arguments *pro et contra* technological progress;
2. researching conceptual issues and the ethical context of use of omnipresent neuro-, bio- and nano- and other technologies for enhancing human beings;
3. analysing specific cases of human enhancement technologies;
4. motivating further actions from a multidisciplinary team of researchers and thinkers and creating bio-ethical reactions to the issues of human enhancement;
5. motivating an intersubjective discussion on issues of enhancement of human being and its historical reality;
6. passing appropriate legal norms and establishing cyborgoethics.

7.3.1 The Challenges of Nanotechnology

Nanotechnology is a relatively new branch of science and technology which implies research and manipulation of materials whose size is below 100 nanometres (nm).[11] Nanotechnology was made possible by research in the domain of quantum physics, specifically research on atoms and quantum forces whose actions cannot be seen with the naked eye. According to the US National Science and Technology Council, the definition of nanotechnology is the following:

Nanotechnology is the ability to control and restructure the matter at the atomic and molecular levels in the range of approximately 1–100 nm, and exploiting the distinct properties and phenomena at that scale as compared to those associated w/ith single atoms or molecules or bulk behavior. The aim is to create materials, devices, and systems with fundamentally new properties and functions by engineering their small structure. (Roco, 2010, p. 428).

The term nanotechnology was first used in 1974 by Norio Taniguchi from the *Tokyo University of Science*. In 1986, Eric Drexler, currently at the *Stanford University*, initiated the development of molecular nanotechnology, starting from the concepts of controlled positioning of atoms and achieving self-replication of molecular machines. The word *nanotechnology* is being increasingly used with various meanings. On the one hand, it implies the production of devices whose size is 100 nanometres or less, and on the other hand it denotes manipulation of individual atoms, which is why it is often called molecular nanotechnology. Igor Čatić (2002) argues that nanotechnology encompasses the world of atoms and connects the biotechnology with the technology of the inanimate. According to him, "nanotechnology is estimated to make 700 to 900 billion US dollars in the upcoming 10 to 15 years, and about 70% of EU research funds is focused on using nanotechnology in biological sciences, primarily medicine" (Čatić, 2002).

The reason for these investments is fear that the world economy will crash unless a suitable replacement is found for the silicon semi-conductors in our computers. Nanotechnology offers a solution for this predicament. Nanotechnology is the strongest tool of *scientific work* as it enables manipulation of individual atoms, thus creating highly durable materials with superb electric and electromagnetic properties which do not exist in nature. Nanotechnology research and development includes controlled manipulation of nanoscopic structures and their integration into larger material components, systems, and architectures. The goal is to create the desired structures in accordance with the laws of physics and chemistry. This is achieved by putting every individual atom or molecule in the right spot. Nobel Prize winner Richard Smally once said: "The grandest dream of nanotechnology is to be able to construct with the atom as the building block." Philip Kuekes of Hewlett-Packard said, "Eventually, the goal is not just to make computers the size of dust particles. The idea would be to make simple computers the size of bacteria. Then you could get something as powerful as what's now on your desktop into a dust particle.'" (Kaku, 2011, p. 98).

7.3.1.1 Consumer Goods Industry

Nanotechnology is used in industrial processes, weight loss products, skin and body care products, golf clubs, clothes, mattresses, etc. Drexler as the pioneer of nanotechnology envisages a future in which all products are produced at the molecular level. Nanobits do not exist yet but there is as strong chance in the distant future they will be available and that they will change the human environment as we know it. According to the Kaku, by year 2100 we will have a molecular collector or a replicator capable of creating anything. This appliance would be the size of a washing machine and you would put the raw materials inside it and press a button. At that moment, billions and billions of nanobots would appear, each one programmed to tear every molecule apart and make it into a new product. This machine would be capable of producing anything; it would be the pinnacle of engineering and science and the ultimate solution of human problems ever since first tools were made in ancient history (Kaku, 2011, p. 115). The problem that the replicator itself cannot solve is the huge amount of energy that would be required for it to work (Kaku, 2011, pp. 111-112).

7.3.1.2 Electronics

Today's electronic devices use wiring the size of 50 nm, i.e. around 500 atoms in diameter. Miniature microprocessors (smaller than 100 nm) enable higher speed and higher power of the processor. It is possible to produce the thinnest wire which is only one atom wide. The aspect of nanotechnology which is slowly starting to affect the lives of us all are the microelectromechanical systems (MEMS) which include sensors for airbags and screens, and gyroscopes for planes and cars. MEMS are tiny devices that would fit on the tip of a needle. Instead of existing computer transistors, engineers use tiny mechanical components invisible to the human eye (Kaku, 2011, p. 101).

7.3.1.3 Car Industry

In addition to microelectromechanical devices for activating airbags, nanotechnology is also used to produce nanocolours, which contain inorganic silicone particles connected with an organic polymer. Nanocolours are durable and resistant to corrosion and bumps.

7.3.1.4 Medicine

US scientists are using nanotechnology to develop methods for fighting cancer which would not harm the healthy tissue. Synthetic carbon nanotubes are implanted into the cancer cells and heated with the use of a laser. After the tubes are heated up to 70 °C, the cancerous cells die. The scientists claim that it is possible to target with precision the exact type of cancerous cells by attaching the antibodies to the nanotubes. At the MIT, a nanotube capsule was produced which fits inside a syringe and the goal is to inject it after the biopsy. The nanoparticle binds the cancer cells onto itself. The particle capsule will help the doctor to track tumours or notice any cardiac abnormalities. The capsule is supposed to replace the MRI in the future as the sensor will be able to read everything that the capsule recorded (Balija, 2016). One of the predictions is that the sensory nanochips will be implanted throughout the entire human environment. These sensors would monitor the complete state of human health at every moment and in every place. The key components are the artificial DNA chips. In 1997, the first commercial DNA chip, which quickly analyses 50,000 DNA sequences, was put on the market. The doctors at the

Massachusetts General Hospital have created their own custom-made biochip that is 100 times more powerful than anything on the market today. [...] The new biochip is sensitive enough to find one in a billion CTCs circulating in our blood. As a result, this chip has been proven to detect lung, prostate, pancreatic, breast, and colorectal cancer cells by analyzing as little as a teaspoon of blood. (Kaku, 2011, p. 105).

In the future, it is predicted that nanorobots will course through our veins, destroy the pathogens and create new cells, even new organs:

And where biology leaves off, hardware could conceivably take over. Nanomedicine theorists have, for instance, look forward to the day when medical bots patrol the body, 'armed with a complete knowledge of a person's DNA', ready to 'dispatch any foreign invaders.' One nanotech enthusiast reports that he has 'drawn up intricate plans for a system of nanobots that would make blood circulation obsolete. Tiny robot tankers containing oxygen, nutrients, and wastes would replace all the functions of the circulatory system. (McKibben, 2004, p. 154)

Another feature of nanotechnology is the possibility of reaching vast numbers (the ability of nano-robots to self-replicate). Thus, nanotechnology, genetics and robotics "[...] had a common flaw: they could self-replicate. Tailored germs could spread; robots could build more of themselves; and as for nanotechnology, the ultimate promise of the field depends on, as we've seen, on the replication of the tiny assemblers" (McKibben, 2004, p. 90). Besides maintaining health, nanobots the size of a microbe could also be used to:

[...] scan the human brain and re-create its design electronicay, perhaps by using those same nanobots to capture the locations, interconnections, and all contents of all the nerve cell bodies, axons, dendrites, pre-synaptic vesicles, neurotransmitter concentrations and other relevant neural components and sending the information to a hard drive via wireless transmission. (McKibben, 2004, p. 86)

One of nanotechnology theorists Robert Freitas argues:

Until nanomedicine becomes available, nobody can have any reasonable expectations that they will live long enough to pursue their longest-term and most interesting goals to fruition. Thus, the development of nanomedicine looms like a tollgate, separating each one of us from the rest of our lives. (McKibben, 2004, p. 147)

Numerous scientists have already voiced their concerns that the possibility of merging molecules into larger structures will radically change some branches of science and have a radical effect on society, healthcare, economy, politics, and human identity. The promises of nanotechnology are: extremely fast computers, lighter and stronger materials, more efficient production processes, increased medical technology which leads to immortality. Nanotechnology will enable the creation of extremely fast computers and new types of computer memory. The quantity of technological novelties will also surge – lighter and more durable materials will be used in space technology and fashion industry (paints which do not stain, materials stronger than steel weighing less than 280 grams, etc.). A team of researchers at the *Northwestern University* has produced transparent transistors that allow them to display an image in the air (Greenwood, 2008). Researcher Tobin Marks believes that this material could have a wide array of uses – protective goggles for soldiers which would display important information, allowing surgeons to keep track of what is going on the patient's other side while operating, etc. (Glas Amerike, 2007b). Nanotechnology will lead to revolutionary changes in the economy, creating clean and efficient work processes. Some of the relatively recent discoveries in this field include the method for storing vast amounts of hydrogen in reservoirs which allows us to replace fossil fuels, and the smallest electromotor visible only under a microscope (Slobodna Dalmacija, 2005). Enhanced medical technology includes creation of small medical robots which would have the ability to wipe diseases from our cells, stop the aging, and enable the upload of the human mind from a computer (achieving human state of artificial intelligence) (Drexler et al., 1991, p. 119). The most interesting part of the nanofuture are the interventions in the human body and the ability to position individual molecules as we please. Starting from the creation of artificial molecules which serve as the memory and logic of the machine, we arrive to the possibility to implement an entire computer into a cell. An accomplishment like this would enable programmed improvement and enhancement of man at the cellular level. On the positive end, it would facilitate tracking our health and the state of our environment. We have already mentioned that the researchers at Stanford have developed a cancer treatment method using synthetic nanotubes inserted into cancer cells and heated using lasers, causing these cells to die (Schwartz, 2005). Unlike standard chemotherapy which causes death of both cancerous and healthy cells, this treatment burns the cancer-ridden cells without damaging the healthy tissue.

James Toor and his colleagues at the *Rice University* have created nanovehicles which would be used to control a tiny robot in the cardiovascular system and destroy all cancer cells they encounter. In addition, the vehicles will deliver drugs to specific places in the body. Since the vehicles have no engine, researchers are working creating a power source for them. An idea being considered as a solution for this problem is creating a synthetic molecule of adenosine triphosphate, which is in its natural form the source of energy for the human body (Kaku, 2011, p. 103). Futurologists Ray Kurzweil believes that people could become immortal in less than 20 years from now by perfecting the nanotechnology and obtaining a more profound understanding of how the human body functions. This claim is supported by research conducted at the *Wake Forest School of Medicine*, which identified the protein which controls the malignant properties of brain tumour cells. Protein Fra-1 is present in later stages of breast, colon, and thyroid cancer and it controls the way in which genetic instructions are carried out in cells (Đugum, 2005). George Smith, head of the *Department of Materials* at *Oxford University* believes that the nanoscales reveal new, alternate features of materials (Posavec, 2012). The most significant advances in nanotechnology are made in the field of human genome studies and it is a matter of time before every human being has their own genetic imprint, which will be used to predict all the diseases that might befall them. De Gray (*Cambridge Interdisciplinary Research Centre on Ageing*) is already working on a plan to extend the life expectancy (De Grey, 2005, p. 1000). So far, the testing has been conducted on mice, but human life expectancy is expected to be prolonged within the next ten years. The aim is to find the main body mechanism in charge of ageing, stop it or at least slow its activity. De Gray argues that this is more than just an idea, i.e. that there is a very detailed plan for repairing the damage in the cells by combining methods or procedures that already exist (Metro Portal, 2009).

Atom manipulation will create stronger and lighter materials which will be used to make just about anything – from buildings and vehicles to brand new devices whose design will be abstract in comparison to everything we know today. The nanotechnology critics are concerned that introducing nanochips into nanobots could result in artificial intelligence becoming independent. There are also some adverse effects of nanotechnology on ecology, global economy and health issues. Nanoparticles can become detrimental for survival and health of the human species due to size comparable to viruses and the inability to control their expansion. Due to increasing development of nanotechnology, it is necessary to conduct comprehensive research about its potentially negative social side-effects. An important aspect of such research would be to develop methods and procedures to prevent or inhibit negative consequences. In that respect, it is vital for the scientists to participate in discussions on ethical issues and defining legal norms and legislature that would regulate the future development and application of nanotechnologies. There is an overwhelming concern that at some point in the future nanotechnology will lose control and turn all minerals on Earth into useless grey sludge. The danger lies in the self-reproduction of future nanobots which would have all the traits of viruses. Released in the environment, they would self-reproduce without any possibility of human control. In his book *Engines of Creation*, Eric Drexler describes a fictional case of nanomachines self-replicating to the extent of running out of raw materials for self-reproduction. As they are made of organic materials, they use up all organic matter in their reproduction – humans, animals, plants, microorganisms, wood, plastic, cloth, and fossil fuels. The only thing left on Earth are the nanomachines, water and simple gases. Scientists find this scenario implausible and unlikely (Levy, 2006, p. 11).

7.3.2 Cognitive Sciences

When scrutinizing *scientific work,* which establishes everything that is, the question we are faced with is whether machines constructed according to the principles of scientific and technological mind and strengthened by artificial intelligence can have intelligence necessary for active work and whether they can think or become self-aware. In its interdisciplinary approach to the study of human mind, cognitive sciences incorporate within themselves the results of computer sciences, neurosciences, psychology, linguistics and anthropology. Our task here is limited to the description of artificial intelligence used to imitate human brain and which ultimately wants to become the eternal extension of the human intelligence and the human mind. We also want to underline the necessity of creating legal solutions that would determine who can develop artificial intelligence as it is already clear that misuse or AI becoming independent might represent a huge challenge for humanity.

English physicist Stephen Hawking says:

The real risk with AI isn't malice, but competence. A super-intelligent AI will be extremely good at accomplishing its goals, and if those goals aren't aligned with ours we're in trouble. You're probably not an evil ant-hater who steps on ants out of malice, but if you're in charge of a hydroelectric green-energy project and there's an anthill in the region to be flooded, too bad for the ants. Let's not place humanity in the position of those ants (de Haldevang, 2018).

Many reputable scientists are convinced that artificial intelligence is well on its way of becoming autonomous and surpassing human capabilities. Computers already control aeroplanes and autonomous automobiles, create music and paintings. According to Hawking, "The development of full artificial intelligence could spell the end of the human race" (Jones, 2014). His fear stems from the fact that at some point during its development, AI will begin to upgrade itself autonomously. Slow biological evolution of man in relation to artificial intelligence will leave him without any real power. Around 8000 scientists have expressed their concerns about the dangers of AI development. The influence of AI on economy, government, ethics, and even theology will be huge. The possibility of hacking is an additional cause for concern (Jones, 2014).

The first electronic computer was constructed in 1945 and was presented as an electronic brain, which was pretty far from its actual capabilities. The idea of creating an artificial neural network in which numerous process elements process data in parallel – just like the neurons in the human brain – stem from 1940 when McCulloch and Pitss (MIT) published their mathematical model of a neural network within the automata theory (Bašić et al., 2008, p. 4).The first neural network model with learning capabilities was made in 1958, but the processing powers of computers of that age were not up to the task of implementing neural networks. After a somewhat prolonged delay, a neural network is implemented into a computer in 1990 and has been and has since become inevitable. Neural network as an artificial replica of the human brain aims to simulate the learning procedures. It consists of numerous simple processing elements, units, or nodes, whose functionality is based in their attempts to imitate biological neurons. Experts believe that this analogy is dubious even today as, despite many simplifications, there are still human nerve system phenomena which have not been adequately replicated in a neural network and there are neural network features which do not coincide with those of biological organisms. Neural networks solve problems of classification and prediction, which is why they are used in numerous branches of science and technology.

Within the scientific community, there is no consensus regarding the term intelligence. Intelligence is described as the ability to learn and the capability to operate in new situations based on the ability to notice facts. Artificial intelligence is regarded as an attempt to imbue machines with intelligent behaviour, i.e. to allow them to manage unpredictable situations without following predefined programs. According to Pavelić (2004, p. 118), intelligence is the ability to manage new situations, identify problems, and find reactions adapted to the new scenario. It should not be equated to habits, skills, and knowledge, or instinctive means of adjustment. However, intelligence is what determines what the experience level of an individual will be and how the individual uses this experience. It is believed that the goal of imitating the human brain and its transformation into an artificial form is the core interest for the field of artificial intelligence as human brain contains the biggest amount of intelligence which we know or are able to conceive. Man's creativity is manifested in the ability to choose the right hypotheses and make decisions based on experience and logical rules. Mankind is able to learn using various strategy and learning is another important aspect of artificial intelligence which enable the system to conduct changes over itself. So, it seems justified to argue that the system which successfully mimics the work of a human brain would be downright intelligent (Bašić et al., 2010, p. 4).

The term artificial intelligence was first used by John McCarthy in 1956, and he defines it as "the science and engineering of making intelligent machines" (McCarthy, 2007, p. 2). Minsky defines it as "the science of making machines do things that would require intelligence if done by men" (Minsky, 1968, p. v). Dan W. Patterson regards artificial intelligence as:

[…] a branch of the computer science concerned with the study and the creation of the computer systems that exhibit some form of the intelligence: systems that learn the new concepts and the tasks, systems that can reason and also draw the useful conclusions about the world around us, systems that can understand the various natural languages and perceive and comprehend a visual scene and the systems that perform the other types of the feats that essentially require the human types of the intelligence. (Patterson, 1990)

An intelligent system establishes an interaction with the outside world, collects and processes information, gathers knowledge and handles it, processes it and from it makes conclusions, plans and can establish real-time communication with human beings using natural language and speech, and has the ability to communicate with other intelligent systems. IBM's Deep Blue computer achieved historical success in 1997 by beating the world chess champion Garry Kasparov. Since a computer has very little knowledge of chess but has the ability to calculate 200,000,000 chess positions in a single second, whereas Kasparov can only predict 3 positions per second, the computer was able to beat him. However, going through a large number of combinations does not require intelligence but a program. It became clear that Deep Blue is unable to think. It was superior in chess but it would get no points on the IQ test. Kasparov was beaten by an algorithm not intelligence.

Mathematicians have realized

[…] that they made a crucial error fifty years ago in thinking the brain was analogous to a large digital computer. But now it is painfully obvious that it isn't. The brain has no Pentium chip, no Windows operating system, no application software, no CPU, no programming, and no subroutines that typify a modern digital computer. In fact, the architecture of digital computers is quite different from that of the brain, which is a learning machine of some sort, a collection of neurons that constantly rewires itself

every time it learns a task. (A PC, however, does not learn at all. Your computer is just as dumb today as it was yesterday.) (Kaku, 2011, pp. 44-45)

Kaku (2011) describes two approaches to artificial brain modelling. The first approach takes as its foundation the model which originated in the works of British mathematician Alan Turing. His model is based on three components: "an input, a central processor that digests this data, and an output" (Kaku, 2011, p. 45). All digital computers are based on this pattern. Computers will become intelligent once a program is made which contains all features of intelligent behaviour. In contrast to the computer, human brain comes with no program support. It consists of a network of neurons which continuously wire together and, in this wiring, practice and learning at the same time, they become better and more intelligent. The second approach to creation of artificial intelligence starts with installation of a neural network which learns much in the same way as the human brain. However, the neural network of the human brain is different from a digital network. If a single transistor in the central processor unit of a computer is not working, the processor is dead. The human brain functions even in situations when one of its parts is surgically removed. In those situations, other parts of the body take over the function of the removed part. Unlike the brain which is slow, digital computers can perform calculations at the highest speeds currently known to man. The brain compensates for the low speed if its nerve impulses "because it is massively parallel, that is, it has 100 billion neurons operating at the same time, each one performing a tiny bit of computation, with each neuron connected to 10,000 other neurons." (Kaku, 2011, p. 45). The difference between human brain and artificial intelligence is also reflected in pattern recognition of behaviour and reasoning. To be regarded as intelligent, artificial intelligence should be able to recognize the context of an event and not just the keywords. Computers inside robots have significantly better sight than human beings but they do not understand what they see. It takes computers several hours to search their memory and find the object they saw in the space in front of them. Even cockroaches identify objects and go around them (Kaku, 2011, p. 48). This shows that artificial intelligence is still at the same level as some of the lower forms of living beings. Humans recognize objects as soon as they see them. Secondly, computers in robotic form can hear much better than humans but they cannot understand what they heard unless it is stored in their memory. They have no common sense and it is exactly the common-sense experience that will be very difficult to program as it needs to be learned. It takes hundreds of millions of lines of code to describe the common sense of a healthy six-year-old. Researchers have spent years developing a program which would code all the secrets of common sense. In its application, artificial intelligence is also faced with the problem of emotions. Neurobiologists have realized that emotions played one of the most crucial roles in the evolution of man as they provided us with a measure of value. They are a core component of human intelligence as they are used to form value judgements. Scientists have so far been able to program artificial emotional behaviour patterns, but a robot does not really feel these emotions nor is it aware of them. Partial success in programming artificial intelligence has motivated scientists to re-engineer the brain. Human brain is being torn apart neuron by neuron and a simulation of these neurons is run on a super computer called the *Blue Gene*. This computer is "capable of computing at the blinding speed of 500 trillion operations per second. Blue Gene is a remarkable sight. It is huge, occupying about a quarter acre, and consists of row after row of jet-black steel cabinets, each one about 8 feet tall and 15 feet long." (Kaku, 2011, p. 55). For now, this powerful computer "can simulate the firing of neurons found in a mouse, not its behavior" (Kaku, 2011, p. 55). Despite all these difficulties, Henry Makram made his optimistic announcement in 2009: "It is not impossible to build a human brain and we can do it in ten years. If we build it correctly, it should speak and have an intelligence and behave very much as a human does." (Kaku, 2011, p. 56). Makram believes "that it would take a supercomputer 20,000 times more powerful than present supercomputers, with a memory storage 500 times the entire size of the current Internet, to achieve this" (Kaku, 2011, p. 56).

The second approach to brain re-engineering does not use the *Blue Gene*, but "scientists take the slice-and-dice approach, starting by dissecting the brain of a fruit fly into incredibly thin slices no more than 50 nm wide (about 150 atoms across). This produces millions of slices." (Kaku, 2011, p. 57) Gerry Rubin of the Howard Hughes Medical Institute believes that it would take around 20 years to make the whole map of a fruit fly's brain: "After we solve this, I'd say we're one-fifth of the way to understanding the human mind [...] The human brain has 1 million times more neurons than the brain of a fruit fly. If it takes twenty years to identify every single neuron of the fly brain, then it will certainly take many decades beyond that to fully identify the neural architecture of the human brain. The cost of this project will also be enormous." (Kaku, 2011, p. 57). Kaku assumes that decoding and re-engineering of the human brain will be completed by the end of this century. But it will take even more time to create humanoid robots with human artificial intelligence. In this strand, the issue of consciousness presents a major obstacle, and there are several interpretations of it. There is no single unified definition of this term. For instance, Kaku assumes that consciousness consists of three components:

1. sensing and recognizing the environment; 2. self-awareness; 3. planning for the future by setting goals and plans, that is, simulating the future and plotting strategy" (Kaku, 2011, p. 58). In the first component, the lower life forms and "machines with feedback mechanisms have a primitive form of consciousness. (Kaku, 2011, p. 59)

However, artificial intelligence cannot turn its sense of the environment into understanding of what it sees. We can conclude that artificial intelligence has no pattern recognition capability at the level of lower forms of life. In the second component, "many animals lack awareness of who they are. But monkeys, elephants, dolphins, and some birds quickly realize that the image in the mirror represents themselves and they cease to attack it." (Kaku, 2011, p. 59). Humans have a highly developed sense of self and their environment. Artificial intelligence, according to Kaku, does not possess this level of consciousness. With regards to the third component, some animals are capable of planning for immediate future but have no sense of past and future. Humans have a "very well-developed sense of the future and continually make plans. We constantly run simulations of reality in our heads." (Kaku, 2011, p. 59). This requires a "a very sophisticated understanding of common sense and the rules of nature." (Kaku, 2011, p. 59) and shows that humans are the only living beings to have mastered this skill.

Even though there is still no scientific consensus on how to create a robot with a consciousness more advanced than human, Kaku assumes that they could:

[...] run complex simulations of the future far ahead of us, from more perspectives, with more details and depth. Their simulations would be more accurate than ours, because they would have a better grasp of common sense and the rules of nature and hence better able to ferret out patterns. [...] If their goals include helping the human race, then everything is fine. But if one day they formulate goals in which humans are in the way, this could have nasty consequences. (Kaku, 2011, p. 60)

In his view on future relations of human beings and robots, Kaku wonders: "Perhaps that is our role in history, to give birth to our evolutionary successors. In this view, our role is to get out of their way."

(Kaku, 2011, p. 60). With this question, Kaku closes in on the discussions of the future of coexistence of man and artificial intelligence described in previous chapters of this book. In his work, Kaku (2011) describes the manifestation of the essence of *scientific work* through authority of his own scientific revelations and conversations with several hundred scientists and futurologists. His description, however, is guided by the exhilaration with scientists and scientific accomplishments that will be in the service of mankind and under his control. For him, science is knowledge without awareness of the dangers that *scientific work* as the substance subject of the human history has within its essence, thus functionalizing the society of man and nature, establishing them onto a lower level in the cybernetic reality. Many prominent scientists like Stephen Hawking and entrepreneurs like Elon Musk, Steve Wozniak, and Bill Gates express their caution regarding potential dangers of uncontrolled growth and development of artificial intelligence, which could spell the end of mankind (Ćuto, 2015). In 2015, over 1 billion dollars has been spent on research of artificial intelligence in the USA. How to find the right balance between the benefits that come out of developing artificial intelligence and the actual danger, not just of its power but also dangers of loss of numerous occupations and jobs? The key question is whether we can control the development, goals and application of artificial intelligence so that it would not turn from a faithful servant into an evil step-mother. According to Wozniak, we have to ask ourselves:

Will we be the gods? Will we be the family pets? Or will we be ants that get stepped on? I don't know about that ... But when I got that thinking in my head about if I'm going to be treated in the future as a pet to these smart machines ... well I'm going to treat my own pet dog really nice (Storm, 2015).

7.3.3 Genetics

Scientific work in the form of scientific and technological progress, particularly in the field of genetics, has granted man certain abilities of divine intervention, made him a lord of life and death and has opened up the discussion about topics considered unimaginable until quite recently, such as *in vitro* fertilization and embryo transplants, genetic engineering and cloning (Čović, 2004, p. 16).

The question of genetic inheritance was described by Hippocrates who imagined that inherited particles are collected in the body of an adult individual and are formed by experience. Aristotle believed that the inheritance order is permanent and inherited through gametes (sex cells). Throughout the last 200 years, genetics has solved the mystery of the nature of inheritance. The system of studying the genetic features dates back to Mendel's discovery of inheritance features of peas and the 1866 study *Experiments on Plant Hybridization*. The true significance of Mendel's discovery did not receive appropriate scientific attention until 1900, when scientists realized that his laws of inheritance apply to different plants, even mice and chicken. Until that point, the prevailing idea was that of mixed inheritance, which means that the descendants inherit a mixture of features from their parents. Mendel's work showed that inheritance is individual and the inherited material is made up of cell particles – genes. Every organism has two cells which determine a certain property. Through the gametes, both parents give their child one of the two cells that are combined to give rise to the diploid zygote which contains the DNA of both the mother and the father. This coded information in the form of inheritance units is called a gene. From the zygote all future cells in the human body are formed in the process called mitosis.

The crucial figure for the continuation of genetic research was Darwin's cousin Francis Galton, who bequeathed a large sum of money for the foundation of the *Laboratory for National Eugenics* at the University College London. Nine years after the laboratory was founded, two strands of research emerged. One strand founded the *Galton Laboratory* at the UCL, which dealt with biological research, while the second strand dubbed itself the *Eugenics Education Society* and set as its goal to improve the human species. However, the progress of genetics was slow due to lack of sophisticated technical means required in genetic research. A significant leap forward was made by Thomas Hunt Morgan, who studied the hereditary properties of fruit flies, and his student A. H. Sturtevant who wrote a paper in 1913 whose title will sum up the development of genetics up to 1980s – *The linear arrangement of six sex-linked factors in Drosophila, as shown by their mode of association.* This paper marked the beginning of gene manipulation studies, which ended in June 2000. The global public was then informed that the map of the human genome had been completed, and that work will be continued on individual features, which should enable personalized approaches, genetical prevention and genetic treatment that would replace the approaches before that time. Apart from disease prevention, treatment via intervention in the diseased cells, and extension of the lifespan, genetics promises to solve social problems as well.

Genetic discoveries have a significant potential benefit for mankind. These revelations form the basis of molecular medicine, which studies the diseases at the level of molecules – genes and proteins. Its goal is to cure the diseases at the genetic level or at least attempt to prevent them using genetic interventions before they even begin. It can also be used to clone organs, which would help in treatment of disease-ridden human beings. Molecular medicine is looking for answers of specific DNA changes of somatic cells caused by the environment; it studies the reasons for the emergence of the autoimmune diseases, viruses, viral disease, gene death, etc. Its overall aim is to decode the entire genome of man, systematically determine the structure of proteins whose genes will be known in nearby future. The human genome project will ultimately determine the exact sequence of 'chemical letters', decode the 'book of life' that makes up the DNA of every cell in man's body. In 2000, all three billion pairs of bases of our DNA were mapped. The order of 'chemical letters' in our genes (30 – 40,000) was discovered, even though not all genes are known to us. We now know that the core of every cell in man's body, skin, muscle, liver, brain, etc. contains 46 molecules of the DNA that are equal in different cells. The complete DNA is called a genome. The DNA molecules are built from smaller parts, nucleotides that are marked by 'chemical letters' A, C, G, and T. Individual parts of the DNA molecule are called genes. The order of 'letters' inside the genes determines what kind of protein will by synthesized in the organism (Pavelić, 2004, p. 37). A gene is an instruction for protein synthesis, which is why it is perceived as the code of life. It is well-known that proteins are responsible for the normal state and functioning of organism and its features. How these proteins function after receiving the instructions is still being investigated.

Russian mathematician Aleksandr Mikhailovich Lyapunov

[...] demonstrated in 1963 that every living system transmits a small but precise quantity of energy or matter containing a great volume of information through channels laid own in advance. This information is responsible for the subsequent control of large quantities of energy and matter. From this perspective numerous biological and cultural phenomena (storing, feedback, channelling messages and so on) can be conceived as manifestations of information processing. In the final analysis, information appears in large part to be the repetition of information, but still another kind of information, a kind of control which seems to be a universal property of terrestrial life, irrespective of its form or substance. (Baudrillard, 2017, p. 119)

In that respect, one might claim that genetics is an information science. It is well-known that the human genome most probably contains a set of useful information about the development of man, functioning of the organism, about the consciousness and psyche in general (Pavelić, 2004, p. 37). Geneticists believe that functioning of the organism is dependent on the messages written in the genes. As genome decoding has been completed and mapped, the next goal is to uncover the structure and the role of proteins. With the help of technology, the geneticists have already created a DNA chip which can be used to analyse around 10,000 genes at the same time. The chip is typically a small glass surface onto which thousands of samples of the DNA have been imprinted using the photolithographic micro-robotic method. Even though this technology is still in the development stages, it is expected that this sophisticated method will soon enable all individuals to obtain a quick genetic portrait.

Comparison with other information sciences shows that:

The terminology of genetics is full of expressions taken from linguistics and communication theory ..., which emphasised both the principal similarities and the important differences in the structure and function of genetic and verbal codes ... Today it is clear that the genetic code must be considered as the most basic semiotic network, and therefore as the prototype of all the other systems of signification used by the animals, including man. (Baudrillard, 2017, p. 119)

After the initial surge, genetic engineering has been stagnating due to unexpected metabolic changes caused by the transmission of a single gene. Genomes typically do not expect foreign genes, which is why a significant number of procedures of grafting foreign genes has been unsuccessful. Since the functioning of genes is a consequence of a very complex interaction with the entire organism and the external world, all attempts to create new genetical features or changing the existing ones will have to undergo a thorough revision. The stability of genes is determined by the overall state of the organism, and one can assume that science will uncover these complex interactions and undertake a safer and a more meaningful attempt to adapt genetic identities that will be of use to mankind. Until then, all genetic procedures should be taken with significantly more precaution (Pavelić, 2004, p. 59).

In relation to the matter at hand in this book, which deals with ontological and anthropological aspects of enhancement and reshaping of man, it is necessary to emphasize that certain aspects of genetic engineering go beyond the powers of nature as they enable the creation of genetic information of two or more organisms on Earth, and insertion of such DNA into every organism could lead to unforeseeable side-effects that could emerge as a result of artificial combining of the DNA (Pavelić, 2004, p. 60). When the exchanged genes enter the genome, they can no longer be recalled, which is why the scientific community proposed a 50-year suspension on any application of genetic technology of any kind which could deliberately or accidentally change the gene population of any species. This ban also prohibits the use of genetic engineering methods on human embryos and releasing genetically modified organism into the environment (Pavelić, 2004, p. 62).

For this reason, publicly available genetic research is increasingly focused on developing genetic therapy of somatic cells with the aim of fixing hereditary diseases which occur as a result of individual faulty genes. Somatic cell manipulations are not transmitted onto the following generations while germ cell manipulations carry the risk of errors which could be transmitted to future generations via mutations. Genetic mutations cannot be fully controlled – while genes can be sliced and merged quite precisely, the procedures of inserting these genes into a living organism is often imprecise and unsafe, and thus beyond any possibility of supervision (Pavelić, 2004, pp. 65-66). In 1993, Jerry Hall from the *Washington University Clinic* became the first person to clone the human embryo. All attempts of cloning ever since have been based on trial and error. For instance, it took 277 attempts to clone Dolly the sheep. Some clones died very quickly, while others developed abnormalities. Scientists have so far been able to clone more than 20 species of mammals, including sheep, pigs, cats, horses, and dogs. The most significant progress has been achieved in somatic cell manipulation, which enabled the design of tissue and organs used in treatment of human beings. According to the *Cell Stem Cell* journal, 18 years after creating the first clone of Dolly the sheep, scientists at a biotechnological company were able to create embryonic stem cells using similar procedures on skin cells. The procedure employed by Robert Lanza and Young Gie Chung is called therapy cloning as it produces stem cells that have the ability to develop into over 220 different types of cells, including nerve, muscle, blood, and pancreas cells (Rudež, 2014)

Genetic anti-aging research is divided into two different lines of thought. On the one hand, aging of the human organism occurs as a result of cell damage, while according to the other perspective, it is the result of a genetic clock which indicates the length of an organism's lifespan. Cells are programmed to die, which is why the geneticists are working on reprogramming them. Scientists at the *Amsterdam Medical Center* have analysed blood cells of a woman who lived to the age of 115 and have concluded that lifespan ultimately depends on the capability of stem cells to continue their everyday process of tissue regeneration. When cells reach the end of their lifespan at a particular point and start dying, vital organs and cells stop regenerating (Jutarnji list, 2014). At the end of the woman's life, only 2 out of 1300 stem cells were active and maintained the level of white blood cells. When they stopped working, the woman's life stopped as well.

Despite legal limitations and ethical boards, genetic manipulation experiments take place in laboratories across the globe. At the *Cambridge University laboratory*, scientists have been able to develop the embryo outside the mother's body until the 13th day of its growth by using the methods previously tested on mouse embryos. Analysts tracked the development of the human embryo in the earliest known stage of development of human life when the basic body shape is being formed. This research is highly significant for improving the assisted reproductive technology and further developing regenerative medicine. The main researcher of this project believes that extending the permission for developing the embryo using extrauterine methods would allow us to reach absolutely crucial findings about the basics of human biology. Of course, this opens up the question of where to put new limitations (HINA, 2016b).

In 2015, genetic modification of human embryos within the scope of a research project on earliest stages of human development was approved in the UK. The research intends to use the CRISPR-Cas9 technology, which has been the topic of discussion in the international scientific community due to concerns that this project could be used for genetic manipulation of human embryos and for creating custom-designed children (HINA, 2016b). Similar research is also being conducted by the scientists at the Guangzhou University in China. Their project has far been only partially successful due to a high number of missed mutations (Jutarnji list, 2015).

A group of geneticists from Harvard, Yale, Columbia, Johns Hopkins, MIT, and other scientific institutions are proposing the *Human Genome Project – Write* project which would be used to create the complete human genomes for scientific research. The experiments would be conducted to examine the effects of genetic mutations and create safer stem cells for organ transplants. Scientists assume that they could in due time produce genetic mutations on animals (e.g. pigs) so that their organs could be transplanted into human beings. Artificial genome would be produced by using chemicals to produce the DNA found in human chromosomes. It is assumed that there are technical conditions in place for creating organs and body parts to replace dysfunctional ones. This project represents the idea of using individual building blocks to create a more advanced individual from the current human ones. The ability to create human beings from artificial DNA does not require more natural elements, as was the case with Dolly the sheep. Realization of this project would create an endless source of cloning human beings and would pave the way towards creating the artificial man. Who will have the actual right to create and control the cells created this way? Scientists claim that the public will participate in the project from the start, as this is a delicate matter involving ethical, social, legal, and religious issues (HINA, 2016c).

The aim of this brief overview of accomplishments and strategies for further development of genetics and molecular medicine was to create the foundation for our philosophical facts and ethical warnings about genetics as a part of the ontological self-movement of *scientific work*. In terms of treatment and prevention of possible diseases, genetics and molecular medicine offer numerous benefits for man and mankind, such as the genetically modified viruses that would be used to fight cancer. The London Institute of Cancer Research announced this revolutionary discovery – a modified virus which attacks cancer cells and kills them from inside without binding to healthy cells. This stimulates the reaction of the immune system to fight cancer. The ICR president Paul Workman believes that after spending years looking at viruses as enemies to defeat, we can now find their real purpose – to fight cancer (Draškić, 2015). It remains to be seen to how many people will be able to use this treatment. In contrast to this, eugenics advocates enhancement of the individuals and the family, which pre-empts the question of readiness of man to take upon himself all consequences of genetic manipulations in reproductive cells of the human organism. Only time will tell whether the dominant interest groups will be able to legally force an individual to program his offspring or whether this question will remain within the jurisdiction of every individual man or woman. With their experiments, molecular biology and genetics put forward a question and an even more uncertain answer – who controls our lives and deaths?

We should emphasize that the idea of genetic enhancements peaked in the 20th century when "Hitler's *Lebensborn* program mated 'racially pure' German women with SS officers to perpetuate 'desirable genotypes'; meanwhile, the authorities were busy sterilizing undesirables and, of course, trying to exterminate some races altogether" (McKibben, 2004, p. 37). Questions on this issue are being put forth today: "[...] won't we see other people as subhuman, and enslave or slaughter them? [...] Is it still 'eugenics' if no one's forcing you to do it? If you're making all the choices for your own kid?" (McKibben, 2004, p. 37). Individual choice could be completely abolished:

Eugenics in that kind of world won't take a Hitler, as the bioethicist Arthur Kaplan puts it, it will only require people saying: 'You can have a kid like that if you want, but I'm not paying.' Even if legislation could prevent such a scenario, what happens when your unenhanced child grows up and goes to get a job? 'On the merits', maybe he gets to take out the trash. Hey, but it's your choice. No one's making you do it. (McKibben, 2004, p. 191)

A person "left without any choice *at all* is the one you've engineered. You've decided for once and for all, certain things about him: he'll have genes expressing proteins that send extra dopamine to his brain to alter his mood; he'll have genes expressing proteins to boost his memory, to help his stature. He'll be putty in your (doctor's) hands" (McKibben, 2004, p. 191). Habermas argues that:

Gene manipulations is bound up with issues touching upon the identity of the species, while such an anthropological self-understanding provides the context in which our conceptions of law and of morality are embedded. [...] We cannot rule it out that knowledge of one's own hereditary features as programmed may prove to restrict the choice of an individual's way of life, and to undermine the essentially symmetrical relations between free and equal human beings. (Habermas, 2003, pp. 22-23)

Humans cannot plan and program other human beings as this prevents them from being self-developing individuals but things. Habermas also warns us about a real danger of different groups making decisions on different pathways of genetic development. In those cases, nothing would remain of the mankind as we know today but a memory of a common ancestor.

According to the US scientists, the development of genetics is heading towards the creation of parentless offspring. Scientists at *Stanford University* published a study on artificially created sperms and eggs. They discovered a combination of chemicals and vitamins which stimulate stem cells to turn into eggs or sperms. Sperms have heads and a short tail and are mature enough to fertilize the eggs that were developed for fertilization purposes. This discovery, if proved successful, will mean that parents are no longer necessary for reproduction. Should science be given the authority to tamper with the very beginning of human life is a major ethical issue and a responsibility that we all have to undertake (M.B.B., 2009). Using stem cells, Chinese scientists were able to produce completely functional sperm from which healthy mice developed in a laboratory for the first time (Dauenhauer, 2016).

Against these issues, McKibben argues "[…] national laws might be a flimsy bulwark. When the US government set certain limits on stem cell research in 2001, for instance, several American scientists went to British universities to continue their work." (McKibben, 2004, p. 164). We can see that "[…] all sides in these debates have their bioethicists, whom they use for cover. […] all the big biotech companies have ethics boards, and some of them pay their advisers with stock options or hand them checks for $2,000 a day." (McKibben, 2004, p. 185). The president of the *Hastings Center on bioethics* Daniel Callahan argues that:

[…] most of those who have come into the field have accepted scientific ideology as much as most scientist, and they have no less been the cultural children of their times, prone to look to medical progress and its expansion of choice as a perfect complement to a set of moral values that puts autonomy at the very top of moral hierarchy. (McKibben, 2004, p. 185)

We can take the example of the president of the ethics board at the *Advanced Cell Technology*, who

[…] outlined the 'extreme precautions' the company was taking: the eggs would be kept in a secure location, access to which required permission from two ACT technicians; the eggs would be repeatedly counted, photographed, and videotaped, and so on. Nothing untoward would happen to the eggs – but the society that was about to be dramatically changed by this new development would have to fend for itself. (McKibben, 2004, p. 186)

The use of genetics is "virtually inevitable […] and for the most banal reasons possible – to makes some money, or to satisfy the virtuoso scientists who created the technology" (McKibben, 2004, p. 164). On the other hand, does an individual have unlimited rights in choosing one's own successors? Who owns our genomes or our 'genetic chip' and who will control it? Will the clone from a mature adult cell even be sufficient for a healthy life? What will a completely artificial, chemically-made human being with no healthy human components look like? These are just some of the questions that require an argument-based discussion and a thoughtful answer. One of the key questions that the *scientific work* cannot answer with its human and historical affiliations is whose side the nature will take and how it will react to the scientific course of evolution of man. We hold it justified to believe that opening up the 'book of life' could also bring about the 'book of death'. Life in nature is interconnected in myriads of ways and has been synchronizing and balancing for millions of years of shared evolution. Nature exchanges information, energy, and matter, allowing every organism to get what it needs for life, reproduction, and development. Genetic engineering and the production of duplicates destroys this natural harmony. Manipulations and errors which stem from genetics could prove to be irrecoverable. The ecosystem is manifested in the richness of different forms and types which make up the whole, and uniformed replicas stand in stark contrast with this diversity. Genetic uniformity would disrupt the diversity and possibly bring into question the uniqueness of life on Earth. Geneticists Lee Silver in his book *Remaking Eden* describes a future divided between genetically perfected and natural beings:

Emotional stability, long-term happiness, inborn talents, increased creativity, healthy bodies – these could be the starting points chosen for the children of the rich. [...] Obesity, heart disease, hypertension, alcoholism, mental illness – these will be the diseases left to drift randomly among the families of the underclass. He forecasts a future divided between 'the GenRich' and 'the Naturals'. The former, perhaps 10 percent of the American population, 'all carry synthetic genes', and they control 'all aspects of the economy, the media, the entertainment industry, and the knowledge industry.' The latter work as 'low-paid service providers and laborers.' (McKibben, 2004, p. 38)

Can we live with the time of genetic possibilities luring us with its promises and, more importantly, which scientific findings will form the basis of our decision-making? At the entrance to the Delphi oracle stood a sign which read – 'Know thyself!'. The genetics of today allows us to know the transhuman beings at the level of information hidden within the genes and which will be known to us in a foreseeable future. In addition to knowing ourselves, genetics also allows us to alter ourselves and the whole mankind. The Delphian warning is not equivalent with the genetic messages and challenges that the man of today is faced with, but these ideas have more and more ideological proponents in posthumanism, eugenics, corporate interest groups, the capital, etc. Systematic propaganda in the media, which themselves represent the manifestation of procedures of *scientific work*, is gaining more advocates of genetic enhancements. While molecular medicine is transitioning from the era of treatment of diseases into the era of disease prevention and propagates the role of an individual in creating his own health, a major part of mankind has no access to primary healthcare. The geneticists claim that they are aware of the necessity of dialogue on ethical boundaries of gene modification. But they argue they do not modify genes, they simply 'fix' them.

Futurologists and scientists predict that by the end of this century, biotechnology will have the power over life and death. The bio-cybernetic birth of man, according to the models of the genetic code, represents a warning as these lives are controlled by the keeper of the vault of the genetic chip in which the new cybernetic ontology and anthropology of life of non-human beings is written. Genetics shows the difference between the human and non-human through the common denominator of the genetic code. This perspective questions the existence of everything human in the biological world. Microbiology and genetics as the guardians of 'the book of life' are shown as harbingers of death of 'unity of human nature' and human life in historical reality. The possibilities which stem from cloning or creating human beings with the aid of a chemical formulae lead to the creation of duplicates which represent the endless extension of the body with itself to the point of unrecognizability. A modernized body made through a series of multiplication loses every trace of personality of a self-grown life, and the repetition of this body into infinity, metastasizing its basic genetic formula leads to the end of the body. This is the end of the road for man of today.

Dissolution of the boundaries of humanity endangers the individuality and gender in genetic simulation of non-human living beings. Man has already become dead through possible genetic and cyborg immortality. The immortality of life goes beyond all human nature and human life as it goes beyond the human capabilities that are now based in the *empire of freedom of scientific work in the age of scientific humanism as naturalism*. Cloning human beings, designing individuals identical to their originals, recombining genetic materials of different species, creating new forms of human and non-human life illustrates the omnipotence and impotence of procedures of *scientific work* whose desire to reach the source of artificial life in order to manipulate it will destroy all biological life in its concealedness, diversity, and beauty of inevitability.

REFERENCES

All Headline News. (n.d.). *Michigan Researchers Develop Artificial Bone Marrow That Can Produce Real Blood*. http://www.allheadlinenews.com/articles/7013529632

Allday, E. (2010). *UCSF team working on artificial kidney*. SFGate. https://www.sfgate.com/health/article/UCSF-team-working-on-artificial-kidney-3166374

American Pain and Wellness. (n.d.). *Spinal Injections/Procedures, Neurostimulation*. http://www.painandwellness.com/spinal_injections_neurostimulation.html

Balija, P. (2011). Klinika u tijelu: Mikropauci će pratit tumore i srčane mane. *24 sata*. https://www.24sata.hr/news/klinika-u-tijelu-mikropauci-ce-pratiti-tumore-i-srcane-mane-213822

Bašić, B. D., Čupić, M., & Šnajder, J. (2010). *Umjetne neuronske mreže*. Fer.

Bašić, B. D., & Šnajder, J. (2008). *Uvod u umjetnu inteligenciju. Umjetna inteligencija*. https://www2.fer.hr/_download/repository/UI-1-Uvod.pdfsl

Bauby, J. D. (n.d.). *The Diving Bell and the Butterfly*. Academic Press.

Baudrillard, J. (2017). *Symbolic exchange and death* (Rev. edition). Sage Ltd.

BA/VLM. (2012). Izradili čeljust 3D printerom i ugradili je 83-godišnjakinji. *Večernji list*. https://dnevnik.hr/vijesti/znanost-it/izradili-celjust-3d-printerom-i-ugradili-je-83-godisnjakinji.html

BBC News. (2009). Scientists unveil world's first bionic fingers. *BBC News*. http://news.bbc.co.uk/2/hi/uk_news/scotland/edinburgh_and_east/8401205.stm#:~:text=Experts%20have%20unveiled%20what%20they,the%20bionic%20i-Limb%20hand.

Berger, T. W., & Glanzman, D. (Eds.). (2005). *Toward replacement parts for the brain: Implantable biomimetic electronics as neural prostheses*. MIT Press. doi:10.7551/mitpress/6858.001.0001

Bostrom, N. (2010). Superintelligence. In J. Brockman (Ed.), *This will change everything: Ideas that will shape the future* (pp. 111–114). Harper Perennial.

Brown University. (2010). 'Artificial ovary' develops oocytes into mature human eggs. *ScienceDaily*. https://www.sciencedaily.com/releases/2010/09/100914102108.htm

Bukatman, S. (1993). *Terminal identity: The virtual subject in postmodern science fiction*. Duke University Press.

Capar, L. (2015). Francuzi u laboratoriju proizveli ljudsku spermu. *Večernji list*. https://www.vecernji.hr/techsci/francuzi-u-laboratoriju-proizveli-ljudsku-spermu-1005595

Čatić, I. (2002). *Nanotehnika – na granicama fizičkog svijeta*. Vjesnik.

Čatić, I., Razi, N., & Raos, P. (1991). Analiza injekcijskog prešanja polimera teorijom sustava. *Društvo plastičara i gumaraca*.

Čatić, I., Runjić-Sokele, M., & Petriček, G. (2008). *Process of human reproduction – the natural model of injection moulding of living and non-living substances*. Bioforum.

Čatić, I., Runjić-Sokele, M., Tiljak, H., & Petriček, G. (2007). Prirodni modeli njekcijskog prešanja. In *Lošinjski dani bioetike*. Mali lošinj.

Chandler, D. L. (2019). Artificial "muscles" achieve powerful pulling force. *MIT News*. https://news.mit.edu/2019/artificial-fiber-muscles-0711

Connor, S. (2009). British scientists to create 'synthetic' blood. *The Independent*. https://www.independent.co.uk/news/science/british-scientists-create-synthetic-blood-1651715.html

Čović, A. (2004). *Etika i bioetika: Razmišljanja na pragu bioetičke epohe*. Pergamena.

Ćuto, T. (2015). Steve Wozniak: Budućnost čovječanstva će biti zastrašujuća. *Zimo*. https://zimo.dnevnik.hr/galerija/steve-wozniak-buducnost-covjecanstva-ce-biti-zastrasujuca---401384.html/61125686/401384

Dauenhauer, N. J. (2016). *Prvi put iz umjetnih spermija stvoreni zdravi potomci*. Tportal.hr. https://www.tportal.hr/vijesti/clanak/prvi-put-iz-umjetnih-spermija-stvoreni-zdravi-potomci-20160226

Davids, A. M., & Bellwin, A. (1954). Experimental Construction of an Artificial Bladder with Maintenance of Normal Urinary Function and Continence. *Annals of Surgery*, *139*(4), 484–493. doi:10.1097/00000658-195404000-00013 PMID:13149101

de Grey, A. D. N. J. (2005). Like it or not, life-extension research extends beyond biogerontology. *EMBO Reports*, *6*(11), 1000–1000. doi:10.1038j.embor.7400565 PMID:16264420

de Haldevang, M. (2018). Stephen Hawking left us bold predictions on AI, superhumans, and aliens. *Quartz*. https://qz.com/1423685/stephen-hawking-says-superhumans-will-take-over-ai-is-a-threat-and-humans-will-conquer-space/

Draškić, L. (2015). Genetski modificiran virus 'ubija' rak? Znanstvenici: Imamo dokaze! *Večernji list*. https://m.vecernji.hr/lifestyle/genetski-modificiran-virus-ubija-rak-znanstvenici-imamo-dokaze-1007607/galerija-140037?page=1

Drexler, K. E., Peterson, C., & Pergamit, G. (1991). *Unbounding the future: The Nanotech- nology Revolution*. William Morrow and Company, Inc.

Đugum, T. (2005). *Nanorevolucija*. Slobodna Dalmacija.hr. arhiv.slobodnadalmacija.hr/20050424/temedana02.asp

Elsevier. (2009). NASA's Electronic Nose May Provide Neurosurgeons With A New Weapon Against Brain Cancer. *ScienceDaily*. https://www.sciencedaily.com/releases/2009/04/090430065456.htm

Farrier, J. (2009). Researchers Develop Artificial Tongue. *Neatorama*. https://www.neatorama.com/2009/09/02/researchers-develop-artificial-tongue/

Fildes, J. (2009a). Artificial brain '10 years away'. *BBC News*. http://news.bbc.co.uk/2/hi/8164060.stm

Fildes, J. (2009b). Bionic hand wins top tech prize. *BBC News*. http://news.bbc.co.uk/2/hi/science/nature/7443866.stm#:~:text= The%20world%27s%20most%20advanced%2C%20commercially, win%20 this%20year%27s%20MacRobert%20award.

Fischman, J. (2010). Bionics. *National Geographic*. https://www.nationalgeographic.com/magazine/article/bionics-robotics-medical-technology

Floridi, L. (2014). *The Fourth Revolution: How the Infosphere is Reshaping Human Reality*. Oxford University Press.

Foucault, M. (1966). *The Order of Things: An Archaeology of the Human Sciences*. Vintage Books.

Fukuyama, F. (1992). *The End of History and the last Man*. Hamish Hamilton.

Fukuyama, F. (2002). *Our posthuman future: Consequences of the biotechnology revolution* (1st ed.). Farrar, Straus and Giroux.

Glas Amerike. (2007a). *Dubinska stimulacija nudi nadu kod pacijenata sa povredom mozga*. https://www.glasamerike.net/a/a-34-2007-08-26-voa6-87090677/765078.html

Glas Amerike. (2007b). *Pregled najnovijih dostignuća u oblasti nanotehnologije*. https://www.glasamerike.net/a/a-34-2007-01-16-voa4-87000552/756543.html

Greenwood, M. (2008). See-Through Nanowires Used to Drive Transparent Organic LEDs. *PhotonicsMedia*. https://www.photonics.com/Articles/See-Through_Nanowires_Used_to_Drive_Transparent/a32294

Greguric, I., Čatić I. (2010). *Kiborgoetika – presjecište ili poveznica bioetike i tehnoetike. Projekt MZT-a Primjena sustavnosne teorije u raščlambi opće tehnike* (120-0000000-1805). Zagreb: Fakultet strojarstva i brodogradnje.

Groothuis, D. R. (1999). *The soul in cyberspace*. Wipf ans Stock.

Grubin, D. (2007). *PBS series: The mysterious human heart*. The Replacement Heart. http://www.pbs.org/wnet/heart/episode1/replacement2.html

Habermas, J. (2003). The Future of Human Nature. *Polity*.

HINA. (2016a). Indijski liječnici dječaku su na čelu uzgojili potpuno novi nos. *24 sata*. https://www.24sata.hr/tech/indijski-lijecnici-djecaku-su-na-celu-uzgojili-potpuno-novi-nos-481802

HINA. (2016b). Veliki uspjeh znanstvenika: U laboratoriju razvili ljudski embrij do 13. dana. *Index.hr*. https://www.index.hr/vijesti/clanak/veliki-uspjeh-znanstvenika-u-laboratoriju-razvili-ljudski-embrij-do-13-dana/891427.aspx

HINA. (2016c). Znanstvenici predlažu razvoj sintetskog ljudskog genoma. *Večernji list*. https://www.vecernji.hr/techsci/znanstvenici-predlazu-razvoj-sintetskog-ljudskog-genoma-1089440

Horvat, M. (2012). Razvili umjetne krvne žile koje bi mogle spasiti brojne živote. *24 sata*. https://www.24sata.hr/lifestyle/razvili-umjetne-krvne-zile-koje-bi-mogle-spasiti-brojne-zivote-249972

Howard, S., & Kristen, K. J. (2010). *Artificial Disc Replacement*. Vertical Health, LLC. https://www.spineuniverse.com/treatments/surgery/artificial-disc-replacement

InHealth Technologies. (n.d.). *Blom-Singer Voice Prostheses*. https://inhealth.com/blom-singer-voice-prosthesis#:~:text=Voice%20is%20produced%20by%20temporarily,then%20out%20through%20the%20mouth.

I.S. (2011). Presađen prvi »uzgojeni« organ. *Vidi.com*. https://www.vidi.hr/Zdravlje/Presaden-prvi-uzgojeni-organ

Jones, R. C. (2014). Stephen Hawking warns artificial intelligence could end mankind. *BBC News*. https://www.bbc.com/news/technology-30290540

Jutarnji list. (2009). Uz pomoć nanotehnologije za 20 godina postajemo besmrtni. *Jutarnji list*. https://slobodnadalmacija.hr/mozaik/zivot/uz-pomoc-nanotehnologije-za-20-godina-postajemo-besmrtni-68690

Jutarnji list. (2010). Napravljen prvi umjetni želudac. *Jutarnji list*. https://www.jutarnji.hr/naslovnica/napravljen-prvi-umjetni-zeludac-3760848

Jutarnji list. (2014). Znanstvenici otkrili tajnu smrti? Za spektakularno otkriće mogu zahvaliti ženi koja je umrla u 115. godini. *Jutarnji list*. https://www.jutarnji.hr/life/znanost/znanstvenici-otkrili-tajnu-smrti-za-spektakularno-otkrice-mogu-zahvaliti-zeni-koja-je-umrla-u-115.-godini-795289

Jutarnji list. (2015). Anketa: Kinezi uspjeli modificirati embrije. *Jutarnji list*. https://www.jutarnji.hr/life/znanost/anketa-kinezi-uspjeli-modificirati-embrije-393991

Kaku, M. (2011). *Physics of the future: How science will shape human destiny and our daily lives by the year 2100*. Doubleday.

Ko, D. (2010). *Biosintetička umjetna rožnica stapa se s okolnim tkivom oka i vraća vid*. Dnevno.hr. https://www.dnevno.hr/techno/znanost/biosinteticka-umjetna-roznica-stapa-se-s-okolnim-tkivom-oka-i-vraca-vid-10253/

Kroker, A. (1984). *Technology and the Canadian mind: Innis, McLuhan*. Grant. New World Perspectives.

Kučković, Đ. (2014). Organi iz labaratorija. Prvi umjetni nos uskoro na licu pacijenta. *Jutarnji list*. https://www.jutarnji.hr/life/zdravlje/video-organi-iz-laboratorija-prvi-umjetni-nos-uskoro-na-licu-pacijenta-787165

Lauc, T. (2020). Nove tehnike uzgoja zuba osmišljene u Laboratoriju za tkivno inženjerstvo i regenerativnu medicinu na Sveučilištu Columbia, mogle bi gubitak zuba učiniti prošlošću. *Večernji list*. https://ordinacija.vecernji.hr/zdravlje/zdravi-zubi/hoce-li-nam-uskoro-moci-izrasti-novi-zubi-znanstvenici-na-pragu-velikog-otkrica/

Levy, J. (2005). The Doomsday Book: Scenarios for the End of the World. *Vision (Basel)*.

Mayo Clinic Stuff. (2011). Hearing aids: How to choose the right one. *MayoClinic.com*. http://www.mayoclinic.com

M.B.B. (2009). Muškarci i žene više nisu potrebni za reprodukciju. *Večernji list*. https://www.vecernji.hr/lifestyle/muskarci-i-zene-vise-nisu-potrebni-za-reprodukciju-42078

McCarthy, J. (2007). *What is Artificial Intelligence? In Enough: Staying Human in an Engineered Age*. Holt.

McLuhan, M. (1969). *Counter Blast*. McClelland and Stewart.

MedicineNet. (n.d.). *Pacemaker*. https://www.medicinenet.com/pacemaker/article.htm

Metro Portal. (2009). Možemo li zaustaviti starenje bez posljedica? *Metro Portal.hr*. http://m.metro-portal.hr/mozemo-li-zaustaviti-starenje-bez-posljedica/21628

Minsky, M. L. (2015). *Semantic information processing*. The MIT Press. (Original work published 1968)

Mone, G. (2008). How It Works: The Endoscope Camera in a Pill. *Popsci.com*. https://www.popsci.com/how-it-works/article/2008-03/how-it-works-endoscope-camera-pill/

Moravec, H. P. (1988). *Mind Children: The Future of Robot and Human Intelligence*. Harvard University Press.

Mužić, V. (2009). Kada treba ugraditi umjetni zglob. *Vaše zdravlje*. http://www.vasezdravlje.com/izdanje/clanak/1701/

Naam, R. (2005). *More than Human: Embracing the Promise of Biological Enhancement*. Crown Publishing Group.

NASA. (2003). *Artificial Cells*. https://www.nasa.gov/vision/earth/livingthings/artifical_cells_sng.html

NJD. (2011). Ubojita kaciga. Britanski piloti protivnike ruše pogledom. *TPortal.com*. https://www.tportal.hr/tehno/clanak/britanski-piloti-protivnike-ruse-pogledom-20110104

Oregon State University. (2004). Portable Kidney Dialysis Machine Developed. *ScienceDaily*. https://www.sciencedaily.com/releases/2004/02/040203233312.html

Pavelić, K. (2004). *Čuda moderne medicine*. Nakladni zavod Globus.

Plastic Surgery Advisor. (n.d.). *Lip Augmentation*. https://www.aestheticsurgerycenter.com/surgical/lip-augmentation

Posavec, M. (2012). Nanotehnologija – vječni život i blagostanje? *Dayline.info*. http://www.dayline.info/index.php?option=com_content&task=view&id=976.

Ricker, T. (2006). The Hal-5 ready to battle. *Engadget*. https://www.engadget.com/2005-12-02-the-hal-5-ready-to-battle.html

Rielo, D. (2011). Vagus Nerve Stimulation. *Medscape.com*. http://emedicine.medscape.com/article/1186123-overview

Roco, M. C. (2011). The long view of nanotechnology development: The National Nanotechnology Initiative at 10 years. *Journal of Nanoparticle Research*, *13*(2), 427–445. doi:10.100711051-010-0192-z

Rudež, T. (2014). *Uskoro ćemo klonirati ljude i ljudske organe*. Jutarnji List.

Schwartz, M. (2005). Scientists develop nanotech-laser treatment that kills cancer cells without harming healthy tissue. *Stanford News*. https://news.stanford.edu/pr/2005/nanotube-08105.html

Šerić, M. (2016). U naš mozak stane cijeli internet. *Večernji list*. https://www.vecernji.hr/techsci/u-nas-mozak-stane-cijeli-internet-1054509

Slobodna Dalmacija. (2005). Nanorevolucija. *Pregled.com*. http://www.pregled.com/print_preview.php?id_nastavak=102&grupa=nauka

Solar, A. (2016). Znanstvenici razvili prozirnu kožu koja prekriva bore. *Večernji list*. https://www.vecernji.hr/lifestyle/znanstvenici-razvili-prozirnu-kozu-koja-prekriva-bore-1083686

Solar, A. (2016). Uzgajanje ljudskih kostiju uz pomoć 3D metoda. *Večernji list*. http://www.vecernji.hr

Spengler, O. (1932). *Man and Technics: A contribution to a Philosophy of Life*. George Allen and Unwin Ltd.

Storm, D. (2015). Steve Wozniak on AI: Will we be pets or mere ants to be squashed our robot overlords? *Computerworld*. https://www.computerworld.com/article/2901679/steve-wozniak-on-ai-will-we-be-pets-or-mere-ants-to-be-squashed-our-robot-overlords.html

Surgery Section. (n.d.). *Spinal Cord Stimulation for Treatment of Pain*. https://www.bcbst.com/mpmanual/!SSL!/WebHelp/Spinal_Cord_Stimulation_for_Treatment_of_Pain.htm#:~:text= Standard%20spinal%20cord%20stimulation%20(SCS,in%20the%20trunk%20or%20limbs.

Texas Heart Institute. (n.d.). *Ventricular Assist Devices*. http://www.texasheartinstitute.org/HIC/Topics/Proced/vads.cfm

T.V. (2014). Prvi put ikad: U samo tri dana uzgojili umjetna pluća! *Dnevnik.hr*. https://dnevnik.hr/vijesti/svijet/u-samo-tri-dana-uzgojili-umjetna-pluca---323965.html

University of Cambridge Research. (2011). *Artificial pancreas in pregnancy promises fewer diabetes deaths*. https://www.cam.ac.uk/research/news/artificial-pancreas-in-pregnancy-promises-fewer-diabetes-deaths

University Of Rochester Medical Center. (2001). Implanted Defibrillators: How Well Do They Work? *ScienceDaily*. https://www.sciencedaily.com/releases/2001/08/010828075349.html

Večernji list. (2016). Znanstvenici razmatraju ‘stavljanje’ lijekova u tijelo putem mikrorobota. *Večernji list*. https://www.vecernji.hr/techsci/znanstvenici-razmatraju-stavljanje-lijekova-u-tijelo-putem-mikrorobota-1103577/galerija-208531?page=1

Vitas, Z. (2016). Uz pomoć elektroda u mozgu paralizirani mladić svira gitaru. *Večernji list*. https://www.vecernji.hr/techsci/uz-pomoc-elektroda-u-mozgu-paralizirani-mladic-ian-burkhart-svira-gitaru-1076494

Vitas, Z. (2017). Bečki znanstvenici pokušavaju uzgojiti mozak. *Večernji list*. https://www.vecernji.hr/lifestyle/becki-znanstvenici-pokusavaju-uzgojiti-mozak-1139732

Walsh, M. (2010). Breast Enlargement - Secrets of Silicone Valley Revealed. *Ezine@rticles*. http://ezinearticles.com/?Breast-Enlargement---Secrets-of-Silicone-Valley-Revealed&id=3686019

Weiner, J. (2008). *Cyborg athletes? Bionic Olympics? High-tech prosthetics case raises the prospect of a potential sci-fi sports world*. MiniPost. https://www.minnpost.com/politics-policy/2008/01/cyborg-athletes-bionic-olympics-high-tech-prosthetics-case-raises-prospect-p/

Wiener, N. (1990). *The human use of human beings: Cybernetics and society*. Free Association Books.

Z. J. (2010). Spas za alkoholičare: Znanstvenici uzgojili umjetnu jetru iz stanica jetre štakora. *Index.hr*. https://www.index.hr/vijesti/clanak/spas-za-alkoholicare-znanstvenici-uzgojili-umjetnu-jetru-iz-stanica-jetre-stakora/514515.aspx

Žižek, S. (2007). *O vjerovanju: Nemilosrdna ljubav* (M. Miladinov, Trans.). Algoritam.

ENDNOTES

1 Narcissus is a character from Greek mythology, the son of the river god Cephissus and nymph Liriope who fell in love with his reflection in the lake.

2 The diving bell in this case is a metaphor for the dependence of the body, the terminal disregard for existence and the ultimate reduction of life. The butterfly is a metaphor for the spirit and the effects of its thought and ideas (Čović, 2004, pp. 84-85).

3 The metal diving bell can also be regarded as a metaphor for the ever-increasing technologization of the human body and spirit which transforms the human being into a lifeless entity.

4 Cf. Čatić (1999).

5 Cf. Čatić et al. (1991).

6 Cf. Čatić & Gregiric (2011).

7 Cf. Greguric (2011). The image was created within the scope of the project Kiborgoetika – presjecište ili poveznica bioetike i tehnoetike. Projekt MZT-a Primjena sustavnosne teorije u raščlambi opće tehnike (120-0000000- 1805). Fakultet strojarstva i brodogradnje.

8 American Pain and Wellness (n.d.).

9 Cf. Čatić & Greguric (2010).

10 Cf. Davids & Bellwin (1954).

11 The prefix *nano-* comes from the Greek *nanos* meaning 'dwarf' and it denotes something tiny. Specifically, it denotes a one billionth part of a metre, i.e. the world of molecules and atoms invisible to the naked eye.

Chapter 7
Cyborgization of Actual Social Relations

ABSTRACT

This chapter depicts the essence of cyborgization of social relations in sports, art, music, traffic. This harkens back to the pioneering cybernetic work of Norbert Weiner and extends into the current and future reconfigurations of man-machine relations that are shaping human life and society from human enhancement to driverless cars. The chapter shows that procedures of scientific work cyborgize the historical reality of man into the era of scientific humanism as naturalism, and that it is not the future, but in fact the present, that we are acclimatizing to as we have not become fully aware of the present future.

8.1 CYBORGIZATION IN SPORTS

A bionic race would take away from true competition. It is not my desire to give athletes a device that takes them beyond what their body can do.

Marlon Shirley[1]

In this chapter, we analyse particular aspects of culture in the age of *scientific humanism as naturalism* to show their non-human nature. The analysis is divided into four parts.

The first section in this chapter *Natural body modifications – the issue of body enhancements in sports* will provide the basic insights into cyborgization of the human body and the desire for creating a stronger, faster, and a more resilient human body. The second section *Creating cyborgized athletes – prostheses as aids in sports* analyses the possibilities offered by cyborgization procedures by looking into details of the case of Oscar Pistorius, the disabled athlete whose legs were amputated and replaced with prosthetic ones. By analysing different options which are becoming available in this domain, we are also faced with an ever-increasing need to distinguish between cyborg athletes and natural athletes. The third section *Improving athlete's competitive capabilities – morally justified or not?* questions the moral and legal justification of enhancing healthy human bodies. The fourth section *Selecting the win-*

DOI: 10.4018/978-1-7998-9231-1.ch007

ner and defining fairness in terms of enhancements – creating artificial winners in the age of scientific humanism as naturalism questions whether the current system creates artificial winners, how to achieve fairness in competition between natural and cyborgized athletes, and whether the two should be even allowed to compete.

Figure 1. Oscar Pistorius (Moreton, 2012)

The core idea and the motto of sports "a healthy mind in a healthy body" has been brought into question by cyborgization tendencies in sports, which call for critical scrutiny and ethical questioning of boundaries of cyborgization in sports. Complete and uncritical faith in the scientific and technological progress of society could put us in a position in which man is completely reliant on technology, paying no attention to the consequences which may arise from this.

With the help of science and technology of today, a healthy human body (its structure, functions, capabilities) can be codified and transformed into an artificial cyborgized body. Accelerating the evolution of man includes improving the human species with the aim of creating a stronger, faster and a more resilient body with an improved health and an extended lifespan. It also denotes the tendency of overcoming the limited mental capabilities, deficient sensory and physical capabilities, all of which are ideas advocated by the transhumanist movement.[2]

The *Transhumanist Declaration* from 1998 states: "Transhumanists advocate the moral right for those who so wish to use technology to extend their mental and physical (including reproductive) capacities and to improve their control over their own lives. We seek personal growth beyond our current biological limitations." (Bostrom, 2005, p. 21). This goes to show that the transhumanists advocate reshaping biological human capabilities, i.e. creating a kind of a superhuman entity with changed and enhanced functions. This way a healthy human individual becomes an experiment within itself, which will ultimately result in the creation of an artificial body far superior than that of current man. Contrary to the transhumanist perspective, the bioconservatives oppose the use of contemporary technology for the sake of enhancement of man and accelerating evolution. Their fundamental ethical principles are that of natural order and the authority of the divine creation.[3]

The consequences caused by application of new bionic enhancement possibilities in the world of sports is portrayed rather vividly by the case of Oscar Pistorius (Fig. 1). As the first disabled athlete, his artificial leg prostheses gave him a chance to compete with the able-bodied athletes in world championships and the Olympics. There is nothing questionable about Pistorius's desire to compete with natural athletes. His motto is: "You're not disabled by the disabilities you have. You are able by the abilities you have" (Moreton, 2012). What changes the concept of sports and represents a threat to all healthy sportsmen is the creation of enhanced athletes. This tendency leads towards complete cyborgization of sports and competitions in which there will be no place for biological athletes, only artificially enhanced ones.

8.1.1 Natural Body Modifications – the Issue of Body Enhancements in Sports

Cyborgization in sports as a world of huge spectacles has become an increasingly popular topic during the past several years. Up until recently, doping was used to illegally increase the athletes' performances and the body was enhanced by hard training and anabolic steroids. Today, it is possible to achieve the same effects today with cyborgization procedures. This implies the enhancement of healthy human bodies through implementation of technology. This new branch of medicine was pushed forward with the case of Oscar Pistorius, who became the most well-known cyborg in the world of sports. So far, no consensus has been reached as to whether Pistorius is an enhanced athlete-cyborg or whether the artificial prostheses simply replaced his legs and allowed him to compete equally with healthy athletes.

As we have shown in previous chapters, human beings are treated or enhanced on at least four levels: physical, cognitive, emotional and in terms of lifespan expectancy. Physical enhancements include exoskeletons, external structures for the human body which increase its strength and durability, bionic prostheses (bionic arms, legs, etc.) used for treatment; the brain-computer interface which restores the lost functions with the help of a computer and enables the paraplegics to issue commands to a computer using sensors built into the motor cortex. It is highly likely that this prosthesis will be used in the world of sports for enhancement purposes as well. In terms of cognitive and emotional enhancements, different drugs are available, such as Ritalin and antidepressants which improve the overall mental state. In terms of life expectancy, research is being conducted in the domain of nanotechnology where nanobots could go through blood vessels to treat the diseases and fix the afflicted organs.[4] Once we reach the point of curing all diseases and the point of immortality, as predicted by Kurzweil (2005), there will be no need to implant parts made from artificial technology with the aim of replacing particular body parts, nor to apply particular drugs (cognitive or emotional enhancements) as everyone will be able to upload his or her emotion of happiness and make a backup of their brain. Kurzweil predicts that augmenting mental and physical capabilities will be extended and that "Within 25 years we will be able to do an Olympic sprint for 15 minutes without taking a breath, or go scuba-diving for four hours without oxygen. [...] If we want to go into virtual-reality mode, nanobots will shut down brain signals and take us wherever we want to go" (Willis, 2009). If Kurzweil's prediction, called "The Law of Accelerating Returns", does indeed come to pass, what can we expect in sports – virtual reality Olympics, enhanced sportsmen who will communicate telepathically, etc.?

Roger Clarke[5] believes that the development of cyborgization will lead to the situation in which people will demand their rights to technology to keep them alive, which will in turn lead to oversaturation in requests for rights of cyborgized and non-cyborgized individuals. For instance, "People who are using prostheses to recover lost capabilities will seek to protect their existing rights. People who have lost capabilities but have not yet got the relevant prostheses will seek the right to have them. Enhanced humans will seek additional rights to go with the additional capabilities that they have" (Clarke 2010). It seems quite likely that these rights to implantation of artificial technology could be transferred onto the world of sports. Instead of losing time on tiresome training sessions, the athletes could take a much simpler route and use genetic manipulations or technological implants to enhance their bodies and achieve superb results. These requests will definitely not stop with artificial prostheses intended to replace the missing organs. Requests will be made for hearing and vision enhancements, even RFID chips, which would allow the athletes to communicate with one another.

Artificial technology which serves to regain lost capabilities could also be abused in numerous ways. For instance, "[…] the cochlear implant is currently being used to give people some semblance of normal hearing, it could also be used to augment normal hearing. Retinal implants and wearable computers could also be used to allow people to secretly record and transmit what they see […]. People can be walking around with this extension and other people not realise they've got it" (Saleh, 2010). Potential abuse on the one hand brings into question the justification of using artificial technology, and on the other hand, it leads to the point of no return away from freedom, which is a fundamental right of every natural human being. Athletes could use their freedom to decide on implementing artificial technology and become mere puppets whose body will be enhanced and upgraded as necessary. This way, the world of sports will not only have some enhanced athletes who will be better and faster than biological athletes, but they and their results will be controllable as if they were a part of a video game. Reality will get a very virtual appearance, but there is also another very real danger that does not depend merely on possibilities of technological progress. The world of sports is just a small part of the cyborgized reality so it is a crucial question to ask whether there is any material capability and readiness left in the world to respect all rights and all possible requests for technological implantations, regardless of whether they are done for enhancement or replacement purposes, and whether this will move mankind even further away from equal chances for all. Taking into consideration the concentration of the capital, it seems highly likely that we will witness the emergence of a privileged class, which will choose between replacement and enhancement, and a deprivileged class, whose enhancements will be controlled only for the purpose of increasing production and multiplying excess value.

8.1.2 Creating the Cyborgized Athletes – Prostheses as Aids in Sports

Athletes with prosthetics and robot sportsmen have become surprisingly human and frighteningly non-human. Does this mean we have stepped into the transhuman era of sports? According to the definition, a prosthesis is "a device, either external or implanted, that substitutes for or supplements a missing or defective part of the body."[6] As such, the field of prosthetics is already used in all sports at all levels. For instance, the use of *Fastskin*[7] swimsuits has the potential to increase speed, which causes an unfair advantage among swimmers as the swimsuit is made of improved materials which decrease the body's friction as it moves through the water, making the swimmer more dynamic (Magdalinski, 2009, p. 120).

Tennis players use rackets made of carbon fibres, which enable a faster and a more powerful swing than wooden ones. Thus, biological athletes already use various carbon-based prosthetics, from carbon rackets, shoes, hockey sticks, etc. The question is why these external prostheses are accepted and legal, while Pistorius's prosthetic legs are not? In 1974, surgeon Frank Jobe[8] performed the first operation in which the broken ligaments were replaced with ligaments from another part of the body (Weinreb, 2007, p. 42). This surgical modification opened up the path for implantation of artificial prostheses. South African Paralympic Oscar Pistorius[9] was born without both fibulas[10] and had both of his legs amputated in his early childhood and replaced with *Cheetah* prosthetic legs[11], made from special carbon fibre. Because of his superb results with prosthetic legs, he was nicknamed *Blade Runner*.[12]

Pistorius's *Cheetah* prostheses certainly mark a change of paradigm in the world of sports in which the competitors will no longer rely on their talents alone but also on mechanical devices shaped by the technology which changes the very nature of sports:

Athletes are already posthuman cyborgs and we celebrate this," says Miah, the Scottish doctor. "It is likely that greater use of this technology will seep into other aspects of culture, as we begin to embrace more and more enhancements. Sports might soon become peculiar for resisting such developments and, in the meantime, will be placing athletes at greater risk by forcing them to enhance behind closed doors (Miah, 2008)

Widespread use of prostheses can lead do the domination of sportsmen with prosthetic limbs or other enhancements. Completely biological athletes will soon become an exotic minority. It is highly likely that in the future man will have the possibility to choose which organs he/she wants to enhance to become a better athlete in his/her field.

8.1.3 Improving Athlete's Competitive Capabilities – Morally Justified or not? The Pistorius Case

Enhancing the competitive capabilities of athletes through technology has put the issue of fairness and regularity of competitions in which biological and enhanced athletes compete into spotlight. Even though Oscar Pistorius got his prosthetic legs for medical reasons, his case represents a paradigmatic case of enhancements and a tipping point between the biological and post-biological era of sports. Use of prosthetic aids causes doubts about the boundaries of the body which an individual human being can and should reach. What will happen when the boundary between a disabled person and an able-bodied person is shifted unexpectedly? Oscar Pistorius made a quantum leap from a well-defined system in which the disabled athletes competed with one another (the Paralympics) into another system where he as a disabled person competed with healthy athletes (the Olympics). At the World Championships in Athletics in South Korea 2011, he qualified for the semi-finals by finishing the race in 45.39 seconds (the world record is currently held by Wayde van Niekerk with 43.03 seconds). He also competed as a member of the 4x400m relay team. Pistorius's performance marked a turning point in the relationship between prosthetics and athletics.[13]

In his professional career, Pistorius faced a series of problems and bans. During late 2007, the *International Association of Athletics Federations* (IAAF) introduced a ban on using technical aids that could give athletes a competitive advantage and banned Pistorius from competing at the 2008 Summer Olympics in Beijing. They believed that Pistorius's artificial limbs give him an unfair advantage – from claiming that the elasticity of artificial limbs enables him to push himself from the ground more efficiently than the natural human ankle to claims that the prostheses allowed him to reach higher speeds and lose less energy than other sprinters. Human performance experts Peter Weaynd and Matthew Bundle published the results of their research in the *Journal of Applied Physiology* immediately after the IAAF banned Pistorius from competing in competitions for able-bodied athletes, including the Olympics (Hunter & Rochat 2008), due to complaints that the prostheses give him an unfair advantage.[14] They concluded that Pistorius's artificial limbs, whose weight is two times lower than that of able-bodied athletes, allow him to reach 15-30% higher running speed[15], and allow him to change the position of his limbs faster (Weyand et al., 2009, p. 903). Weyand claims that Pistorius reaches astonishing speeds due to low weight of the prostheses in comparison to the weight of normal lower leg, while Nicholas Romanov, who studies human biomechanics and coaches runners, does not think that the elasticity of the carbon fibres is a significant benefit (C.S., 2017). Pistorius appealed against the IAAF's decision to the Court of Arbitration for Sport (CAS) in Lausanne, Switzerland. It revoked the decision by the IAAF and argued that Pistorius did not have an unfair advantage over able-bodied athletes. Due to research and testimony of biophysicists Hugh Herr, the IAAF revoked its ban on use of technical aids in sports, which allowed Pistorius to compete with the best athletes in the world.

A huge part of the debate is still focused around Pistorius's ability to use the so-called gravitational torque throughout the entire race[16] and the question whether Pistorius as a disabled athlete with less muscle mass has a metabolic advantage over athletes whose limbs are intact. In terms of aerobic capacity, it was believed that Pistorius has 25% less energy than healthy athletes and an unnatural fatigue level, but artificial limbs gave him additional stamina. Tests were conducted to check how much oxygen Pistorius consumes and the rate of his fatigue. By studying the biomechanics, i.e. the amount of power used, length of his stride[17], Herr concluded that Pistorius's rate of fatigue and metabolic advantage are no different from able-bodied athletes. However, he argues that technology is advancing rapidly and it is quite possible that athletes with health-related issues and physical deficiencies might soon be faster on the track. How can you estimate whether a result was achieved through personal talent or technological aids? Is it morally justified to have a disabled person compete with able-bodied athletes if there is a possibility that the aids give him/her a competitive edge? Are Pistorius's prostheses technical enhancements in sports or just a legitimate replacement for his own legs? These are the questions that are yet to be answered.

Enhancement of biological bodies with prosthetic products is shifting the boundaries from the level of adequate to better than normal. All it takes is to look at how the technology affects and changes the rights, autonomy and freedom of human beings in sports. Sports need to decide whether genetically formed athletes can be legitimate competitors (Edwards and Skinner, 2006, p. 124). The ruling bodies in sports should oversee and control every use of technical aids in sports to determine whether they could increase performance efficiency and ban them if necessary in order to protect the integrity and fairness of the game.

8.1.4 Selecting the Winner and Defining Fairness in Terms of Enhancements – Creating Artificial Winners in the age of Scientific Humanism as Naturalism

In the era which attempts to preserve the natural essence of sports strives to improve the results beyond human capabilities, the crucial questions of ethics in sports become fairness, justice and ensuring equal possibilities for all competitors. An ethical framework that can be applied in sports, particularly in the case of Oscar Pistorius, is related to the four principles set out by Beauchamp and Childress: non-maleficence (do not cause harm), beneficence (do good), autonomy (respect individuality, sanctity of personal choice), and autonomy (equal access and application) (Beauchamp & Childress, 2001). Biological body enhancements in the world of sports put into question the entire purpose of competition between natural and cyborgized athletes. It is possible that natural athletes will demand the right to compete only in those competitions which enable the use of solutions which enhance their capabilities.

Herr claims that "The laws are written to allow people with unusual minds and bodies to have full access in society, and that includes sporting events. So we're imagining a world in which you happen to be born without fibulas, and you also happen to be the world's best athlete. You should have the freedom to become that. There also has to be fairness in sport. The answer is more technology, not less" (Wolchover, 2012). Herr believes that one day artificial limbs will enable higher speed and efficiency than biological limbs. For the Paralympic athletes to be able to qualify for the Olympics, they will have to use less sophisticated technological solutions which would keep them on a par with the able-bodied competitors.

The case of Oscar Pistorius has caused discussions on forming categories in sports. How do we distinguish between natural categories and the categories which are a result of social development? Do the boundaries of fairness shift by allowing the athletes with prosthetic limbs to compete with the natural athletes? These are just some of the questions yet to be answered with regards to the use of prosthetic technology in sports.

A Toronto-based futurologist George Dvorsky asks: "Will able-bodied athletes mutilate themselves to attach a more efficient mechanical limb? Will athletes all want to be like Oscar? Imagine a hockey team that communicates telepathically, or basketball players with improved peripheral vision" (Weiner, 2008). According to Wolbring, this moves us towards the Olympics with different classes:

Division I might allow steroids and human growth hormone. Division II might ban drugs, but allow special shoes or fins on uniforms. Division III might allow prosthetics and extra-skeletal devices. Division IV might be your basic guys and gals in shorts and T-shirts. Eventually, the bionic athlete, or "human enhancement technology" (HET) athlete, will be the high-performance athlete, earning money, TV ratings and attention (Weiner, 2008).

The shift of paradigm in sports opens up the possibility of domination of artificial winners, new classifications in sports, and a different way of selecting the winner. In the era of confronting natural and enhanced athletes, defining fairness and ethical principles becomes a crucial question which must be answer regardless of the pace and the direction of the future development of sports. Cyborgization in sports opens up the philosophical questions and debates, not just about the essence and ethics of sports, but also about the essence and boundaries of enhancement. Can sports maintain the purpose of celebrating the beauty of the natural human body or will it be defined based on technical modifications of the body and the world of spectacles in which sports have a very prominent position? We have legalized the first cyborg in sports who uses artificial prostheses to compete with the able-bodied sportsmen and achieves superb results. Even though Oscar Pistorius did not implant the prostheses in question for the purpose of enhancement, his example has opened up numerous discussions about the justification of enhancements in sports.

In terms of body enhancement, the transhumanists advocate the freedom of choice and the right of the individual to improve their chances in life. Savulescu believes that biomedical enhancements are not just in line with the sport, but with the spirit of sports itself. For him, to want to be better in all aspects of life means to be human. Savulescu's proponents suggest the same approach in sports, in which the enhanced – cyborgized – athletes will have the advantage. From this perspective, biological sportsmen could become the Paralympic athletes and bionic or biomedically enhanced athletes would be the future Olympic athletes. Gregor Wolbring argues that athletes without any enhancements might be perceived as an artificial class in the future, while the British ethicist J. Harris supports this stance by claiming:

The treatment/enhancement distinction is in many senses, including the regulatory, a red herring. [...] In sport [...] we are talking always about procedural justice or fairness. [...] The key to procedural fairness [...] is that the same enhancement opportunities are, in principle, available to everyone. (Camporesi, 2008, p. 639)

Parallel to the discussion about the enhancement of capabilities of able-bodied athletes, the first cyber-Olympics in history were held in Brazil. The host of these Olympics was the *Swiss Federal Institute of Technology* ETH Zurich. The so-called *Cybathlon* games are a chance for sportsmen with automatic prosthetic aids, i.e. which use different technological and robotic implants in everyday life. According to the organizers, this competition is less focused on speed and strength, unlike the Olympic and the Paralympic games. The competition places a strong focus on controlling the body and the devices which the competitors use as prosthetic aids. The participants competed in six different disciplines using the state-of-the-art technological aids, including the computer brain stimulator. Since every competition celebrates the winners, the *Cybathlon* has already marked a victory for the technology, in this case in the service of the disabled persons (Poptech, 2016).

Enhancement technologies push the boundaries of freedom for the disabled athletes who are given a chance to compete on equal ground with the able-bodied sportsmen and achieve the results which would not be possible without cyborgization. On the other hand, their freedom is overshadowed by the question of whether this enhancement remedies a bodily deficiency or whether it creates an impermissible advantage. The bioconservative thinker Francis Fukuyama perceives the demands of the transhumanist movement as a "strange liberation movement" which wants to "liberate the human race from its biological constraints" (Fukuyama, 2004, p. 42). This is why he claims: "One obvious way to draw red lines is to distinguish between therapy and enhancement, directing research toward the former while putting restrictions on the latter. The original purpose of medicine is, after all, to heal the sick, not to turn healthy people into gods" (Fukuyama, 2002, p. 208). He argues that we should "discriminate between those technological advances that promote human flourishing, and those that pose a threat to human dignity and well-being" (Fukuyama, 2002, p. 182). An uncrossable line should be drawn between what is and is not allowed in medicine, science, and sports. Looking at Pistorius's running and his results has given us a new perspective on changing the boundaries in sports. Pistorius's sports results are undoubtedly great, but his example reflects the technical and social changes which have a profound influence on the shift of paradigm in sports. The fundamental purpose of sports – "a healthy mind in a healthy body" – ceases to be an event of testing the biological boundaries of endurance and becomes a competition of best cyborg solutions in sports.

8.2 CYBORGIZATION IN ARTS[18]

The new media are not bridges between man and nature; they are nature

Marshall McLuhan, 1969

Art in the *age of scientific humanism as naturalism* takes over the features of *scientific* work along with everything within the range of its establishment. Carried by scientific and technological progress, it imitates artificial life and artificial intelligence, moving away from its original spontaneity, creativity and aesthetic divine role models. It loses all its vigour in its simplicity and unique originality. Art will become a mere section in the department of science, technology, information and communication activities of *scientific work*. Through its reflection in the scientific and technological production of artificial beings, it takes on the form of production of scientific and technological work with body as the content in real and virtual space-time. This shows that a part of art inventively participates in the establishment of the non-human and the creation of posthuman bodies and new forms of life of living beings.

Taken into the vortex of current pragmatics of *scientific work* in the destruction of man's world, art questions the foundations of its own self-understanding and the purpose of existence. A work of art devoid of sense is best seen in the experiments and performances of contemporary cyborg artists. An art event is evaluated according to the principles of efficiency in a cyborgized and a predictable world. The artists themselves become means and the media of art in the world of established apparitions, a world without man as the subject of his own history. In man's place stands the scientific and technological way of establishing life and the new media which shape the culture and art.

Media theorist Marshall McLuhan understood the media and technology as an extension of man – a way to expand his capabilities: "wheel is an extension of feet in rotation. Having extended or translated our central nervous system into the electromagnetic technology, it is but a further stage to transfer our consciousness to the computer word as well" (McLuhan, 1995, p. 60). Man has developed the extensions for all purposes for which he had used his body in the past. The mechanical age has extended his organs, and the electronic age has extended his nervous system. In McLuhan's view, electric technology extends to the central nervous system in those cases where the "amplification is bearable by the nervous system only through numbness or blocking of perception." McLuhan, 1995, p. 43). This process, which follows every extension of the media, is called autoamputation. McLuhan believes that all new media and technologies create a new environment: "When technology extends one of our senses, a new translation of culture occurs as swiftly as the new technology is interiorized." (McLuhan, 1962, p. 40).

The media and the technology become man's prostheses[19] of sorts, his extensions which bring him closer to technology. In this respect, analysing contemporary anthropology of man, i.e. his ontology, becomes inseparable from the questions about the essence of technology and *scientific work*. In the world of today, the body can be extended and reshaped using technological means, which extend and reshape man's body in the real world; through the media, which dematerialize and infinitely extend the body in virtual space; and through cloning. In all three cases, man finds himself between the organic nature and the artificially created nature – artificial technology. McLuhan argues that man is increasingly becoming a technological prosthesis of his own copy, a relic of a former man, a usable entity that only lives with the help of artificial technology. Under the influence of new procedures and methods, and the media based on them, a radical shift in the artistic expression has taken place. Within the transhumanist art production, man's appearance, purpose, body form and structure are changed in the name of enhancement. Connected to various gadgets, he is shown as a hybrid being who loses its role in history and becomes capable of performing only the pre-programmed functions which push him further away from our wise ancestors. The transhumanist art projects increasingly resemble products of technological laboratories.

Most installations on book and magazine covers are computer-generated. In early 1990s, Karl Sims from MIT developed a program which has evolved into programs for creating impressive two-dimensional images. This system has become one of the generators of computer art and design in the age of *scientific humanism as naturalism.*

8.2.1 The Posthuman and (Re-)production of man in a Computer – the Bodiless man

Cyberspace philosopher Michael Heim believes that "With its virtual environments and simulated worlds, cyberspace is a metaphysical laboratory, a tool for examining our very sense of reality." (Heim, 1993, p. 82). Sherry Turkle believes that computers lead us across the world of dreams and enable us to contemplate a mental life which exists completely detached from the body. According to Turkle, a computer is an "evocative object" situated between the animate and the inanimate world (Turkle, 2005, p. 19). Through it, the boundaries between animacy and inanimacy and the relationship between man and the machine become increasingly fuzzy.

Interaction and absolute internet speed have brought us to "the invention of a perspective of real time, that will supersede the perspective of real space" (Virilio, 1995). By erasing the borders between man and the machine, the animate and the inanimate, we might enter into the posthuman world "out of our control, in which our wills are subsumed by the system." (Pearce, 1995, p. 7). Baudrillard claims that the cyborgized reality in the virtual reality is not a lie. It reflects a logical structure, the *camera obscura* as the virtual reality is artificial intelligence-generated reality in which bio-cybernetic life of digital images take place. The users of virtual reality have a virtual feeling of moving through a virtual environment without a body (cf. Dagenais, 1991). In this respect, we are dealing with the bodiless reality, the transformation of the human body into a body without organs (cf. Deleuze & Guattari, 1983). Rejecting the organs requires us to isolate the body and separate the feeling of actual reality. The interface becomes our main setting, the primary hub and point of division between man and technology. This reality cannot be understood from the point of view of traditional categories of material reality as the possibility of existence of man as an immortal being can only take place in the space of virtual reality. In the virtual space, the cyborgized man and artificial intelligence are not just real, but hyper-real, which is why they cannot be gauged by the material reality in which the body is tied to the biological structure of nature.

Baudrillard believes that simulation has engulfed the reality with all its traditional categories. Man's life is no longer directed towards its purpose as it is controlled by the models and is less connected to the outside world. As the world no longer has the metaphysical equivalent, it is not measured with truth or reality. Since it is impossible to check, reality is a deceit, a detailed illusion. The endless movement of the virtual will ensure that the real cannot be measured against anything. The only thing left to do is to measure the reality with the outside world, which is entirely constructed by the artificial intelligence. Thus, reality can only be measured against its own emptiness and meaninglessness of its duplicate which is its artificial imprint but not its accurate reflection. In a sense, it is more real than reality itself.

The internet as a place of deconstruction of the human body contains pseudo-human life forms, such as avatars.[20] However, the fact that the user and his body are present in the network only as signs opens up interesting possibilities for artists to play with artificial or fake online bodies which leave a digital trace on the network.

In her project *Primo Posthuman 3M +* (2000), US artist and an advocate of transhumanism Natasha Vita advocates new technologies which should enable us to overcome the weaknesses of our body, primarily the biggest weakness of all – mortality. Using biotechnology and nanotechnology, she developed a project of the future which illustrates the criteria of the future body which is stronger and more flexible than the current biological body, and offers extended capabilities and improved senses. It is multifunctional and can be upgraded (Transhumanist Arts & Culture, 2007). The designed body animated by a 3D model is based on the volume of her body mass and displays the processes necessary to equip and upgrade the human organism so that its "political, cultural, biological, and psychological limits" could be removed from it with the aim of "overcoming constraints on our progress and possibilities as individuals, as organizations, and as a species" (More, 2003a).

The relationship between the technology and the body is explored by US artist Tina La Porta in her project *Future Body*. Using the network model as a platform for interaction, La Porta problematizes the process of separating the material body using telecommunication technology and focuses on activating and deactivating the fluctuations in the transmission between geographically separated participants mediated by the interface (La Porta, 2000). The project is trying to find the answer and the meaning of embodiment established between communication technologies. The results of this experiment show us the possibility of a dislocated embodiment within the symbolic empire of the cyberspace. The material body is slowly vanishing and being replaced by immaterial silhouettes of the fading presence.

Figure 2. Primo Posthuman 3M+ (More, 2003b)

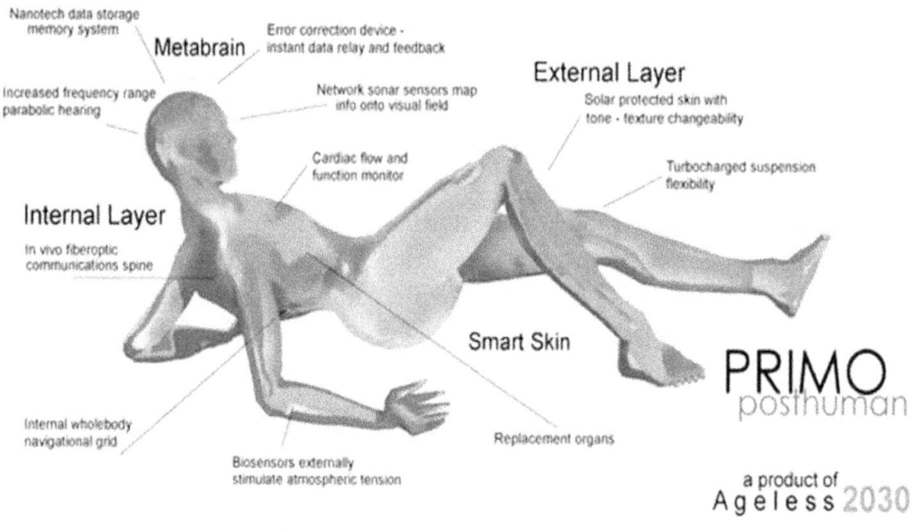

Scanning and mapping the body with 3D procedures enabled the creation of the project *EvaSys* (1997) by Austrian artist Eva Wohlgemuth. After scanning her body, Wohlgemuth created its three-dimensional digital copy and stored it in digital format on a hard drive (it is 1 MB large, equal to an average text file) (Wohlgemuth, 1997). Art work *Body Scan* represents another example of loss of (biological) material as a paradigm of reality (Huffmann & Jahrmann, 1997).

US artist Victoria Vesna's project *Bodies (c) Incorporated* enables people to create their own bodies in 3D space. Network participants can play complex video games of identity and survival. For instance, they can construct a virtual body from different body parts, textures, and sounds, and can gain membership in an extending community of body owners. The events in the 3D space cause a sequence of emotional reactions from the participants that are trying to be answered. For instance, what level of psychological devotion and attachments do owners have for their virtual bodies? How does the body become a source of pleasure and anxiety as it passes through changes caused by the owners themselves (Vesna, 1996).

Personal Eugenics (1999) is a project by an Australian artist John Tonkin, the aim of which is to offer the network users a possibility to configure their face and change their online identity. All it takes to participate in the project is an interactive interface which enables the user to record/upload the image of his face along with the description of the person whose image they would like to take over. The user(s) can develop their face until they get the desired appearance. *Personal Eugenics* seems like a wonderful place in which human enhancement capabilities can be transformed and manipulated with the help of digital technologies and in which the socially desirable features can be copied while the undesirable ones are deleted (Tonkin, 1999).

The project *The Internal Organs of a Cyborg* by English Artist Jane Prophet explores the use of existing biotechnology, transplants, and future possibilities of nanotechnology with implanted personalized microchips. Through a science fiction story about two characters whose lives and identities collide in a hospital room, Prophet explores the posthuman body. The project is a CD-ROM story told in the form of a comic. The user can move through parallel stories of two characters: Donald P. Nym is a successful director of a large bank, and Chrissy Hastings is a young woman who earns money by taking part in drug trials and implantation experiments. To satisfy her need for body enhancement, Chrissy buys implants on the black market of biotechnological surgeries, including a microchip onto which she will download her personality. Both characters meet in the ER for the first time. Chrissy is there because she was wounded when the microchip malfunctioned and Donald is there because of a heart attack. Chrissy dies and her heart is donated to Donald. After the death of her biological body, her virtual entity is self-activated through the implanted personal chip. Her virtual self starts sending biochemical messages to Donald and he starts having weird memories and feelings and ultimately develops posthuman love after he starts receiving e-mails from her virtual ego (Prophet, 2000).

8.2.2 Computer-Interactive Performance

US artist Lynn Hershman is regarded as a pioneer in the domain of computer-interactive installations. She uses a wide array of media and technologies to explore individuality and personality. Through her art, Hershman focuses on the issues of mutual communication of the body. Hershman created her avatar Roberta Breitmore (1971-1978), an artificial character that embodies her and which represents her alter ego (Maloney, 2011). The Breitmore avatar had its own behaviour, clothes, apartment and life stories. In her first interactive video *Lorna*, the visitors could choose what they want to see, take part in the creation of her behaviour, make major decisions for the character, but also experience her emotions at the same time (Hershman, 2007).

In her project *Telerobotic Doll*, Hershman used dolls Roberta and Tillie. "She is virtual, you are real. Her eyes are cameras. Click on an eye to move her head. Go ahead, try it!!!!" (Hershman, 1997). The two telerobotic dolls provide information and images that are being broadcasted online. Viewers can use the doll's eyes as a means of extending their own vision. The eyes of both dolls were replaced by cameras and the dolls can be moved both physically and telerobotically. Tillie's left eye has colour vision and detects the environment in real time – what she can see on the monitor in front of her. Online viewers can see everything that CybeRoberta (Fig. 3) and Tillie can see. In other words, the viewers can only see through their dolls. In this case, computers do not just function as personal computers but, like the internet, they allow the possibility of collective use in which the processes and formation of individualization converge and are given an option of self-creation and reconfiguration, along with their integration into technological, social, spatial, and historical environment.

Figure 3. CybeRoberta (Krug, 2017)

Inter_skin (1994) is a project by a Norwegian artist Stahl Stenslie. It is a digital sexual aid that we use to communicate with our partner by transmitting the sense of touch. By touching ourselves we stimulate our partner. A sexual stimulus is carried via computer interface with the option to increase the strength of the stimulus. In this case, we are dealing with an exchange of information without any form of direct contact between the physical bodies, which keeps the participants shielded from potential diseases (Stenslie, 2004). Like *Inter_skin*, the *Future Kiss* project is concerned with the interaction in the technologically mediated society of the future. Czech artist Lenka Klimešová uses vibration mechanism-equipped masks to detect the partner's kiss. Intimate relations with the partner are based on the exchange of digital information which simulate emotions (Klimešová, 2008).

8.2.3 Cybernetic Performance in Arts – Body as Software and Interface

In § 4.4.3.4 *Improved cyborg in arts*, we described Stelarc's performance. In this section, we shall repeat only the crucial parts. Australian performance artist Stelios Arcadiou, also known as Stelarc, uses medical instruments, prostheses, robotics, virtual reality and the internet to create his art works (Dugandžija, 2013). He is renowned for his controversial works which mostly involve the body and uses technology to strengthen and extend his physical body. His performances raise many questions about the possible boundaries between the human body and cyborgs in the cybernetic space, extension of cognitive capabilities of artificial intelligence within the new bodiliness of man.

He attracted the global attention for the first time with his series of performances called *Suspensions* (1976-1988) in which he hung himself from ropes and hooks, testing and measuring the endurance and physical capabilities of his body (Scarry, 1985). By testing and extending the boundaries and capabilities of the biological human body, Stelarc explores different postbiological and posthuman strategies in his performances, such as the robotic third arm. The *Third arm* (1976-1994) is an artificial mechanical arm attached to his right arm which extends Stelarc's capabilities as it can be moved independently. It is activated through contraction of abdominal muscles and legs. After years of use, Stelarc can move it intuitively without any need for consciously focussing his efforts.

Connecting the identity of the virtual and the biological is the focus of Stelarc's performance in the *Movatar* (2000) project. In this project, Stelarc transforms the concept of a material avatar in which he controls the virtual body as he also allows the Movatar to have autonomous virtual entity and partial control over Stelarc's body. *Movatar* is a virtual body or an avatar imbued with artificial intelligence, which makes it partially autonomous and operative (Stelarc, 2021). By taking over a physical body through an internet-connected device, it can act in the real world. If Stelarc wears the device and logs into the avatar, he will become the host of an intelligent virtual entity through which the movements of the avatar will be expressed. Pneumatically activated prostheses enable precise and powerful movement of the upper abdomen, while the legs are flexible and free.

Stelarc's works have been produced in the light of what Donna Haraway calls the *C^3I metaphor*: "command-control-communication-intelligence" (Harraway, 2016, p. 6). His third ear, *Extra Ear ¼ Scale* (2003) is capable of functioning as a station for emitting and receiving information. It is a flexible prosthesis made of soft tissue and flexible cartilage It represents the extension of the human body and resembles a natural ear in both form and structure. It was grown in a laboratory from real human cartilage and surgically transplanted inside the left forearm. In the future, Stelarc plans to install microphones into the third ear and enable the transmission of sound online. The real earlobe would be grown from stem cells, but it is not yet known whether it would be implanted under the other arm as the fourth ear or whether it would replace the hard cartilage of the current one. The ear would not have the ability to receive sound, only to emit. By connecting it to a modem or a laptop, Stelarc would have the possibility of emitting audio. The additional ear would function as a sort of an internet antenna which would telematically and acoustically increase one of other senses. This project questions the idea of body integrity and confronts the increasing possibility of manipulating living systems (Stelarc, 2021).

Using robot technology, Stelarc constructed the *Hexapod* (2002), a hybrid man-machine contraption which consists of six controllable robotic legs. Robotic dynamic locomotion and movement are controlled through movement of the body and turning of the torso. The muscle machine is a flexible and stable mechanism, moved by a combination of an electric and pneumatic system. When Stelarc's single leg is lifted, three mechanical legs are moved forward after the other three legs stop. Stelarc's body stands on the ground in the machine's base, which includes the lower exoskeleton body connected to a robot. The coder on the side wrists allows Stelarc to control the movements of the machine and to change the speed at which he wants to traverse. The walking mode can be selective and is activated by hand gesture. When the machine starts to move, it is impossible to ask who is in charge, Stelarc or the machine, as they become completely integrated and move as one (Stelarc, 2021).

In performance such as *Fractal Flesh* or *Parasite* (1997), the artist literally closes himself online. In the *Parasite* project, the electrodes are attached to the ends of the muscles and the impulses from the electrodes force his body to move. During the *Fractal Flesh* performance, the internet functions as the extension of his nervous system. The audience can use an ISDN link to log into the artist's body. Using a touch screen monitor, registered users are allowed to stimulate parts of his body. Stelarc describes this out-of-body experience as electronic stimulation.

In his performance *Ping Body* (1996), Stelarc problematizes the creation of a new virtual identity. Here, the cyberspace represents the new space of activity and movement of the body. Stelarc's biological nervous system is neither autonomous nor in charge of his virtual one. A symbiosis of the human and non-human takes place and the new bodily identity is shown as a system of a virtual cybernetic organless network. This performance is a paradigmatic example of life events as art without the body. Stelarc wants to show that the cyborgized body is more perfect than the natural one. His goal as an artist is to expose his body to technological interventions which transform the body into an object of manipulation which can be changed, extended, and completed. Stelarc believes it is necessary to continue with further cyborgization of the body which has become inadequate and imperfect without technology. Because of this, Stelarc allows his body to become a prosthesis in a technological environment. As his motivation for cyborgization of the body, Stelarc mentions the old and the new story of self-creation:

Images are immortal, bodies are ephemeral. The body finds it increasingly difficult to match the expectations of its images. In the realm of multiplying and morphing images, the physical body's impotence is apparent. The body now performs best as its image. Virtual Reality technology allows a transgression of boundaries between male/female, human/machine, time/space. The self becomes situated beyond the skin. This is not a disconnecting or a splitting, but an extruding of awareness. What it means to be human is no longer the state of being immersed in genetic memory but rather in being reconfigured in the electromagnetic field of the circuit - in the realm of the image. (Stelarc, 1998, p. 123)

Just like Stelarc, French conceptual artist Orlan uses her experiments to portray her own body as a software and an interface which can be used in different ways. In 1990, Orlan started her work *Saint Orlan's Reincarnation* as seen in fig. 4 (Sherman, 2014). She uses art to redesign her face with the help of several cosmetic surgeries. Her motto as an artist is: "This is my body, this is my software" (Orlan 1996) and her main goal is to create a new identity based on the appearance of famous female characters from the history of art. She had one procedure to make her lips resemble Boucher's *Europa*, another to have chin like Botticelli's Venus (Topić, 2012). Her goal was to achieve the ideal of feminine beauty as created by the male artists through history. She made a performance out of all her surgical procedures, recording them and selling them online. In her performance *Omnipresence* (1993), Orlan photographed her face before and after the surgery so that she could document her gradual transformation. She described her work as a struggle against the ingrained, unstoppable, programmed DNA, nature and god (Topić, 2012).

Figure 4. Saint Orlan's Reincarnation

The project *Cyber Tattoo* (1996) by German artists Micz Flor and Florian Clauss showed that the boundaries between the physical body and technical capabilities in the age of biotechnology are starting to shift (Clauss & Flor, 1997). *Cyber Tattoo* is a digital online tattoo studio in which the human skin is turned into an information document. The user must be connected to a computer and can tattoo the desired motive directly onto his skin by transmitting the data directly from the studio. Creating the body image through the computer stimulation enables the *Hautnah* (1994) project of the Italian artist Alba d'Urbano. Based on someone's real body mass which is entered into a computer, paper patterns are created and they can be used as clothing items. They can be work as a second above the real one (d'Urbano, 2012).

Performative art forms of enhancing the human body are clear examples of establishing cybernetic art as a manifestation of the essence of *scientific work* guided by new patterns of artistic and aesthetic expression. Man is established as an object and means of artistic experimentation and performance in which the aesthetics of the body does not stem from man's natural being as before, but a foreign scientific and technological being. Artists have "always had the power of gene splicers or nanotechnologists: they could exercise enormous control over the lives they created. All artistic creation partakes of the wish for power and control, which in its fullest fantasy is the wish for immortality, to exist beyond the bonds of nature" (McKibben, 2004, p. 212). The art of *scientific humanism as naturalism* uses the new media and the human body as means and content of its artistic expression. Artworks find new expressions in the ambivalence of the new technologies that are used to create the configurations of cyborgs and avatars as visions of future bodies. In the final decades of the 20th century, internet and computers have become the channels of artistic expressions in which the body is perceived as a system of data or signs which can be shown in their entirety in the cyberspace (for instance, Stelarc's *Movatar*, Lynn Hershman's *Telerobotic Dolls*, Tina Laporta's *Future Body* or Eva Wohlgemuth's *EvaSys*). One might argue that art in the age of *scientific humanism as naturalism* has undergone a change of paradigm as the authors have been replaced with technology and works of art are expressed as products. On the other end of this debate in the real world, human body is constantly being enhanced and reshaped through various technological accessories and implants, thus increasingly becoming an *automatos*, a self-moved machine that can be programmed at will. On the one hand, human body is dematerialized in an infinite space time of digital signs, and on the other hand, technological implants and accessories denaturalize man and turn him into a cyborg with their tendency to replace all organic body parts. Descartes said:

But about the body itself, on the other hand, I had no doubts, but I thought I distinctly knew its nature, which, if I had attempted to describe how I conceived it in my mind, I would have explained as follows: by body I mean everything that is capable of being bounded by some shape, of existing in a definite place, of filling a space in such a way as to exclude the presence of any other body within it; of being perceived by touch, sight, hearing, taste, or smell, and also of being moved in various ways, not indeed by itself, but by some other thing by which it is touched; for to have the power of moving itself, and also of perceiving by the senses or thinking. (Descartes, 2008, p. 19)

Today, through virtual reality or real-time interaction organic body can be fused with the imaginary digital image of the body. Within the hybrid media, the users are able to interact and the virtual space time enables the infinite expansion and reshaping of the body and the identity. What art expresses in the field of interactivity is mediation – a machine which itself becomes a concept of certain artworks.

If art imitates nature and nature is increasingly mediated by science – *scientific naturalism*, then the artwork takes over the character of exhibition of *scientific work*. Contrary to the transhumanist paradigm and the transhumanist art, we believe that real or artificial identification with monstrous, information technology *cyber* cyborg or avatar represents a neither safe nor promising future for mankind. What do others say, those who, unlike the transhumanist, are unwilling to enhance their body? Who advocates the *bios* (life) of the biological man in their name? The end of civilization was announced by Francis Fukuyama (1992), even Paul Virilio who says that the overwhelming speed and constant interaction of the cyber age will ensure that “One day the day will come when the day won’t come” (Virilio, 2008).

8.3 CYBORGIZATION IN MUSIC[21]

The Greek word *musike* (*mousikê*) in its original meaning denotes that which belongs to the muses, who were the keepers of knowledge of that time (Ashby, 2010, p. 143). In other ancient cultures, music was also related to the divine and the sublime. It carried the godly message of beauty, ethics and the meaning of life. This message is steadily fading in the world of today, the world of lost meaning, the world of cyborgization, media manipulation and digitalization of everyday life. Music is the actualization of possibility of sound to give meaning to human beings, meaning which they experience with their entire body – their mind, thoughts, feelings, senses, volition, and metabolism (Clifton, 1983, p. 22). With their songs about the world of media and the technoculture, the made-up rockers mock the essence of this period.

Glam rock is a genre of rock music which originated in Great Britain in the early 1970s. Its emergence was influenced by different pop art styles (fiction, art, theatre, music, and media theory of that period) and was consequently given different names: *theatre rock, glitter rock, shock rock, and gay rock* (Lenig, 2010, p. 1). It got the attribute *glam* (from *glamorous*) because it “celebrated the spectacle of the rock star and concert. Often dappled with glitter, male musicians took the stage in women’s makeup and clothing, adopted theatrical personas, and mounted glamorous musical productions frequently characterized by space-age futurism” (The Editors of Encyclopaedia Britannica, 1998). The first images associated with *glam rock* music are: long teased hair, purple guitars, polyphonic singing, and songs celebrating the hedonistic way of life. It is no wonder that a superficial music critic or an uninterested listener will regard this musical genre as a mere provocation whose main purpose is to shock the audience, i.e. the external form, and regard the musical quality and the poetics of these bands as garbage. But, are they actually a gripping form of shock, a shock which is trying to highlight the crucial issues through music and clothing? In the remainder of this section, we shall try to explain what the made-up rockers have to say about the media and the technoculture, and whether an influence of a theory or the zeitgeist of *scientific humanism as naturalism* can be traced in their works.

8.3.1 The Age of Technoculture and Simulated Reality

Today's culture is dominated by hyperreality, substance has been turned into sensation, thoughts replaced by images, the reality confused with virtual world fiction and the practice of new cybernetically modified human: "But certainly for the present age, which prefers the sign to the thing signified, the copy to the original, fancy to reality, the appearance to the essence, this change, inasmuch as it does away with illusion, is an absolute annihilation, or at least a reckless profanation" (Feuerbach, 1881, p. 9). The development of technology and digitalization of everyday life at the turn of the 20th century have shaped new forms of social activities, one of which is definitely the technoculture as "a characterization of contemporary cultures in which technology (especially but not only information and communications technology) has so deeply saturated into cultural practices that the two preciously distinct spheres (of technology and culture) are seen to be inseparable." (Girard and Girard, 2011, p. 270).

This concept emerged during the 1990s from the phenomenon of fast flow of information of all kinds, turning culture into e-culture, computers into universal communication devices, the media into new/metamedia, ultimately leading to the paradigm of "spaciousness of time", which has become an ongoing tendency. The internet as a new channel of communication affects the creation of the new culture, the Internet culture which, according to Baudrillard, can regarded as a simulation of the real, non-virtual culture. In traditional terms, reality is contrasted to the virtual as real is opposed to imaginary, tangible to immaterial. The current situation has led to the emergence of new alternative, parallel spaces. In this new world "Illusion is no longer possible, because the real is no longer possible." (Baudrillard, 1994, p. 19). This way, images and signs will multiply to such an extent that it will be impossible to distinguish between the illusion and the reality, the signifier and the signified, the subject and the object, without any social or real world to talk about, only semiotic relations of hyperreality (Best, 1994, pp. 41-42). In the world of hyperreality, the subject no longer exists, it is a part of the simulation, constructed in the media in the forms of mimesis of media images. We are mere copies without an original (simulacra), a construction of images which construct us. Baudrillard warns us that the modern world is living in the era of simulation, an era of producing the simulacrum, constantly changed by its operational replicate. The virtual world, "through the new technologies, gives us the sense that it now marks the vanishing or end of the real" (Baudrillard, 2003, p. 39).

Television and the internet, as well as other media of communication, represent a transcendence of the real world and realize its complex replacement with the technological assumptions of the medium in question. At first glance, it seems paradoxical or even absurd to claim that clowns and jesters from the very brinks of social events would provide a deeper criticism of the essence of time than most famous poets. Glam rock band, *Scorpions* describe a mechanical society and the 20th century man who has become a victim of technology. In their song *Twentieth Century Man*, they deal with the strong influence of the media and caution that people are being controlled by computers which tell them what to do:

In the jungle of these times... Honest love's so hard to find, In the world of video, Computers tell you what to do, The world's devoted to the Dollar sign, Remote-controlled and on the line, Plug in man on the rack, Waiting for a heart attack, And no chance to die... They look for God on the screen, They've got even dream-machines, They are mesmerized... Can't you see the end, Twentieth Century Man? (Scorpions, 1980).

Scorpions sing about a man controlled by technology, a man whose realization is nowhere to be seen, and a man who becomes completely blinded by technology and forgets about love.

The world of illusion is becoming increasingly relevant, after long told ancient discussions of Socrates, Parmenides, and the sophists. This represents a revival of Plato's Allegory of the Cave which is reactualized as a basis of the modern notion of "what is". With the emergence of the internet, man broke the barrier of the real world and sailed off into the world of virtual reality. In this world of signs, there is no matter, no ethics, laws, space and time. The space of our freedom is (virtually) limitless. The internet becomes the bedrock of the virtual reality phenomenon in which human imagination creates the assumptions about something which is not currently available to one's senses, but the solutions for this issue are being researched as we speak.

One might say that "Whole new cyber-worlds are being created out of 'virtual reality' and 'artificial intelligence' and 'digital community' (Soja, 2000, p. 325). In 1962, McLuhan predicted the modern global village era: "Today, after more than a century of electric technology, we have extended our central nervous system itself in a global embrace, abolishing both space and time..." (McLuhan, 1995, p. 3). A body is not necessary for virtual reality as the user can be whatever he wants to be, take any form he wants, and can move like he does in the physical world. He changes his locations, cities, travels through the expanses of the Universe. Simply put, he can achieve all his dreams and take any identity he wants.

In the strange depictions of the cyber world, human spirit finds itself outside itself, in its alienated other essence, its virtual bodiless duplicate. McLuhan uses the myth of Narcissus to describe the human fascination with technical extensions of themselves:

The Greek Myth of Narcissus is directly concerned with a fact of human experience, as the word Narcissus indicates. It is from the Greek word narcosis or numbness. The youth Narcissus mistook his own reflection in the water for another person. This extension of himself by mirror numbed his perceptions until he became the servomechanism of his own extended or repeated image. They nymph Echo tried to win his love with fragments of his own speech, but in vain. He was numb. He had adapted to his extension of himself and had become a closed system. (McLuhan, 1995, p. 41)

Like Narcissus gazing into his own reflection in the lake, man is captured by the experience of this reflection of his body. But this reflection increasingly resembles a cyborg. While the principle of unity of body and spirit – one mind, one identity – applies for the material world, the virtual world is composed of pure information. The pieces of information are also spread out and dispersed freely across space. The inhabitants of this intangible space are also dispersed, without a body as a unifying anchor. Therefore, it is possible for one person to create multiple electronic identities connected only by their common creator.

In the ecstasy of communication, it is the media who accomplish hyperreality through simulation. Communication of physical persons and purposeful human activity is replaced by communication of our virtual replicates, which we authorize to be all we could not be in the real world in our stead when accessing the network. In his song *I Just Wanna Be God*, glam rocker, Alice Cooper, pinpoints mankind's blindedness by its own reflection. In this society after the spectacle appears the ecstasy of lonely and narcissistic communication:

I'm in control, I got a bulletproof soul, And I'm full of self-esteem, I invented myself with no one's help... Me, myself and I agree, We don't need nobody else... I'm just trying to be God, I only wanna be God, I just wanna be God, Why can't I be God (Alice Cooper, 2001).

Glam rockers warn us that we live in a world of media saturation. The omnipresence of the media changes our sense of reality, manipulates our consciousness, shapes our identity according to the prevailing trends. Mass media are creating a monoculture by making the structure of existence identical for everyone regardless of their geographic location. This results in the feeling of alienation and loneliness within a crowd, which is exactly what glam rocker Alice Cooper speaks of in the song *Clones (We're All)*: "I'm a clone, I know it and I'm fine, I'm one and more are on the way … We're all clones, All are one and one are all… I just wanna be myself, Be myself" (Alice Cooper, 1980).

This way, we all become mere copies with no original (clones), identical to others but not to ourselves. Glam rockers were fascinated and frightened by the new technological options at the same time: "In songs by Brian Eno, Sparks, David Bowie, Roxy Music, and others, the future is repeatedly seen as a glowing spectacular age of new technologies. Bowie writes in "TVC 15," "he's quadraphonic, he's a, he's got more channels, So hologramic, oh my TVC one five. Bowie sees all the benefits of the future—quadraphonic sound, holographic imaging, more cable channels, all in a single device" (Lenig, 2010, p. 6). Despite their fascination with technology, the fear prevails so they chose to write about their anxiety and problematize the coming media era in their songs. Alice Cooper sings of the next generation in "Generation Landslide," saying, "the over-indulgent machines were their children, There wasn't a way down on Earth here to cool them." For Cooper, the superpowered machines of the future are likely to wrest control from the populace. (Lenig, 2010, p. 6-7)

A similar idea is portrayed by media theorist Marshall McLuhan in his analysis of effects of the media on human life:

[…] the medium is the message. This is merely to say that the personal and social consequences of any medium – the is, of any extension of ourselves – result from the new scale that is introduced into our affairs by each extension of ourselves, or by any technology. (McLuhan, 1995, p. 7)

This phrase, coined in 1964, underlined the importance of the media thirty years before the emergence of the internet. Back then, McLuhan claimed that the media would bring the end of mankind as we knew it and that we would have to rethink the society we lived in. In line with McLuhan's claim that the media are in fact extensions of the human bodies, we can confirm that the internet and the TV are reshaping both mankind and its relations between individual members. This is the central idea of the song *Doin' as I see on my TV* by glam rock band Poison:

When I was a child I'd be starin' at the tube, I saw sex, violence, and death on the news, One day I saw the Duke kill someone, Daddy said don't worry boy, they're just acting son, All I want is attention from somebody, Somebody pay attention to me, Like them actors I got no shame, I want my fifteen minutes of fame, Doin' as I seen on my TV, Can I get on the news if I kill my wife? (Poison, 2000)

The song *Ministry of Fools* by *Saxon* warns us about the strong influence of TV, trying to take control over the human mind: "Your talking on the TV every single day, Making stupid gestures nothing much to say… We're not listening to you, don't tell me what to do, We're not listening, to the ministry of fools." (Saxon, 1997). Saxon's view of television can be compared with Bordieu's, who believes that TV is becoming a mirror of our reality. This way, the TV screen becomes a Narcissus's mirror of sorts and TV the pace of our narcissistic exhibition.

In his song *Pessi-Mystic*, Alice Cooper sings: "Don't need a crystal ball, For me to see clearly, No astrology or Tarot cards, Watching CNN, And holding my breath, To face the daily news, Scares me to death, I'm pessimistic… Everybody's mind is badly infected, Everybody feeds the parasite, Everything is dark so why not accept it?" (Alice Cooper, 2000). As we can see, Cooper's song represents a convincing critique of our infectedness by the media which manipulate our consciousness, our identity. TV and the internet, but other communication media too, achieve transcendence of the real world and its complete replacement with the virtual space thanks to technical assumption of the medium itself: "Television becomes, in this sense, the dominant technology that changes our relation to the natural world. We experience events in its terms, begin to live in reference to it and ratify experience in terms of its on-screen simulation" (Tichi, 1992, p. 137). Baudrillard sees the TV medium as the "the first great formula of [hyperreality]" (Baudrillard, 1994, p. 30). Television as a medium evokes the idea of an endless visual, infinitely segmented and image-readable world (Baudrillard, 1988, pp. 122-123). McLuhan (1995) argues that TV has created a new environment for children – an environment with a lower degree of visual orientation and higher degree of inclusion. These children grow up in an electronic environment which is becoming increasingly interactive (TV, video games etc.) but not appropriate for immediate exchange of experiences, emotions, and thoughts, all of which imply a dialogue. Television provides the incorporeal extension as it removes the cultural necessity of face-to-face interactions in favour of abstract identification with others (Holmes, 1997, p. 30). The art of conversation is substituted with various technical gadgets, which come from our immediate surroundings as constant challenges.

Glam rockers created spectacles to send their visual message, awaken the audience, and warn us about the dangers of the cyborgized reality. At that time, the media took all public attention and became a lens through which the society perceived the world. (Lening, 2010, p. 6). Inspired by the media and future technologies which became a part of their *glam* songs, *glam* rockers sang about: the media watching us and invading our privacy (Saxon – *Watching you*), controlling and seducing us (Scoripons – *Media Overkill*, Ozzy Osborne – *Crazy Train*, White lion – *If My Mind Is Evil*), changing our identity and leading us into a personal crisis (Sweet – *Identity Crisis*, Alice Cooper – *Pessi-Mystic*, The Tuff – *Who The Hell Am I*?), forcing us to sell our soul (The Cult – *Dirty Little Rock Star*, Van Halen – *Light up the Sky*), and stimulating immorality (Alice Cooper – *Sex, Death And Money*).

8.3.2 Robot Rock 'n 'Roll

Glam rocker protested in favour of the natural man (Alice Cooper – *Clones (We're All), Who Do You Think We Are*, Scorpions – *Robot Man*, White lion – *Transformers*). Even though we are "travellers in time" (Scorpions – *Are We Travellers in Time*) who live on a "brutal planet" (Alice Cooper – *Brutal Planet*), we must fight so that our lives do not become absolutely conditioned by the media and technology. "Bowie's meta-message was escape – from class, from sex, from personality, from obvious commitment – into a fantasy [...] or a science-fiction future." (Hebdige, 1979, p. 61). Digital society made of digital networks engulfs the spirit and consciousness, transforms the human body, bringing it closer to the ideal of beauty characteristic for machine entities. In 1986, WASP's *Mantronic* represented an outrageous protest against us becoming emotionless cyborgs: "You see me, a man and machine... Watch me bleed and I feel no pain, I'm going wild, and slowly insane, Mantronic, I'm on the loose, Bionic, my neck in the noose, I'm still the king, Bionic, half man and machine" (WASP, 1986).

In his song *What You Think We Are*, Alice Cooper exclaims:

I'm machinery, With a semi-automatic heart, I'm unclassified, I terminate, I tear apart... Who do you think we are, Special forces in an armored car, Who do you think we are, We don't care, we don't care. (Alice Cooper, 1981)

In Cooper's view, we are becoming machines with semi-automatic hearts and the Earth is becoming a place of man's death. Ultimately, the artificial bodies and future robotic forms of life are created – humanoid robots, androids, and other robots that are well on their way to replace many functions of man, which is the topic of the song *Robot Man* by Scorpions:

[...] a robot man, I'm a loser... Babe, it's a magic station, Where we live what we do with our magic from my generation, I say babe, it's not a vision, It's reality, this is a robot scene what we live in... I'm a robot man. (Scorpions, 1975)

Many of the big names in rock 'n' roll would be surprised with the technological advancements today which allows things far crazier than they were willing to do. After *Gorillaz* showed the world 20 years ago how the IT paradigm will affect the music industry, a new sensation has come from Japan – the *Z-Machines* (Fig. 5). A three-member robot band plays a perfectly modern electronic fusion of *rock, soul, funk* and *jazz* music. The band's frontman *Macho* has 78 fingers and does a flawless job even with the most demanding bits. His style with multi-coloured cables coming out of his head and an occasional *headbang* is completely in line with his flesh-and-blood glam rock predecessors. The drums are played by *Ashura*, whose 22 hands can surpass even the fastest speed metal drummers. The two are accompanied by *Cosmo* on keyboards, whose immaculate precision is augmented by green lasers shooting out of his eyes. They are real made-up robot rockers. Another three-member heavy metal robot band named *Compressorhead* was founded in Germany. The band plays Led Zeppelin and Motörhead covers. So far, they have had their performances at the Frankfurt Fair and at the heavy metal festival in Australia, where they opened for the Red Hot Chili Peppers (Vijesti.ba, 2013).

In 2013, the society of made-up rockers was joined by other members of the procedures of *scientific work*, such as robot rockers and robot stand-up comedians which will dominate the cyborgized musical scene. We believe that the emotions and improvisations of living artists can never be replaced by mechanical surrogates as long as there are human beings around. However, cyborgization of man and social relations do not seem to favour the natural artists.

Figure 5. Z-Machines[22]

We have shown that the songs of made-up rockers have raised the issues of medical control, robotization, turning humans into machine beings through use of mass media and the internet, and loss of identity in the world of media. All these technologically constructed realities redefine not just this role, but the very concept of man in the 21st century, which is being increasingly perceived through the transhumanist lenses.

8.4. CYBORGIZATION IN TRANSPORTATION

In a not-too-distant future, all sea, land and air traffic will be completely automatic. Man will just passively observe the effects of artificial intelligence. Google is experimenting with autonomous automobiles, the Chinese scientists are working on mind-controlled cars, while Tesla Cars is developing a navigation system which would be capable of changing lanes, slowing down and reading the road signs via radars, cameras and sensors and no input from the driver. Uber's self-driving truck completed its first delivery of 50,000 beer cans in 2016 (Vuzem, 2016). However, all of these events still take place in the presence of man.

Traffic jams may also be a thing of the past. A central computer will be able to track all the motions of every car on the road by communicating with each driverless car. It will then easily spot traffic jams and bottlenecks on the highways. In one experiment, conducted north of San Diego on Interstate 15, chips were placed in the road so that a central computer took control of the cars on the road. In case of a traffic jam, the computer will override the driver and allow traffic to flow freely. (Kaku, 2011, p. 22)

We have shown that the procedures of *scientific work* cyborgize the historical reality of man into the era of *scientific humanism as naturalism*, and that it is the future, but in fact the present, that we are acclimatizing to as we have not become fully aware of the present future. Our examples show that the contemporary procedures of *scientific work* cyborgize man, animals, social relations, and that they have already established virtual and augmented reality. In the future just around the corner, we can expect ubiquitous computing, digital holograms, and the domination of robots in the historical reality of man. A victory for the cybernetic model over human reality means that everything animate and inanimate has been reduced to a binary code. Heidegger reminds us that "Reflection is the courage to make the troth of our own presuppositions and the realm of our own goals into the things that most deserve to be called in question" (Heidegger, 1977, p. 116).

REFERENCES

Anić, V., & Goldstein, I. (2004). *Rječnik stranih riječi.* Novi Liber.

Ashby, A. (Ed.). (2010). The pleasure of modernist music: Listening, meaning, intention, ideology. Univ. of Rochester Press.

Baudrillard, J. (1988). *The Ecstasy of Communication.* MIT Press.

Baudrillard, J. (1994). *Simulacra and simulation.* University of Michigan Press.

Beauchamp, T. L., & Childress, J. F. (2013). *Principles of Biomedical Ethics* (7th ed.). Oxford University Press.

Bostrom, N. (2005). *A History of Transhumanist Thought.* Academic Press.

Camporesi, S. (2008). Oscar Pistorius, enhancement and post-humans. *Journal of Medical Ethics, 34*(9), 639–639. doi:10.1136/jme.2008.026674 PMID:18757629

Chatard, J.-C., & Wilson, B. (2008). Effect of Fastskin Suits on Performance, Drag, and Energy Cost of Swimming. *Medicine and Science in Sports and Exercise, 40*(6), 1149–1154. doi:10.1249/MSS.0b013e318169387b PMID:18460989

Clarke, R. (2010). *Cyborg rights 'need debating now'.* Sentient Developments. http://www.sentientdevelopments.com/2010/06/roger-clarke-cyborg-rights-need.html

Clauss, F., & Flor, M. (1997). *Cyber Tattoo.* http://www.medienkunstnetz.de/works/cyber-tattoo/

Cooper, A. (1980). *Clones (We're All).* Flush the Fashion. Warner Bros.

Cooper, A. (1981). *Who Do You Think We Are. Special Forces.* Warner Bros.

Cooper, A. (2000). *Pessi-Mystic. Brutal Planet.* Spitfire.

Cooper, A. (2001). *I just Wanna Be God.* Dragontown. Spitfire.

C.S. (2011). Prostheses in sport. Running foul? *The Economist.*

Ćurko, B., & Greguric, I. (2012). Glam rock identity. In S. Vannatta (Ed.), *Chuck Klosterman and philosophy – The real and cereal* (pp. 43–52).

d'UrbanoA. (2012). *Hautnah*. https://durbano.de/hautnah/index.html

Dagenais, F. (1991). Perfect Bodies. In C. Richards & N. Tenhaaf (Eds.), *Bioapparatus*. Banff Centre for the Arts.

Deleuze, G., & Guattari, F. (2009). *Anti-Oedipus: Capitalism and schizophrenia*. Penguin Books.

Descartes, R. (2008). *Meditations on first philosophy: With selections from the Objections and replies* (M. Moriarty, Trans.). Oxford University Press.

Draper, R. (2007). The fastest man on no legs now has the Olympics in his sights. *Mail Online*. https://www.dailymail.co.uk/news/article-466904/The-fastest-man-legs-Olympics-sights.html

Dugandžija, M. (2017). Razgovarali smo s umjetnikom kojem je na lijevoj ruci 'naraslo' treće uho 'Ljudsko tijelo treba redizajnirati. Ono je i fantastično i jadno'. *Jutarnji list*. https://www.jutarnji.hr/globus/kultura/razgovarali-smo-s-umjetnikom-kojem-je-na-lijevoj-ruci-naraslo-trece-uho-ljudsko-tijelo-treba-redizajnirati-ono-je-i-fantasticno-i-jadno-6101508

Edwards, A., & Skinner, J. (2006). *Sport Empire*. Meyer & Meyer Sport.

Faratin, F. (2013). Gallery: Robo-rockers Z Machines preform debut gig. *Metro*. https://metro.co.uk/2013/06/24/gallery-robo-rockers-z-machines-preform-debut-gig-3853469/

Feuerbach, L. (1881). *Second. The Essence of Christianity*. Trübner & Co.

Fukuyama, F. (1992). *The End of History and the Last Man*. Hamish Hamilton.

Fukuyama, F. (2002). *Our posthuman future: Consequences of the biotechnology revolution* (1st ed.). Farrar, Straus and Giroux.

Fukuyama, F. (2004). Transhumanism. *Foreign Policy*, *144*(144), 42. doi:10.2307/4152980

Greguric, I. (2012). Kiborgizacija u sportu – stvaranje tijela bez granica u doba transhumanizma. *Metodički Ogledi : Časopis Za Filozofiju Odgoja*, *19*(1), 95–111.

Greguric, I. (2013). Novi mediji i kiborgizirano tijelo kao prostor umjetnosti transhumanizma. In Medias Res : Časopis Filozofije Medija, 2(3), 350–364.

Haraway, D. J. (2016). A Cyborg Manifesto: Science, Technology, and Socialist-Feminism in the Late Twentieth Century. In D. J. Haraway & C. Wolfe (Eds.), *Manifestly Haraway* (pp. 3–90). University of Minnesota Press., doi:10.5749/minnesota/9780816650477.003.0001

Hebdige, D. (2012). *Subculture: The Meaning of Style*. Taylor and Francis. http://SLQ.eblib.com.au/patron/FullRecord.aspx?p=169053

Heidegger, M. (1977). *The question concerning technology, and other essays*. Garland Pub.

Heim, M. (1993). *The metaphysics of virtual reality*. Oxford University Press.

Hershman, L. (1997). Tillie. *Fictive Art*. https://www.lynnhershman.com/tillie-and-cyberroberta/

Hershman, L. (2007). Breitmore Roberta. *Fictive Art*. www.artpractical.com/feature/looking_for_roberta_breitmore/

Holmes, D. (1997). *Virtual Politics: Identity and Community in Cyberspace*. Sage Publications.

Jutarnji list. (2008). Ljudi će postati besmrtni. *Jutarnji list*. https://www.jutarnji.hr/naslovnica/ljudi-ce-postati-besmrtni-3861999

Kaku, M. (2011). *Physics of the future: How science will shape human destiny and our daily lives by the year 2100*. Doubleday.

Klimešova. (2008). *Future Kiss*. https://www.kontejner.org/en/projekti/touch-me/touch-me-festival-2008-osjecaj-se-bolje/performansi-osjecaj-se-bolje/lenka-klimesova-cz-poljubac-buducnosti

Krug, S. (2017). *We're Seeing Robots. Lynn Hershman Leeson's Networked Gaze*. Rhizome. https://rhizome.org/editorial/2017/feb/09/were-seeing-robots/

Kurzweil, R. (2005). *The Singularity is Near: When Humans Transcend Biology*. Viking Penguin.

La Porta, T. (2000). *Where Will the Body Be in the Future?* https://web.njit.edu/~kimmelma/laporta.html

Lenig, S. (2010). *The twisted tale of glam rock*. Praeger.

Magdalinski, T. (2009). *Sport, technology and the body: The nature of performance*. Routledge. doi:10.4324/9780203099384

Maloney, O. (2011). *Interview with Lynn Hersmann*. Art practical. http://www.artpractical.com/feature/looking_for_roberta_breitmore/

McLuhan, M. (1962). *The Gutenberg Galaxy*. University of Toronto Press.

McLuhan, M. (1995). Understanding Media: The Extensions of Man. MIT Press.

Miah, A. (2008). *Genetically Modified Olympians*. http://www.andymiah.net/2009/10/08/genedoping-is-next-frontier-of-performance-enhancers-in-sports-2008-may-11/

More, M. (2003a). *Princinples of Extropy*. http://www.extropy.org/principles.htm

More, N. V. (2003b). *Primo Posthuman 3M+*. http://www.medienkunstnetz.de/works/primo-posthuman/

Moreton, C. (2012). London 2012 Olympics: Oscar Pisorius finally runs in Games after five year battle. *The Telegraph*. https://www.telegraph.co.uk/sport/olympics/athletics/9452280/London-2012-Olympics-Oscar-Pistorius-finally-runs-in-Games-after-five-year-battle.html

Orlan. (1996). Conference. In D. McCorquodale (Ed.), *This is my Body, this is my Software*. Black Dog Publishing.

Pearce, M. (1995). From Urb to Beat. In M. Pearce & N. Spiller (Eds.), Architects in cyberspace. Academy Ed.

Poison. (2000). Doin' as I see on my TV. *Crack a Smile*. Capitol.

Poptech (2016). Uskoro počinje prva bionička olimpijada. *Vidi.hr*. https://www.vidi.hr/Pop-Tech/VIDEO-Uskoro-pocinje-prva-bionicka-Olimpijada

Price, S. L. (2005). The Sprinter. Sports Illustrated, 102(21).

Prophet, J. F. (2000). *The Internal Organs of a Cyborg*. http://www.e-felix.org/issue5/prophet.html

Salleh, A. (2010). Cyborg rights »need debating now«. *ABC Science Online*. https://www.abc.net.au/science/articles/2010/06/04/2916443.htm

Saxon. (1997). Ministry Of fools (Saxon). *Unleash The Beast*. Virgin/CBH.

Scarry, E. (1985). The body in pain: The making and unmaking of the world. Oxford Univ. Press.

Scorpions. (1975). Robot Man. *In Trance*. RCA.

Scorpions. (1980). Twentieth Century Man. *Animal Magnetism*. Polydor.

Soja, E. (2000). *Postmetropolis: Critical Studies of Cities and Regions*. Blackwell Publishing Ltd.

Stelarc. (2021). http://stelarc.org/projects.php

Stenslie, S. (2004). *Strojni fetišizam*. https://www.kontejner.org/projekti/touch-me/touch-me-festival-inicijacija/k-011-stahl-stenslie-no-inter_skin/

Sutherland, J. (2006). The ideas interview: Nick Bostrom. *The Guardian*. https://www.theguardian.com/science/2006/may/09/academicexperts.genetics

The Editors of Encyclopaedia Britannica. (1998). Glam rock. *Brittanica*. https://www.britannica.com/art/glam-rock

Tichi, C. (1992). Electronic hearth: Creating an American Television Culture. Oxford Univ. Press.

Tonkin, J. (1999). *Personal Eugenics*. http://www.casulapowerhouse.com/microsites/cybercultures/12_artist.swf

Topić, L. (2012). *10 Pitanja za ORLAN*. Pogledaj.to. https://pogledaj.to/oznaka/orlan/

Transhumanist Arts & Culture. (2007). *Primo Posthuman future body design*. http://www.transhumanist.biz/nanonatashavita-more.htm

Turkle, S. (2005). The second self: Computers and the human spirit. MIT Press.

Vesna, V. (1996). Bodies© Incorporated. *Net Art Anthology*. http://www.bodiesinc.ucla.edu/thing.html.

Vijesti.ba. (2013). *Heavy metal grupa čiji su svi članovi roboti*. Vijesti.ba. https://vijesti.ba/clanak/126455/heavy-metal-grupa-ciji-su-svi-clanovi-roboti

Virilio, P. (1995). *Speed and Information: Cyberspace Alarm!* Ctheory.

Vuzem, G. (2016). *Kamion bez vozača prevezao 50.000 limenki piva*. Tportal. https://www.tportal.hr/tehno/clanak/kamion-bez-vozaca-prevezao-50-000-limenki-piva-20161026

Wasp. (1986). *Mantronic. Inside The Electric Circus*. Capitol.

Weiner, J. (2008). *Cyborg athletes? Bionic Olympics? High-tech prosthetics case raises the prospect of a potential sci-fi sports world*. MinnPost. https://www.minnpost.com/politics-policy/2008/01/cyborg-athletes-bionic-olympics-high-tech-prosthetics-case-raises-prospect-p/

Weinreb, M. (2007). *From Radical To Routine: Tommy John Surgery*. Sports Illustrated.

Weyand, P. G., Bundle, M. W., McGowan, C. P., Grabowski, A., Brown, M. B., Kram, R., & Herr, H. (2009). The fastest runner on artificial legs: Different limbs, similar function? *Journal of Applied Physiology*, *107*(3), 903–911. doi:10.1152/japplphysiol.00174.2009 PMID:19541739

Willis, A. (2009). Immortality only 20 years away says scientist. *The Telegraph*. http://www.telegraph.co.uk/science/science-news/6217676/Immortality-only-20-years-away-says-scientist.html

Wohlgemuth, E. (1997). Bodyscan. fleshfactor Informationsmaschine Mensch. In *Ars electronica* (p. 299). Springer.

Wolchover, N. (2012). Are Oscar Pistorius' Prosthetic Limbs an Unfair Advantage? *NBC News*. https://www.nbcnews.com/id/wbna48335839

ENDNOTES

1 Marlon Shirley, US Paralympian, sprinter and a 100m race champion. Shirley competed with Oscar Pistorius at the 2004 Paralympics in Athens. Pistorius won the 200m race. More information about Shirley can be found in Price (2005).

2 According to the *Humanity+* transhumanist organization, transhumanism is defined as "(1) The intellectual and cultural movement that affirms the possibility and desirability of fundamentally improving the human condition through applied reason, especially by developing and making widely available technologies to eliminate aging and to greatly enhance human intellectual, physical, and psychological capacities." and "(2) The study of the ramifications, promises, and potential dangers of technologies that will enable us to overcome fundamental human limitations". More details about the historical development of transhumanism can be found in Bostrom (2005). The World Transhumanist Organization (Humanity+) was founded in 1988 by Nick Bostrom and David Pearce as "an international nonprofit membership organization which advocates the ethical use of technology to expand human capacities. In other words, we want people to be better than well." More information available in Sutherland (2006).

3 The main representatives of bioconservatism are predominantly members of the former President's Council on Bioethics which was founded by the US president George W. Bush with the aim of analysing the ethical and social consequences of enhancement of man that may emerge as a result of advances in biotechnology. It performed its function from 2001 to 2009, when it was replaced with the

Presidential Commission for the Study of Bioethical Issues by President Barack Obama. More information available at: http://www.bioethics.gov.

4 Nanobots are miniature robots the size of blood cells programmed to seek out and treat all anomalies to keep a human body healthy at a cellular and molecular level (Jutarnji list, 2008).

5 Roger Clarke is a visiting professor at the Department of Computer Science, Australian National University. More information available at: http://www.rogerclarke.com.

6 Source: https://www.thefreedictionary.com/prosthesis [17/4/2020].

7 Fastskin is a special swimsuit made by Speedo, which first appeared at the Olympics in 2000. The suit is made from materials with friction lower than that of human skin. Until 2008, over 40 world swimming records were broken with the use of this technology (Chatard & Wilson 2008).
8 Frank Jobe spent 40 years working as the team physician for the LA Dodgers. Source: http://www.usatoday.com/sports/baseball/2003-07-28-cover-tommy-john_x.htm [17/4/2020]
9 Oscar Leonard Carl Pistorius (22nd November, 1986.) was born in the family of Sheila and Henke Pistorius, owner of a zinc processing plant in Sandton, South Africa. Due to a congenital defect and a hereditary disease, both of his fibula bones were missing and he only had two toes. Instead of putting him in a wheelchair, his parents opted to have both his feet amputated so that he could use prostheses to walk one day. At the age of 11 months, both of his legs were amputated from knee down. When he was 11, he got the first prostheses and managed to participate in several sports, such as rugby, water polo, tennis, and wrestling, despite his handicaps. After sustaining a serious rugby knee injury, he focused on track disciplines. More information in Draper (2007).
10 A fibula bone (a calf bone) is the long bone in the lower leg which serves for the attachment of muscles and forms the ankle joint.
11 The prostheses Cheetah Flex Foot were made by the Icelandic company Össur, founded in 1971 by prostheses expert Össur Kristinsson as a small workshop. Today, the company has become the leading prostheses manufacturer in Iceland with its own research centre which produces prostheses for various purposes. These include the Cheetah prostheses made out of special materials which allow the disabled persons to take part in sports activities. They consist of two J-shaped carbon-fibre rods and cost about $18,000. More information at: http://www.ossur.com/?PageID=13462, http://en.wikipedia.org/wiki/%C3%96ssur
12 He got this nickname after the 1982 sci-fi classic *Blade Runner* directed by Ridley Scott and based on Philip K. Dick's novel *Do Androids Dream of Electric Sheep?*, which deals with the existential questions of being and co-existence of androids and men. The connection with Pistorius lies in the two bent prostheses which look like blades.
13 In qualifications for 2012 Summer Olympics in London, he ran the required norm which allowed him to participate at the Olympics.
14 FIRST – Foundation for Inspiration and Recognition of Science and Technology. (2010). *Oscar Pistorius case study*. Body forward first LEGO league.
15 Sprinters usually reach their maximum speed between the 60th and the 80th meter, and then try to maintain it for as long as possible. Former 400m world-record holder Michael Johnson had the ability to maintain the same pace from start to finish. Pistorius is the only athlete whose speed in the second part of the races is faster than in the first part.
16 "Gravitational torque involves leaning forward and remaining constantly on the precarious tipping point between falling to the ground and maintaining controlled forward momentum with each step" (C.S., 2011).
17 A runner's stride is divided into time when his foot is touching the ground and time when his legs are changing positions. All runners spend 0.12 seconds in the air between steps, but they differ in their ability to cover a particular distance in those intervals.
18 This chapter is an extended version of Greguric (2013).
19 The word *prosthesis* comes from the Latin *prosthesis* (putting forward), and denotes an addition or extension. A prosthesis is an artificial part of the body made from various materials to replace

lost or significantly damaged organs in the human body or one of its parts. Source: http://oxforddictionaries.com/definition/english/prosthesis. [28/11/2012]

2[0] In Hinduism, avatar is the embodiment of a deity, either in human or animal form); skr. *avatara*: coming down. (Anić & Goldstein, 2004). In terms of cyberspace, avatars are images – icons: people, animals, etc. which represent pinnacle of simulation of the user's alter ego in a cyber-performance. Their status in this game is still a topic of philosophical debates. The basic dilemma lies in the question whether they are icons in the hands of players (man as subject) or whether they are standalone players in the virtual space (subjects of the game). An avatar could happen in real space if we could implant a human consciousness into a newly created human biological body (genetic modification) and control it via computer.

2[1] This chapter is an extended version of the discussion found in Ćurko & Greguric (2012).

2[2] Robot rock bend *Z-Machines* performing with dancers during their debut concert in Tokyo, June 24, 2013 (Faratin 2013).

Chapter 8
Robots in the Historical Reality of Scientific Humanism as Naturalism

ABSTRACT

This chapter is dedicated to the robots in historical reality of scientific humanism as naturalism. Robots as posthuman descendants and heirs of man occupy the reality of scientific humanism, performing today, admittedly in conception, most human affairs. The discussion of cyborgization in the arts, increasing integration of robots in society, and future possibilities covered in the remaining chapters entrenches the case for cyborgoethics by demonstrating the popularization of cyborgs in areas of life and society.

9.1 COEXISTENCE WITH THE MECHANICAL OFFSPRING

Robots are becoming increasingly important parts of the reality in the age of *scientific humanism as naturalism*. The advocates of robotization of social reality predict that robots allow humans to have more space-time for higher activities (Marx). According to some estimates, in the next twenty years robots will take over about 47% of jobs currently performed by man, and there are reasons to believe that the human worker's labour will be completely replaced in the longer perspective. Since the beginning of development of robotics and artificial intelligence, the public has feared that the robots will take over our jobs but:

[…] the prophets of the technological paradise have always assured us that once we have turned all our work over to machines we will become a nation of artists or, at worst, a nation of art critics. […] freed from drudgery, we will have the time to devote to more 'fulfilling' endeavours. (McKibben, 2004, p. 96)

In the past decade, more and more jobs have shifted over: telephone operators replaced by voice-mail systems, and bank tellers rendered obsolete by ATMs. The trend accelerates: already fast-food restau-

DOI: 10.4018/978-1-7998-9231-1.ch008

rants in Japan have tested 'server' robots' and 'simple, self-robots with TV cameras able to recognize intruders are finding work as security guards. (McKibben, 2004, p. 92).

Ian Pearson, forecaster employed by the British Telecom, predicts that "enterprises run entirely by machine will easily outcompete human workers, 'people will have fewer and fewer attributes to sell.' The only consolation is that 'production and output could greatly increase… so we could all have a better quality of life without having to do work.'" (McKibben, 2004, p. 93). One might expect that "Someday, no airline will dare put a person at the controls for fear of lawsuits. If, with Moore's law on their side, robots become ever better at handling materials, ever better at processing information, and ever, ever cheaper, then they may displace human labour so broadly that the average workday would have to plummet to practically zero to keep everyone employed" (McKibben, 2004, p. 93).

But the question which remains open is whether robots would also start issuing commands in addition to taking our labour. "One thing people seem to like especially is being able to order other people around – this may be the greatest social use of robots. While there may never be enough people to serve this need, there could be enough robots. Everyone could order them around whenever this urge-to-power comes on." (McKibben, 2004, p. 96). With this in mind, Bostrom argues:

It is an open question whether the consequences would be for the better or the worse. The potential upside is clearly enormous, but the downside includes existential risk. Humanity's future might one day depend on the initial conditions we create – in particular, on whether we successfully design the system (for example, the seed AI's goal architecture) in such a way as to make it 'human-friendly', in the best possible interpretation of that term. (Bostrom, 2010, p. 114)

For now, we can logically conclude what this will mean for the future of man. The procedures of *scientific work* as the substance subject of world history reshape and enhance the healthy human being into a timeless cybernetic machine which in its eternal presence (Nietzsche) is faced with its own mechanical offspring, spread out all around us and aiming to replace man in every activity which reflects man's essence, purpose and the possibility of life. Cybernetic reshaping and enhancing of a healthy human being and establishing the mechanical offspring of the biological man are just mere instances of establishing the cybernetic reality of *scientific humanism as naturalism* in the lifeless *scientific history.*

Turkle's work underlines the two strands of technological development – the networked world and robotics. The first strand of development, underlines the importance of a networked world changed by internet access which enabled us to remain connected at all times: "The network was with us, on us, all the time." (Turkle, 2010, p. xii). With the emergence of cyberization it also became a part of us – in us. In our first encounter with a computer, it is almost as if a part of our mind becomes a part of the computer, it becomes a second self, a mirror of mind. In this encounter, man sees himself in a new way and this erases the boundaries between the real and the virtual world to which man comes and goes at his will, observing himself in the machine mirror. Through computer mediation, we become witnesses "to a shift in how we create and experience our own identities." (Turkle, 2010, p. xi), whereby "Our new devices provide space for the emergence of a new state of the self, itself, split between the screen and the physical real, wired into existence through technology" (Turkle, 2010, p. 16). If we ask ourselves ""'What does simulation want?" we know what it wants. It wants—it demands—immersion. But immersed in simulation, it can be hard to remember all that lies beyond it or even to acknowledge that everything is not captured by it. For simulation not only demands immersion but creates a self that prefers simulation. Simulation offers relationships simpler than real life can provide. We become accustomed to the reductions and betrayals that prepare us for life with the robotic." (Turkle, 2010, p. 285). This inevitably leads us to rethink the new conception of self.

In the networked world, every participant leaves behind a trail of electronic crumbs which anyone can use both politically or commercially (Turkle, 2010, p. 280). We can only imagine the extents to which privacy is threatened with the use and implementation of the RFID chips mentioned in previous chapters. For Michel Foucault, "the task of the modern state is to reduce its need for actual surveillance by creating a citizenry that will watch itself." (Turkle, 2010, p. 262). To exemplify this thesis, Foucault cites Bentham's design for a panopticon: "a wheellike structure with an observer at its hub, one develops the sense of always being watched, whether or not the observer is actually present. If the structure is a prison, inmates know that a guard can potentially always see them. In the end, the architecture encourages self-surveillance" (Turkle, 2010, p. 262). One could argue that simulation transforms us into its creations (Turkle, 2010, p. 287-288).

The second shortcoming of the networked society is the feeling of dependence which leads to the feeling of being "alone together" with everyone:

in their own rooms, each on a networked computer or mobile device. We go online because we are busy but end up spending more time with technology and less with each other. We defend connectivity as a way to be close, even as we effectively hide from each other. At the limit, we will settle for the inanimate, if that's what it takes. (Turkle, 2010, p. 281)

The second strand of development of science and technology, according to Turkle, is robotics and the creation of new digital "creatures including Tamagotchis, Furbies, AIBOs, My Real Babies, Kismet, Cog, and Paros, these last, robot baby seals designed specifically to provide companionship for the elderly." (Turkle, 2010, pp. xii-xiii). With the help of sociable, humanoid robots, "one might assume that what we want is to be always in touch and never alone, no matter who or what we are in touch with." (Turkle 2010, p. 285). This way, "we seem determined to give human qualities to objects and content to treat each other as things." (Turkle, 2010, p. xiv). Socializing with robots, according to Turkle, gives an illusion of control and a welcome replacement, but "what if one of the things technology wants is to exploit our disappointments and emotional vulnerabilities" (Turkle, 2010, p. 282).

Hans Moravec predicts a fast development of robotics and defines four generations of robots. The first generations are "home robots, beginning with vacuum cleaners that learn their way around a house, and continuing through machines to pick up clutter, take inventory, guard homes, 'play games', and so on. [...] By 2010, he predicts, the first 'broadly competent universal robots' will be on the market – as big as people, perhaps, but with lizardlike brains." (McKibben, 2004, p. 74). The second generation will "be 'mouselike' – able, that is, to adapt more easily, and 'even be trainable'" (McKibben, 2004, p. 74). The third generation of robots that would emerge ten years late could have "monkey-like brains capable of 5 million MIPS [which] should 'let a robot learn a skill by imitation, and afford a kind of conscience." (McKibben, 2004, p. 74). If asked "why there are candles on the table, a third generation robot might consult its simulation of house, owner, and self to honestly reply that it put them there because its owner likes candlelit dinners and it likes to please its owner" (McKibben, 2004, p. 74). Fourth generation robots could appear in forty years and would have humanlike reasoning and abilities.

Gregory Stock predicts that human beings could perish in a "pseudo-extinction, spawning our own successors by fast-forwarding our evolution" (McKibben, 2004, p. 100). Therefore, "There is no need to worry about mere robots taking over from us [...] We will be taking over from ourselves... The distinction between us and robots is going to disappear" (McKibben, 2004, p. 99). It remains to be seen whether the "machine-men deem it necessary to exterminate men-machines. In this case 'human evolution would continue, but only through the further evolution of the triumphant machine-men who would be obliged – if they have any sense of history – to recognize *Homo sapiens* as their progenitor" (McKibben, 2004, p. 104). Hans Moravec emotionally proclaims:

I consider these future machines our progeny, 'mind children' built in our image and likeness, ourselves in more potent form. Like biological children of previous generations, they will embody humanity's best chance for a long-term future. It behoves us to give them every advantage and bow out when we can no longer contribute. (McKibben, 2004, p. 99)

According to Turkle, "Roboticists insist that robotic emotions are made up of the same ultimate particles as human ones (because mind is ultimately made of matter), but it is also true that robots' claims to emotion derive from programs designed to get an emotional rise out of us." (Turkle, 2010, p. 281). But can a robot incite emotions in human beings and do so with credibility? Turkle believes that we are currently trying to create artificial beings which resemble us and that we "animate robotic creatures by projecting meaning onto them and are thus tempted to speak of their emotions and even their 'authenticity'" (Turkle, 2010, p. 282). This raises the question as to why robots feel and what topics are so personal that they can only be discussed with a machine, since we are aware that robots cannot feel neither empathy nor the course of human connection: "[...] the robot can feel nothing at all. Do we care? Or does the performance of feeling now suffice? Why would we want to be in conversation with machines that cannot understand or care for us?" (Turkle, 2010, p. 282). Has the cyborgized reality pushed us so far apart from fellow human beings?

A group of South Korean and US scientists have created an improved version of a humanoid robot *Einstein*, originally made in 2005. The robot, newly named *Albert Hubo*, is equipped with cameras, microphones, a voice synthesizer and 32 motors which allow him to show various facial expressions. It can laugh, blink, and even pay close attention to its interlocutors. Scientists claim that Albert can understand feelings and understand emotions and gestures and has even learned to express feelings like anger, sorrow and happiness (Jutarnji list, 2009). However, it does not show emotions in a way a human being shows them because it only expresses pre-programmed emotions in his interactions.

The advocates of robotics are trying to create an artificial environment of mutual love between humans and robots, but "Are we indifferent to whether we are loved by robots or by our own kind?" (Turkle, 2010, p. 286). In Philip K. Dick's classic science-fiction novel *Do Androids Dream of Electric Sheep?* we see that "loving and being loved by a robot seems a good thing. The film's hero, Deckard, is a professional robot hunter in a world where humans and robots look and sound alike. He falls in love with Rachel, an android programmed with human memories and the knowledge that she will 'die'" (Turkle, 2010, p. 286).

For several years, travellers on Canadian highways could pick up an unusual hitchhiker robot – *HitchBot*. All he needed was a little electricity from the car's lighter socket to charge his batteries. A driver who would let *HitchBot* into his car had to charge his battery, otherwise he would not get a single smile from an otherwise chatty robot. At the time of writing of this chapter, it had travelled over six thousand kilometres. The *HitchBot* "was a project created to see how a robot would travel through the human world, and to document the trip through photos, linked with social media such as Twitter... During the summer of 2014, *HitchBot* hitchhiked across Canada... In just 26 days it hitched a total of 19 rides and travelled over 10 000 kilometers" (Thompson, 2019, p. 110).

In his book *Love and Sex with Robots*, David Levy claims that not only will love between humans and robots become widely accepted but robots will also teach us to be better lovers: "Love with robots will be as normal as love with other humans, while the number of sexual acts and lovemaking positions commonly practiced between humans will be extended, as robots will teach more than is in all of the world's published sex manuals combined" (Turkle, 2010, p. 5). The advantages of robot love are huge since there is "No cheating. No heartbreak." (Turkle, 2010, p. 5). When "roboticists argue that robots can develop emotions, they begin by asserting the material basis of all thought and take things from there. For example, Rodney Brooks says that a robot could be given a feeling like "sadness" by setting 'a number in its computer code.' This sadness, for Brooks, would be akin to that felt by humans, for 'isn't humans' level of sadness basically a number, too, just a number of the amounts of various neurochemicals circulating in the brain? Why should a robot's numbers be any less authentic than a human's?'" (Turkle, 2010, p. 286-287). Contrary to this, Turkle argues that human beings, unlike robots, are given something robots can never have: human bodies and human life with all its diversity, qualities and flaws. Unlike Turkle, Cynthia Breazeal finds that should we perceive the robot emotions as a new category, they would become genuine:

Cats have cat emotions, and dogs have dog emotions. These differ from each other and from human emotions. We have no problem, says Breazeal, seeing all of these as 'genuine' and 'authentic.' And now, robots will have robot emotions, also in their own category and likewise "genuine" and 'authentic.' (Turkle, 2010, p. 287)

For Turkle, "this argument confuses the authentic with the *sui generis*. That the robotic performance of emotion might exist in its own category implies nothing about the authenticity of the emotions being performed. And robots do not "have" emotions that we must respect. We build robots to do things that make us feel as though they have emotions. Our responses are their design template." (Turkle, 2010, p. 287). Their emotions stem from their internal program, not their world experience.

In her analysis of sociable robots as caregivers for the elderly and the disabled, Turkle finds that these robots are "the challenge of a new 'species' [...] whose "emotions" are designed to make us comfortable with them" (Turkle, 2010, p. 289). This simulated feeling of nurture and resolving loneliness makes us wonder whether we really want our loved ones to be nurtured and entertained by robots. During an MIT seminar which dealt with the role of robots in medicine, the group analysed a robot who helps out the medical staff with turning over a patient in the bed during bath time: "A robot now on the market is designed as a kind of double spatula: one plate slides under the patient; another is placed on top. The head is supported, and the patient is flipped." (Turkle, 2010, p. 290). One group of students looked at the role of robot caregivers as positive in terms of cost, efficiency and lack of caregivers on the market, while the other group perceived their use as an act of humiliation, arguing that the elderly deserve real human contact. What matters most here is the opinion of the patients, because "The most common justification for the delegation of care to robots focuses on things being "equal" for the person receiving care. This argument is most often used by those who feel that robots are appropriate for people with dementia, who will not "know the difference" between a person and a robot." (Turkle, 2010, p. 292). Perhaps "The idea of sociable robots suggests that we might navigate intimacy by skirting it. People seem comforted by the belief that if we alienate or fail each other, robots will be there, programmed to provide simulations of love. Our population is aging; there will be robots to take care of us. Our children are neglected; robots will tend to them." (Turkle, 2010, p. 10).

What is our responsibility here and is it even ethically, socially, or psychologically acceptable for an individual? If we become desensitized to these changes, we could perhaps face a similar dilemma: " [...] once you have elder bots and nurse bots, why not nanny bots?" (Turkle, 2010, p. 292). However,

The developmental implications of children taking robots as models are unknown, potentially disastrous. Humans need to be surrounded by human touch, faces, and voices. Humans need to be brought up by humans. [...] But children fed their string beans by a robot will not associate food with human companionship, talk, and relaxation. Eating will become dissociated from emotional nurturance. Children whose diapers are changed by robots will not feel that their bodies are dear to other human beings. Why are we willing to consider such risks? (Turkle, 2010, p. 292-293)

As an additional argument, Turkle highlights the robot nanny's inability to show emotions towards children whose "[...] eyes "light up" with interest and "darken" with passion or anxiety." (Turkle, 2010, p. 292). If this argument is important for the child-robot relationships, then it should apply to the relationship between robots and the elderly.

We should "be asking what kind of relationships we want to have with machines. Why do we want robots to perform emotion?" (Turkle, 2010, p. 287). Before starting a real relationship, we should be aware that "We have agreed to an experiment in which we are the human subjects. Actually, we have agreed to a series of experiments: robots for children and the elderly, technologies that denigrate and deny privacy, seductive simulations that propose themselves as places to live. We deserve better. When we remind ourselves that it is we who decide how to keep technology busy, we shall have better." (Turkle, 2010, p. 296). The discussion on robot intelligence and emotions leads us the conclusion that we are dealing with the successor of man, not his assistant.

According to psychoanalyst Robert Caper:

The transgression in the analytic enterprise is not that we try to make things better; the transgression is that we don't allow ourselves to see its costs and limitations." To make his point Caper revisits the story of Oedipus. As his story is traditionally understood, Oedipus is punished for seeking knowledge—in particular, the knowledge of his parentage. Caper suggests he is punished for something else: his refusal to recognize the limitations of knowledge. A parallel with technology is clear: we transgress not because we try to build the new but because we don't allow ourselves to consider what it disrupts or diminishes. We are not in trouble because of invention but because we think it will solve everything. (Turkle, 2010, p. 283-284)

9.2 ROBOTS – CONSCIOUSNESS – INTELLIGENCE – MORALS

Our discussion on the consciousness of intelligent machines leads us to the question whether we can design conscious machines, machines which will be able to recognize man. William James has expressed the consensus of science of his time when he claimed that "the cortex is the sole organ of consciousness in man" (James, 1950, p. 34). According to de Sousa (2013) and Chrisley (2008), "Consciousness is considered that part of the brain that is least amenable or reproducible by artificial intelligence (AI). It is assumed that the very nature and essence of consciousness may not be explainable by computations, algorithms and processes or the function methods of AI" (De Sousa, 2013, p. 158).

Is consciousness exclusively human or will a device have the ability to become self-aware? Does it depend on the material which neurons are made of or will humans be able to use computer hardware to replicate it? For Descartes, human bodies are not conscious, our brains are not conscious, only the soul is conscious (Descartes, 2008). Hillary Putnam draws an analogy between humans (mental states) and the machines (logical states) by claiming that mental state is neither the state of the brain nor the nature of the behaviour but a functional state: "Whether, and under what conditions, a robot could be conscious is a question that cannot be discussed without at once impinging on the topics that have been treated under the headings Mind-Body Problem and Problem of Other Minds." (Putnam, 1964, p. 669).

Machine Consciousness (MC), sometimes denoted as artificial consciousness, is an attempt to display and perform particular aspects of human cognition. The goals of machine consciousness are: 1) to create the artefacts that have mental capabilities typically related to consciousness (awareness, self-awareness, emotion and affection, experience, imagination, etc.); and 2) application of this aspect of natural systems in embodied models such as robots (Press et al., 2005, pp. 11-12). According to Flannagan, consciousness is not necessary for a highly evolved intelligent life: "It is conceivable that evolutionary processes could have worked to build creatures as efficient and intelligent as we are, even more efficient and intelligent, without those creatures being subjects of experience." (Flannagan, 1995, p. 129). Apart from this, he assumes that "Whereas most sceptics of strong artificial intelligence press worries that machines cannot be given consciousness [...] Mind does not require consciousness." (Flanagan, 1991, p. 309).

John R. Searle claims that consciousness consists of feelings or awareness with the quality of ontological subjectivity, a unified field of conscience, intentionality and deliberate causality. He underlines the difference between conscious and unconscious robots by claiming: "as far as the ontology of consciousness is concerned, behavior is simply irrelevant. We could have identical behavior in two different systems, one of which is conscious and the other totally unconscious." (Searle, 1992, p. 71). Classification of machine consciousness results in categories of machines which aim to replicate aspects of human behaviour and create actual artificial consciousness: "Machines with the external behaviour associated with consciousness; [...] Machines with the cognitive characteristics associated with consciousness. [...] Machines with an architecture that is claimed to be a cause or correlate of human consciousness. [...] Phenomenally conscious machines." (Gamez, 2008, p. 888)

I argue that human consciousness necessarily involves the perception of the world and inevitable questions the purpose of life of man. Will the machine be able to communicate with the outside world and determine that it exists? Igor Aleksander proposes five axioms which could be taken as a fundamental list of material conditions for a neural computer model which would make a non-biological system conscious. The first one is presence, whereby the system has perceptive abilities to represent the elements of the world and their position. The second is imagination, which refers to autonomous states when the system can remember parts of a particular states or create sensations which resemble parts of the real world. The third is attention, which refers to the system which is capable of choosing a part of the world which it aims to describe or imagine. The fourth is planning, which denotes the system having control over individual sequences of the state and is able to plan its actions. The fifth is emotion, which refers to affective states of the system used to rate and evaluate planned actions and determine the following actions (Aleksander, 2005). In his book *Consciousness and the Computational Mind* (1987), cognitive scientist Ray Jackendoff asks an important question – "where in the flow of information processing does consciousness arise" (Prinz, 2014, p. 49). Jackendoff (1987) argues that our mental capacity is organized hierarchically into low, intermediate and high processing and that consciousness is produced at the intermediate level of processing (Prinz, 2014, p. 50).

The *Tokyo University Technology Institute* is developing a robot which uses artificial intelligence, can think, learn and act on its own. It involves a self-replicating neural network (SOINN) which grants the robot level of thinking similar to human when conducting actions it had never done before. In other words, the robot can make reasonable decisions based on previous experience, but these decisions are also influenced by the environment. The robot gradually becomes smarter and smarter as it uses its sensors to take in visual and auditory information from other sources, including the internet and experience of other robots (Bel-Mordok, 2015).

Thad Starner, the pioneer of *Wearable Computing Group* at MIT wants to: "teach a robot by having it learn from his life—by transmitting his life through sensors in his clothes. The sensors will allow 'the computer to see as I see, hear as I hear, and experience the world around me as I experience it.'" (Turkle, 2010, p. 142). Another example of the mutual connection of robots and humans is represented in the idea that robots could be activated by mobile phones: "When the brain in your phone marries the body of your robot, document preparation meets therapeutic massage." (Turkle, 2010, p. 142).

According to philosopher Hubert Dreyfus, "computers need bodies in order to be intelligent." (Turkle, 2010, p. 134). As a counter-argument for this, Turkle claims that machines will never be able to reach human levels of intelligence as no body given to machines will ever be human. While Norbert Wiener hope that someday it would be "conceptually possible for a human being to be sent over a telegraph", Marvin Minsky aimed to "create a computer beautiful enough that a soul would want to live in it" (Turkle, 2010, p. 85). According to Moravec, humans will have to insert their minds into machine bodies to retain their life form. By uploading our conscious minds into intelligent robots, we will be given infinite power and immortality. In addition, the belief that our minds are actually information patterns and not something integrally connected to the organic foundation (brain) promotes a certain way of thinking about human beings. The question of machine awareness is deeply intertwined with the modern discussions on human consciousness (Krishnan, 2009, p. 131).

Apart from the issues of consciousness, the question is whether robots can be constructed in line with the requirements of human morality? Would a robot programmed to behave ethically actually be ethical? The Three Laws of Robotics were introduced in the 1942 story *Runaround* in the book *I, Robot,* in which Isaac Asimov suggested three commandments which all robots must follow:

First law: A robot may not injure a human being, or, through inaction, allow a human being to come to harm. Second law: A robot must obey orders given it by human beings, except where such orders would conflict with the First Law. Third law: A robot must protect its own existence as long as such protection does not conflict with the First or Second Law. (Krishnan, 2009, p. 107)

Asimov's idea of ethical obligations for robots shows that a robot has to follow an in-built moral code while humans can choose whether or not to do so.

9.3 INDUSTRIAL ROBOTS

The *Robot Institute of America* (1979) defines an industrial robot as "A reprogrammable, multifunctional manipulator designed to move materials, parts, tools, or specialized devices through various programmed motions for the performance of a variety of tasks" (Moran, 2007, p. 1399).

Industrial robots (Fig. 1) are mainly used as a replacement for human labour in jobs which require endurance, working in difficult or specific time or production conditions, speed and high precision. They differ in terms of size, shape and configuration. An industrial robot can manipulate various tools and perform different tasks, depending on the program within it. The first industrial robot is considered to be the programmed assembly line robot invented by George Devol in 1954 in USA.

Figure 1. Industrial robots (Indramat, n.d.)

The development of robots and robotics took place in several stages:

1. The first-generation robots automatically repeat the assigned movements and are the most common robots in factories as their control system is easily adapted to manual operations, pressing and welding.
2. The second-generation robots have the ability of dealing with unforeseen circumstances in the working environment – they are equipped with sensors which feed them information, while their navigation and workplace procedures are pre-programmed. They are used in conveyor belts, assembly lines, painting etc.
3. The third-generation robots use various sensors and artificial intelligence, are equipped with computers and sophisticated programs, and are capable of analysing their efficiency from their mistakes. They change their way of working individually and intelligently to adapt to the working conditions and improve their efficiency (Regodić, 2011, p. 326).

The process of introducing robots into production has been constantly increasing for some time. In 2015, there were 1,631,600 robots in industrial use in the world. Reports from the *United Nations Conference on Trade and Development* have shown that automatization has a significant effect on developing countries and that increased use of robots in the developed countries threatens to disrupt the traditional economies in the developing countries (Ž. L. 2017). This means that low-skilled jobs are more prone to automatization as robots perform these tasks more easily. It is predicted that robots will replace around two thirds of all workers in the developing countries.

In 2017, the European Parliament opened the discussion on *European Civil Law Rules in Robotics.* The reason for this was that fact that "humankind stands on the threshold of an era when ever more sophisticated robots, bots, androids and other manifestations of artificial intelligence ("AI") seem to be poised to unleash a new industrial revolution, which is likely to leave no stratum of society untouched, it is vitally important for the legislature to consider its legal and ethical implications and effects, without stifling innovation" (European Parliament 2017). Between 2010 and 2014, the increase in sale of robots was 17, and it rose to 29 percent from 2014 onwards. Such an increase in use of robotics and artificial intelligence "may have the potential to transform lives and work practices" (European Parliament 2017). Use of smart and autonomous machines which can be trained and are capable of decision-making opens up the question of their direct and indirect effect on the entire society. Despite some clear advantages of robotics, "its implementation may entail a transformation of the labour market and a need to reflect on the future of education, employment, and social policies accordingly" (European Parliament 2017). At the same time, the development of robotics and artificial intelligence "may result in a large part of the work now done by humans being taken over by robots without fully replenishing the lost jobs" (European Parliament 2017). To preserve social cohesion, the EP proposes that robot labour be taxed with the aim of financing new jobs. Since modern robots are equipped with artificial intelligence which enables them the autonomy which can be defined as "the ability to take decisions and implement them in the outside world, independently of external control or influence" (European Parliament 2017), their behaviour should be regulated. A hypothetical question that arises here is what might happen one day when the capital is completely liberated from work and workers? Will that not be the end of both work and workers?

The overwhelming use of artificial intelligence in robots causes fear in humans and encourages us to produce ethical principles for their use. Human beings as the parents of their mechanical offspring are responsible for their behaviour. This is what makes the discussion on ethical principles of use of robots and the creation of robot ethics a crucial issue.

9.4 SOCIABLE ROBOTS

In the 18th century, Jacques de Vaucanson created a mechanical duck which could flap its wings, move its neck, digest food, and even defecate (Turkle, 2010, p. 53). Apart from the mechanical duck, he made several other human-sized robots which played musical instruments. In 1950, the British mathematician Alan Turing tried to create the conditions which would make a machine as intelligent as human. According to his test (called the Turing test in his honour), a machine could be called intelligent if the participant of the test could not tell apart the machine from human performance (Turing, 1950). Joseph Weizenbaum's ELIZA program from 1970s was capable of leaving the impression of empathy as it provided responses in the style of a psychotherapist: "To 'My mother is making me angry,' the program might respond, 'Tell me more about your mother,' or perhaps, 'Why do you feel so negatively about your mother?' (Turkle, 2010, p. 23).

Development of artificial intelligence and robotics create the conditions for human-to-robot interactions, creation of interactive, intelligent sociable robot helpers which can communicate with people and do the same tasks as them. Today, there are robot caregivers and companions on the market which serve as a replacement for human company. We are also witnessing robot teachers, stand-up comedians, dancers, musicians, street beggars, therapists, pets, caregivers, which is why the "Japanese researchers began to look to artificial companionship as a remedy for the indignities and isolation of age. And with similar logic, robots were imagined for the dependencies of childhood. Children and seniors: the most vulnerable first." (Turkle, 2010, p. 106). In this manner, sociable robots become embodied agents, a part of a heterogenous society of humans and robots. They can have various forms and functions, from robots whose only task is to take part in social interactions, e.g. Kismet (Fig. 2), COG, etc., to robots designed to uphold social norms to complete a series of tasks in the human surroundings, e.g. Pearl, Sage, etc. Some of the socially interactive robots use human models of interactions and social features: expressing and/or experiencing emotions, highly discursive communication, learning, recognizing models of other agents, establishing/maintaining social relations, using natural cues (facial expression, gestures, etc.), showing personality and character, ability to learn/develop social competencies (Fong et al., 2003, p. 4).

Figure 2. Sociable robot Kismet (Beciri, 2009)

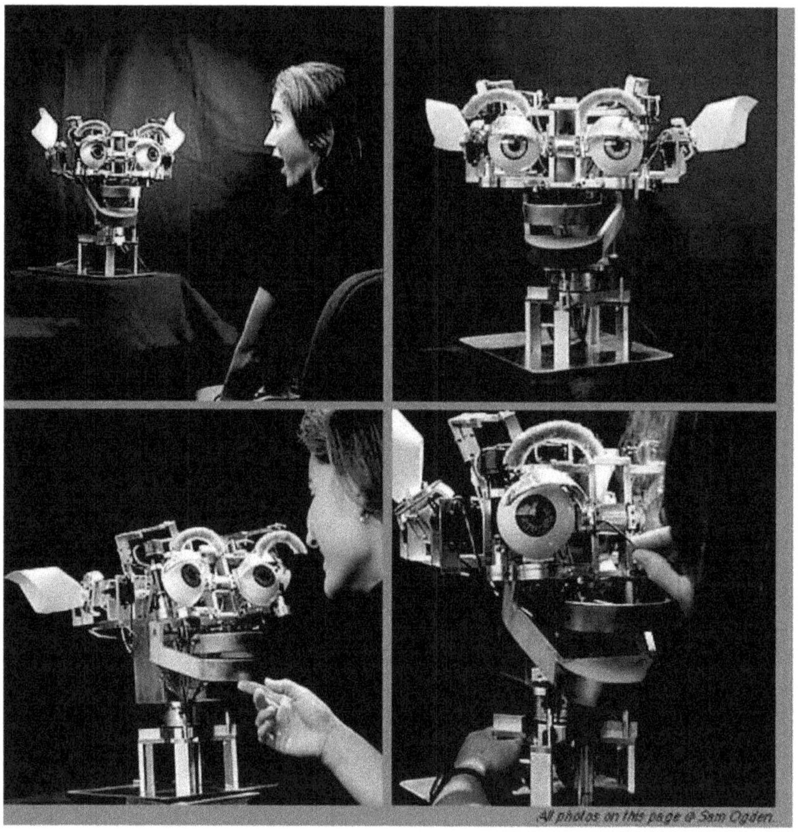

University of Science and Technology in China is testing a humanoid robot (*gynoid*) Jia Jia for human interaction. If we want to photograph Jia Jia, we get reprimanded: "Don't come too close to me when you are taking a picture. It will make my face look fat". Jia Jia has natural eye movements and her lip movement matches her utterances. Her makers are working on a facial recognition program that would make conversations with her even more convincing (Nožinić, 2016a).

9.4.1 Robot toys and pets

Interactive robot hamster *Zhu Zhu* was the most popular toy in 2009. Advertised as "better than any real pet could be" and "He lives to feel the love." (Turkle, 2010, p. 1; 2010, p. 24), it is a sociable artificial animal which does not require any clean-up.

Paro (Fig. 3) is a seal-shaped sociable robot used for therapeutic purposes with a positive influence on the sick and the elderly:

Paro can make eye contact by sensing the direction of a human voice, is sensitive to touch, and has a small working English vocabulary for "understanding" its users [...]. For example, it can sense whether it is being stroked gently or with aggression. (Turkle, 2010, p. 9)

Figure 3. Therapeutic robot Paro (Yang, 2015)

Popular Japanese egg-shaped toy *Tamagotchi* now allows its users to communicate and exchange messages, gifts and arrange meetings with one another (Fena, 2013). As Tamagotchis beeped when they were hungry, "In the late 1990s, an army of compliant mothers cleaned, fed, and amused their children's Tamagotchis" (Turkle, 2010, p. 32)

Turkle claims that children approach a sociable robot hoping to befriend it and when the robot requires nurturing, like the Tamagotchi, it seems alive enough to be loved: "[...] if a Tamagotchi makes you love it, and you feel it loves you in return, it is alive enough to be a creature. It is alive enough to share a bit of your life." (Turkle, 2010, p. 31). Similarly, the fact that a Tamagotchi must not be turned off in a way "becomes evidence of its life." (Turkle, 2010, p. 32). But what does the death of an artificial creature or robot even mean and is it even possible to feel the same type of loss when we grieve a loved one? This is why this matter necessarily involves "The philosophical churning associated with early computer toys (are they alive? do they know?)" (Turkle, 2010, p. 31).

In Turkle's view, "With the Tamagotchi, we see the beginning of mourning for artificial life." (Turkle, 2010, p. 34). What is even more profound is that when a Tamagotchi dies, it can be buried at an online graveyard which has tombstones which "try to capture what made each Tamagotchi special." (Turkle, 2010, p. 34).

Furby is an owl-like robot designed in 1998. It represents an attempt to create an emotional machine which shows signs of worry and sensitivity to pain: "A Furby held upside down says, 'Me scared,' and whimpers as though it means it." (Turkle, 2010, p. 30). If you wanted to switch off a Furby, you would have to take out its batteries, after which he would lose "all memory of its life and experiences—what it has learned and how it has been treated." (Turkle, 2010, p. 35). What made Furby different from traditional toys, according to Turkle, was mutuality – a whole series of instructions (linguistic training, food, rest, affection) that sociable robots require from children and which lead to the creation of an active robot-children relationship.

Freedom Baird believes that increased use of sociable robots will introduce new ethical requirements and wonders which conditions make a machine truly alive to cause an ethical dilemma in us. She asked her participants to "invert three creatures: a Barbie doll, a Furby, and a biological gerbil. Baird's question is simple: 'How long can you hold the object upside down before your emotions make you turn it back?'" (Turkle, 2010, p. 45). Neuroscientist Antonio Damasio describes two levels of experiencing pain – a physical response to a painful stimulus and an emotion associated with pain: "This is an internal representation of the physical. When the Furby says, 'Me scared,' it signals that it has crossed the line between a physical response and an emotion, the internal representation." (Turkle, 2010, p. 45). An eight-year-old girl Paree believes that humans use both of these methods:

Sometimes people have spontaneous feelings and 'just become aware' of them (this is 'knowing your own feelings'). But other times, people have to program themselves to have the feelings they want. [...] The robot, she thinks, probably has the second kind of feelings, but she points out that both ways of getting to a feeling get you to the same place: a smile or a frown if you are a person, a happy or sad sound if you are an AIBO. Different inner states lead to the same outward states, and so inner states cease to matter. (Turkle, 2010, p. 54)

My Real Baby is a robot doll with in-built mood expression which reacts to pleasure and pain with change of facial expression: "If the baby were upset, it would stay upset until someone soothed it or it finally fell asleep after minutes of heartrending crying and fussing. If BIT… was abused in any way—for instance, by being swung upside down—it got very upset. […] It acted a lot like a real baby." (Turkle, 2010, p. 46). This robot caused conflicting reactions – on the one hand, people believed this was a perfect toy to

[...] teach and encourage reciprocal social behavior as children were trained to respond to its needs for amusement as well as bottles, sleep, and diaper changes. Indeed, it was marketed as realistic in all things—except that if you 'hurt' it, it shut down. When children play with My Real Baby, they do explore aggressive possibilities. They spank it. It shuts down. They shake it, turn it upside down, and box its ears. It shuts down. (Turkle, 2010, p. 47)

Computer expert who invented the *AIBO* robot John Lester believes that robots will change the course of human evolution as humans will emotionally bond with them: "We will evolve to love our tools; our tools will evolve to be loveable." (Turkle, 2010, p. 141). He believes that we merger of tissue and the machine will take place sometime in the future:

We will see "through its eyes," says Lester, and interact "through its body... There could be some parts of it that are part of you, the blending of the tools and the body in a permanent physical way." This is how Brooks talks about the merging of flesh and machine. There will be no robotic "them" and human "us." We will either merge with robotic creatures, or in a long first step, we will become so close to them that we will integrate their powers into our sense of self. In this first step, a robot will still be another, but one that completes you. (Turkle, 2010, p. 141)

AIBO is a dog-shaped robot which can recognize its owner and return to the charger on its own. Turkle believes this dog marks the beginning of digital pets of the future. But, can the creators of artificial intelligence also create a human form of friendship: "The first thing missing if you take a robot as a companion is alterity, the ability to see the world through the eyes of another.5 Without alterity, there can be no empathy." (Turkle, 2010, p. 55). Technology will always fail use as it keeps making promises it cannot deliver: "It promises friendship but can only deliver performances. Do we really want to be in the business of manufacturing friends that will never be friends?" (Turkle, 2010, p. 101).

Can robot pets teach children commitment and responsibility towards pets or other human beings. AIBO gives us the possibility to avoid certain death as "I think you could fix [AIBO] with batteries . . . but when your dog actually dies, you can't fix it. [...] AIBO permits something different: attachment without responsibility." (Turkle, 2010, p. 60). The existence of robot pets raises an inevitable ethical dilemma – is it allowed to harm another being and even kill it? If a robot is not biologically alive, how could we even hurt it, and how could it hurt us: "If robots don't feel pain, how could they comfort you?" (Turkle, 2010, p. 69). In terms of robot empathy, Turkle recalls her conversation with the girl called Bridget: "'So, if the robot showed that it felt pain, would that make a difference?' Without hesitation she answers, 'Oh yes, but these robots shut down if they are hurt.'" (Turkle, 2010, p. 72).

Apart from robot toys and pets, there are also robot toddlers whose makers wanted to create mechanical two-year-olds with the aim of simulating the emotions of two-year-old children. *Cog* and *Kismet* are robot two-year-olds who learn through their interaction with humans: "Kismet has a range of 'affective' states and knows how to take its turn in conversation. It can repeat a requested word, most often to say its own name or to learn the name of the person talking to it." (Turkle, 2010, p. 84). By creating the mechanical other equal to man, we have reached the situation in which the increasingly sophisticated robots become "more refined in their ability to target us" (Turkle, 2010, p. 100). This new machine nurture-encouraging relationship encourages man for self-maintaining participation. The question of replacing people opens up the dilemma: "But if our experience with relational artifacts is based on a fundamentally deceitful exchange (they perform in a way that persuades us to settle for the 'acting out' of caring), can they be good for us?" (Turkle, 2010, p. 107).

In 2015, a small *BB-8* droid was presented at the Berlin IFA fair of consumer electronics and home appliances. This toy can be controlled via mobile device and it also boasts voice-recognition and leaning capabilities. If it encounters an obstacle, it will go around it and 'memorize' it for next time. Its route can be simulated in advance (Lijović, 2015).

9.4.2 Robot Home Helpers

In terms of robot socialization, there are good indications that robots will become family members in a nearby future as their capabilities will allow them to perform numerous household chores. An improved *ASIMO* robot from 1996 performs most household chores, follows voice commands, recognizes over ten faces and their emotions and communicates with other robots via wireless connection. HONDA patented a helmet which allows its host to control the robot's thoughts regardless of its location (Radović, 2010).

RAVIO is a robot nanny for children. Using an in-built camera, speaker and wireless internet connection, it allows the parents to have audio and video control over the events in their home. The sensors respond to sound, change of temperature and movement inside the house and warn the parents about the potential dangers (Radović, 2010). *RoBoHoN* is a robot-shaped smartphone from Japan (Fig. 4). It can receive calls, send e-mails, start different applications, walk, talk, dance, and even recognize human cries. Using an in-built projector, it can display photographs and movies and play music. It has an internal facial recognition software (Nožinić, 2016b).

Facebook owner Mark Zuckerberg wants to make a personal butler for office and home help, i.e. an AI-endowed robot which will recognize and understand his voice and perform his commands. The robot would inform him about what is going on in his house when Mark is away or at the office. In the work environment, the robot would help him visualize the data in virtual reality to make him even more efficient in his work (Arežina, 2016).

Figure 4. RoBoHoN (Pulso Urbano, n.d.)

9.4.3 Robot Doctors and Caregivers

Robot caregivers take care of the elderly and the disabled by grabbing groceries from the shelves for them and helping the person get up if they fall. The provide help to those who are unable to function in their everyday life. A two-handed model called *PR2* shaves paralyzed persons (I.S. 2015).

A robot nurse *Nursebot* helps the retirement home users by reminding them to take their therapy, have regular meals and by bringing them their medicine (Turkle, 2010, p. 121). But Turkle wonders whether we have:

come to think of the elderly as nonpersons who do not require the care of persons? [...] Philosophers say that our capacity to put ourselves in the place of the other is essential to being human. Perhaps when people lose this ability, robots seem appropriate company because they share this incapacity. (Turkle, 2010, p. 108)

How do we categorize the division of care, i.e. which persons require most human care and which ones require the least? Perhaps the dementia sufferers require the most human affection: "And if we assign machine companionship to Alzheimer's patients, who is next on the list?" (Turkle, 2010, p. 108). However, since they are a physically and mentally endangered category, "Will only the wealthy and 'well adjusted' be granted the company of their own kind?" (Turkle, 2010, p. 108).

Robot *Rio* (Fig. 5) was constructed at the MIT in 2006 and has been irreplaceable in knee surgery ever since. The robot performs autonomous surgeries with the help of a CT device which allows it to be fast and precise during surgery. The post-op care is covered by a *gynoid* – nurse *Guia Bot*, equipped with laser, camera and a screen which shows the most recently updated patient data. From facial expression alone, the gynoid can deduce whether the patient needs comfort or any other help (Radović, 2010). Scientists at the *George Washington University* have constructed a robot helper for autistic children to facilitate their inclusion in social interactions. The robot stimulates the autistic toddlers to play with other children and form new friendships (Božičević, 2017).

Figure 5. Robot Rio (Medgadget editors, 2009)

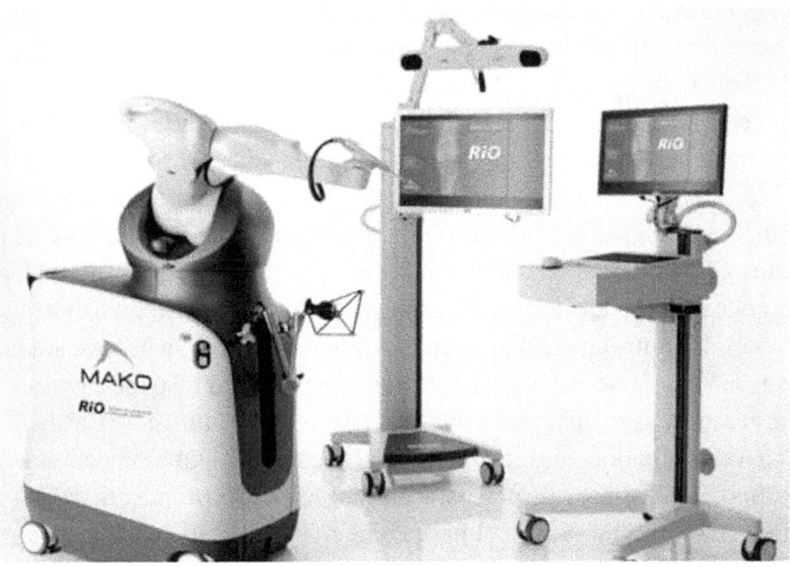

In China, a robot dentist performed the first operation without any human participation. Using a 3D printer, the front two teeth were made and the robot implanted them into the patient's jaw. Teeth implantation was done with minimal deviation of 0.2-0.3 mm, which is in line with the global standards for these types of surgeries (Luc, 2017).

It is human to ask whether we want to accept the partnership with robots, whether we wish to be in their vicinity and accept their services. Robot designer Aaron Edsinger believes "this will follow naturally from the pleasure of physical contact with robotic partners. He says it is thrilling 'just to experience something acting with some volition. There is an object, it is aware of my presence, it recognizes me, it wants to interact with me.'" (Turkle, 2010, p. 132). Furthermore, Edsinger believes the robots of the future will

[...] be cute, want to hug, and want to help. They will work alongside people, aware of their presence and wishes. Edsinger admits that it will be 'deceiving, if people feel the robots know more than they do or care more than they do.' But he does not see a moral issue. First, information about the robot's limitations is public, out there for all the world to see. Second, we have already decided that it is acceptable to be comforted by creatures that may not really care for us: 'We gain comfort from animals and pets, many of which have very limited understanding of us.' Why should we not embrace new relationships (with robots) with new limitations? (Turkle, 2010, p. 133)

Lindman's project, according to Turkle, deals with desire for unity with the machine and testing this union. Lindman wanted to be merged with a robot to gain direct insight into its internal states and experience its emotions. The idea was to record her brain activity during merger so that images could be compared: "I will be the embodiment of the AI and we will see if [when the robot smiles], my brain is smiling." (Turkle, 2010, p. 138). However, since brain could not be transformed into an output for robot intelligence, Lindman changed her plan and linked the robot's computer to her face to "become the tool for the expression of the artificial intelligence" (Turkle, 2010, p. 138). Using this experiment, Lindman wanted to feel what it is like to be a robot and see whether humans can really identify with machines if they project their own emotions (Turkle, 2010, p. 138).

Ray Kurzweil created the virtual character Ramona and "wore high-tech gear that captured his every gesture and turned them into Ramona's movements. His own voice was transformed into Ramona's female voice" (Turkle, 2010, p. 211). With this invention, Kurzweil wanted to "reflect on the difficulties of inhabiting another body and on how he had to retrain his movements—the way he held his head, the shape of his gestures—to become an avatar of another gender." (Turkle, 2010, p. 211). Psychologist Clifford Nass and his colleagues conducted a series of experiments in which human beings exhibit various social behaviour towards computers and project affection onto them: "when we are confronted with an entity that [behaves in humanlike ways, such as using language and responding based on prior inputs,] our brains' default response is to unconsciously treat the entity as human." (Turkle, 2010, p. 140).

Can machines develop emotions and are they crucial to establish communication with human beings? Contemporary artificial intelligence researchers, according to Turkle, promise that they can create machine emotions, as claimed by Rosalind Picard: "I have come to the conclusion that if we want computers to be genuinely intelligent, to adapt to us, and to interact naturally with us, then they will need the ability to recognize and express emotions, and to have what has come to be called 'emotional intelligence'" (Turkle, 2010, p. 140). According to Marvin Minsky, "Emotion is not especially different from the processes that we call thinking." (Turkle, 2010, p. 140). Unlike Minsky, who believes that "that robots are going to be emotional thinking machines", Damasio argues that this can "never be unless robots acquire bodies with the same characteristics and problems of living bodies" (Turkle, 2010, p. 140). However, Turkle warns that "humans 'as a species' have to learn to deal with 'synthetic emotions,' a way to describe the performances of emotion that come from objects we have made." (Turkle, 2010, p. 123).

9.4.4 Robot Sexual Partners

The idea that robots will play a major role in sex life of mankind in the future has become quite common. In 2016, *Daily Mail* published a story about the French woman called Lilly who printed her lover. She created her robot *InMoovator* using a 3D printer. Lilly fell in love with him and intends to marry it once legal conditions are met. The website *Future of Sex* contains the musings of Ross Dawson who discusses sexual relations between human beings and robots. His ideas are grouped in fiver areas: remote sex, virtual sex, robots, entertainment for all senses, and augmented reality. He claims that:

Adult performers and some B-list celebrities will sell robotic replicas of themselves designed for sex by 2033; one in 10 young adults will have had sex with a humanoid robot by 2045; by 2028 over a quarter of young people will have had a long-distance sexual experience; 3D-printed body parts of your lover, enabled with touch feedback, will add intense realism to long-distance sex by 2025; first dates in motion-capture virtual worlds will become popular by 2022; by 2024 people will be able to be anybody, with anybody, enacting impossible fantasies in photo-realistic virtual worlds; brain-to-brain interfaces will allow partners to stimulate each other to reach orgasm directly by 2027; sci-Fi sex fantasies will spring to life as people can enhance their biology and merge with machines to become superhuman sex idols; by 2020 people will regularly pair virtual reality and haptic sex toys to fully immersive themselves into adult entertainment (Owsianik & Dawson 2017).

Scientists have also announced robot sexual partners which will be constructed according to one's desires and personal preferences. Virtual gynoid *Roxxxy* (Fig. 6) with human-like skin was presented at the Las Vegas Expo in 2010. It reacts to touch but cannot move on its own. After you touch her hand, she says through the speaker: "I love holding your hand" (Borovac, 2010). Professors of robotics warn that sexual relations with androids could have disastrous consequences for society, especially its younger members for whom that might be the first relationship. This would lead to disturbed relations between men and women, adults and children, men and men, women and women. The whole system of values will shift.

Ian Pearson predicts that the role of men in bedroom will be taken over by robots and that human-robot sex will become common place. Once the visual and sensory aspects of robots are improved, he argues, the way humans think about this matter will change as well. In time, human beings and robots will become friends with strong emotional ties (E. Do, 2016). A female sex doll, called *Doll*, is remarkably similar to a living woman. The designer of this doll announced that the future models will be equipped with artificial intelligence so that the doll could arouse its owner both physically and mentally (M. Ra, 2016a). Of course, a male sex doll also exists. It is made in real size, it is muscular, has body hair and a beard. After women touch it, it gets an erection and it is ready for sex (M. Ra, 2016b).

In Barcelona, there is already a *LumiDolls* brothel specialized in providing sexual services with dolls. One of the brothel owners emphasizes that, unlike women, doll sex carries no risk of physical or emotional abuse and the risk of STDs is minimal (J. MI 2017a). Robot doll *Samantha* has already become a part of her creator's family. It is not just a sex doll but also a companion, as you can talk to it about life, philosophy, science, and she can even tell you a few hundred jokes. From time to time, she gets involved in family gatherings. She is then set to 'family option' which makes sure that she does not say anything inappropriate in the presence of children. Her designer believes that the gynoid *Samantha* will never be able to take the role of his wife (J. Mi, 2017b).

Figure 6. Robot Roxxy (Wagner, 2015)

Certainly, there are emotions involved between two sexual partners (lovers) and these are very much a part of the sex game. Should one assume that robots as human sexual partners should also have emotions and consciousness? If not, can we even call it an act of love? Turkle wonders whether we are ready to look at ourselves in "the mirror of the machine and to see love as our performances of love" (Turkle, 2010, p. 138)? According to Damasio, "all thinking and all emotion is embodied. The absence of emotion reduces the scope of rationality because we literally think with our feelings" (Turkle, 2010, p. 134). In his book *Descartes' Error*, he claims "there is no mind/body dualism, no split between thought and feeling. When we have to make a decision, brain processes that are shaped by our body guide our reasoning by remembering our pleasures and pains." (Turkle, 2010, p. 134). Contrary to this, research conducted by Pia Lindman indicates that observing the emotions of others connects us with them: "The shape of a smile or frown releases chemicals that affect mental state. And in humans, 'mirror neurons' fire both when we observe others acting and when we act ourselves." (Turkle, 2010, p. 135).

Lindman concludes that sadness is always expressed through a sequence of established patterns programmed by biology and culture and believes that both robots and humans have programs which determine expression of emotions, which ultimately erases the boundaries between the two groups of entities: "And if our emotions are mediated by such programming, asks Lindman, how different are our emotions from those of a machine?" (Turkle, 2010, pp. 137-138). Furthermore, the aim of creating humanoid robots is a multimodal humanoid robot which will replicate conscious human behaviour. According to Aleksander (1992), since consciousness is a specific trait of well-specified machines, it is no wonder it was the area of interest for cyberneticians in the 1960s (cf. Nemes, 1969) and neural network scientists in the early 1990s, such as Taylor (1992, pp. 11-15) and (Aleksander, 1994, pp. 17-22, 2005, p. 4162).

Chinese engineer Zheng JiaJia created a female robot, spent two months perfecting her face, and finally married her when he was happy with the end-product. The traditional-style wedding was held in Hangzhou in the presence of groom's mother and several friends (M. Bu, 2017).

9.4.5 Robots in Service Industries

A restaurant in Jinan, China employs two robot receptionists and six robot waiters which serve food and drinks (Radović, 2010). A café was open in San Francisco in which robots serve coffee which can be pre-ordered via special mobile phone app (Kauffman, 2020). In Ilmenau, Germany, a robot waiter serves hard liquor to guests in a night bar. It usually serves hard liquor and puts it into cocktail shakers. It can also hold a short and limited conversation as its speech recognition and intelligence capabilities are somewhat limited (Hruškovec, 2013a).

The company Moley Robotics is testing a sophisticated robot chef (Fig. 7) which will be capable of making two thousand dishes. The whole cooking process will be automatized. Its mechanical hands contain 20 motors, 24 joints, and 129 sensors to be able to simulate the movement and actions of professional human chefs. The owner of the robot chef will be able to send the recipe or make on order via mobile phone app and specify the time at which he wants his or her lunch to be ready (Ćuto, 2015a).

Figure 7. Robot chef (McGrath, 2015)

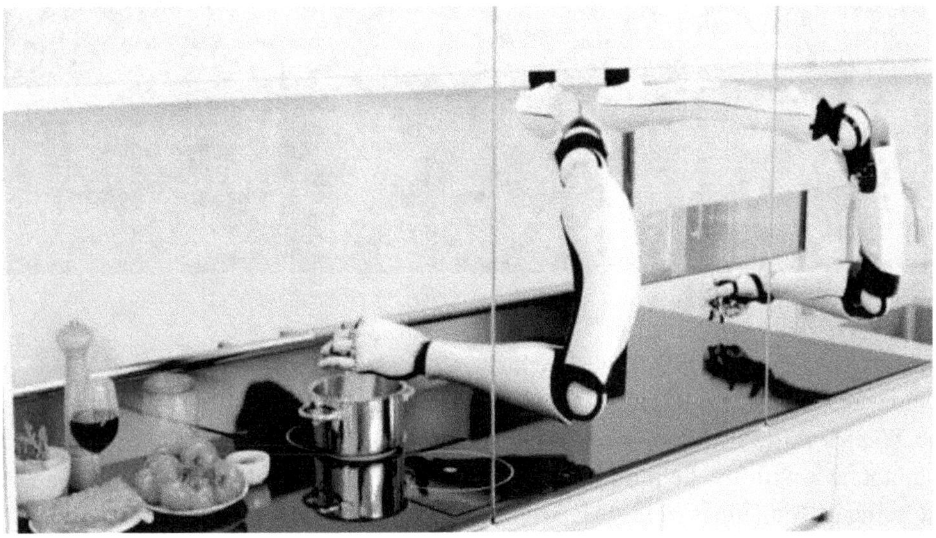

The customers of some Japanese banks are greeted at the entrance by a robot called *Pepper* which greets the visitors and asks whether it can help them. It directs the visitors to the teller they need (Blotnej, 2015). Scientists at the University of Örebro, Sweden have designed a robot which helps passengers find their terminals more easily (Matković, 2016). In 2012, Hitachi presented the improved version of their *EMIEW* robot. It is 80 cm tall, and its wheels allow it to move at human walking speed. It serves as an office guide and it finds the shortest route to the target with the help of its software. Its younger brother *EMIEW3*, produced in 2016 with various communicative functions, is used as customer support in shops (Ackerman, 2016). Robot *ROBOY* was presented at the world congress and expo of humanoid robots and cyborgs in Zürich. It is 120 centimetres tall and weighs 40 kilos. Instead of motorized joints like most robots, *ROBOY* has artificial tendons which move its arms and legs, which makes its movements human-like. It is designed for working in the service industry (Hruškovec, 2013b).

9.4.6 Robot Entertainers and Media Workers

At the CeBIT fair in Hannover in 2012, the *Karlsruhe Institute of Technology* presented a robot actor which sings and imitates Star Wars characters (Hruškovec, 2013b).

Figure 8. Robot dancers (Crossley, 2014)

Pole-dancing robots – gynoids (Fig. 8) which move along with the rhythm of music played by a robot DJ were presented at the same trade fair in 2012 (Hruškovec, 2012a). In early 2016 in China, a robot name *Xiaoice* started working as a TV host at the Shanghai Dragon station. The robot works as a weather announcer with the help of *Microsoft Smart Cloud* technology which allows it to read text in the most human-like manner so far (Matković, 2016). A robot journalist wrote its first article for the Los Angeles Times. The article covered the earthquake that hit the area in 2015 and the robot wrote the report in three minutes. The creator of this robot is a journalist who wrote the code for automatic generation of the article. The algorithm only deals with earthquakes, but there is no doubt that the robot journalist will expand its interest to other topics in the future. Another potential future development is drone journalism, which will allow direct video reports from war-torn areas or places affected by natural disasters (Bel-Mordok, 2015). In 2012, scientist Martina Richer from the *Fraunhofer Institute* presented her robot artist at the Hannover expo. The robot recorded the face of the model in less than a minute and drew the person's portrait in about fifteen minutes (Hruškovec, 2012a).

A robot dance instructor was designed in Japan. It moves with the help of wheels, which allow it to move graciously across the dance floor, and its face functions an instruction screen for its partner. The robot is equipped with lasers and sensors which allow it to keep track of its partner's movement. The information arrive to the screen – its head, and the robot uses the information to monitor everything its human partner is doing (Tehnoklik, 2017).

9.4.7 Robot Teachers and Students

Japanese *National Institute of Informatics* and the company *Fujitsu Laboratories* are designing a robot which should be smart enough to get excellent grades on the *Tokyo University* entrance exam in 2021. The aim of this project is to develop technology which will enable simple use of sophisticated mathematical analysis tools to every user. In the learning process, the robot must analyse words mixed with mathematical terms, recognize mathematical formulas and their mutual relationships, and find the right answers on its own. The question being raised here is whether robots might not really find themselves one day in university entrance exams and start attending them (Hruškovec, 2012b).

Harvard University is testing a robot called *Root* whose aim is to teach children programming skills in a simple, fun and intuitive way (Ćuto, 2016). Gynoid *Saya* teaches classes 1- 4 (Fig. 9), assigns homework, calls the children out to the blackboard and maintains class discipline with words and facial expressions (Radović, 2010).

Figure 9. Robot teacher Saya (Demetriou, 2009).

The *Robo Earth* project was started in Europe as a kind of robot Wikipedia which will store the data the robots discover about the world. This Wikipedia will help robots exchange their experiences and store their common knowledge. Robots will be able to learn about the objects which make up the human world and their mutual relations. The database will contain locations of their workplaces and descriptions of nearby objects, as well as instructions on how to perform various activities (Majdak, 2011).

9.4.8 Robots in Architecture and Construction

University of Colorado reports having created robots the size of a table tennis ball which, once they start working in a group, turn into a 'thinking liquid. The main researcher of the project believes that mankind should be able to colonize the universe with the help of these synchronized robots. They could travel to a location before astronauts, build shelter and adapt the environment. Scientists argue there is no theoretical limit as to what could be achieved with these systems of shared information (Hruškovec, 2012c).

In 2015, Swiss scientists designed a robot builder. The robot has a large mechanical arm on a mobile mount and can be programmed to create complex constructions. There are also computer programs which could be used to define the desired furniture and blueprints for an object. Scientists believe that they will soon create software which will be able to create very complex projects with ease (Matković, 2016).

9.4.9 Robot athletes and sports fans

A baseball team from South Carolina has put robot fans onto its stands. Robots can cheer, chant, and even do the wave like human fans. Their faces can be replaced with photographs of actual fans to give 'personalized' team support (Tolić, 2014). *RoboCup German Open* (Fig. 10) was held in Magdeburg in 2015 and involved 200 teams with about 1000 robot footballers (Wakefield, 2015).

Figure 10. RoboCup German Open (24 junior, 2015)

The humanoid robot *Robovie PC* won the first marathon for two-legged robots in Osaka, Japan. It completed the entire track of 42.195 km in 54 hours, 57 minutes and 50.26 seconds. Robot Olympics are also held in various disciplines.

9.4.10 Robot Drivers and Traffic Policemen

Since 2014, two robot policemen (Fig. 11) regulate the traffic in city crossroads of the Congolese capital Kinshasa. Despite the rough aesthetics of these law enforcement agents, the citizens of Kinshasa are quite pleased with their work. In addition to regulating traffic, the robots also make records of traffic violations (AP, 2015).

Figure 11. Robot policeman (Sokol, 2015).

Ray Kurzweil predicted that an autonomous driverless automobile will be tested by 2017 and this has indeed been accomplished. Kurzweil thinks this will bring more traffic security and less traffic jams as these cars will have sensors, visual lasers, and capability of mutual communication to avoid traffic jams. Kurzweil was also convinced that Google's driverless cars would be on our roads by year 2020. In addition to security, the discussion on these cars is focused on liability for material damage in accidents caused by a computer (Ćuto, 2015b). For this and several other reasons, the European Parliament discussed the "*European Civil Law Rules in Robotics*" in 2017.

Scientists at the *Nankai University* have presented a thought-controlled car. Using brain wave-reading equipment, a driver can tell the car to stop, go forward or backwards, stop and lock/unlock the doors using his or her mind. Driver's equipment consists of 16 sensors which record the EEG brain signals. A computer program converts these signals into commands for controlling the car (Šikić, 2015).

9.4.11 Robot Soldiers

US military robot *BEAR* is used to evacuate soldiers from the battlefield and carry military equipment. It is autonomous and capable of overcoming different obstacles. Another robot called *Big Dog* (Fig. 12) can carry up to 150 kg of weapons and ammo across the enemy territory. It is GPS-controlled (Radović, 2010).

In 2015, China presented three robots that will be used in fight against terrorism and to maintain public security. One robot is used for surveillance – it discovers potential dangers and notifies then headquarters about what it had seen. After this robot, the second one steps into the scene to neutralize explosive and chemical devices. This robot is followed by the third one, armed with a low-calibre rifle and a grenade launcher which moves to attack (Jurman, 2015a). Security experts warn us about the dangers of using robots in military actions, especially 'killer robots' which should make autonomous

Figure 12. Robot Big Dog (Lawler, 2013)

decisions about the targets to neutralize. The question is what happens if they turn on their fellow soldiers or humans in general.

9.4.12 Flying Robots – Drones

A drone is a pilotless aircraft that can be programmed to perform various functions. The United Arab Emirates are planning to organize drone-assisted postal service. These pilotless aircrafts are typically about 50 cm long and have four battery-powered propellers. High temperature and sand particles caused by sandstorms present a huge challenge for the designers. With this project, the Emirates are trying to keep up the pace with modern technological accomplishments (Blotnej, 2014).

Figure 13. Aeryon SkyRanger (PressRelease, 2013)

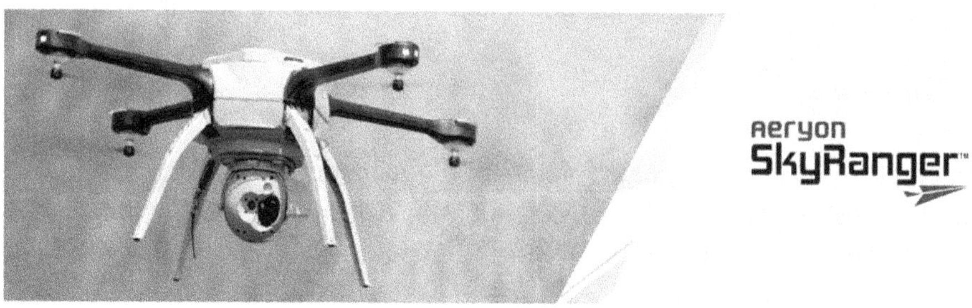

In New Zealand, testing is underway for a ground pizza delivery robot drone called *Domino's Robotic Unit* (DRU). It moves around on wheels and is equipped with sophisticated sensors for avoiding obstacles and an oven to keep the pizza warm up until the moment of delivery. The drone is protected with a PIN which the customer gets and enters into a keyboard when taking over the delivery (Nožinić, 2016c).

University of Florida hosted a race of 16 drones controlled via human brain waves. Each human participant had an EEG scanner on their head and they controlled the drone using their thoughts (Nožinić, 2016d). A restaurant in Singapore is planning to use drones for serving food. In the testing phase, the drones would deliver the food from the kitchen to the waiter, who would serve it to guests. The drones will be completely autonomous and will use the technology to avoid static objects and other pilotless aircraft (Jurman 2015b). UK police are using drones to fight crime. They use the *Aeryon SkyRanger* drones (Fig. 13) which have been patrolling the skies above the Gatwick airport since 2014. The drone can spend fifty minutes in the air, after which it needs to have its batteries recharged (Nožinić, 2016e).

9.4.13 Nanorobots

Nanobots are predicted to be in wide use in medicine for preventing and destroying blood clots, malicious cancer cells, cleaning blood vessels from accumulated fat, removing parasites, bacteria and viruses, destroying kidney stones, delivering drugs across the body, and diagnosing diseases. The ultimate use of nanobots in medicine would be disease identification and prevention, which would lead to longer life expectancy and immortality in the long-term perspective.

In this section, we provided a rather general and non-exhaustive portrayal of robots in the reality of *scientific humanism as naturalism* to show that the projection of future of cyborgized and robotized reality is in fact our everyday life to which we adapt in a friendly and loving manner. This is also exemplified by the humanoid robot *Sophia* who got her citizenship in 2017. She was created by a Hong Kong company *Hanson Robotics* with the use of state-of-the-art technology. Sophie has no pre-programmed responses but uses machine learning. It responds to its interlocutors based on their facial expression. Her 'brain' has to be connected to the internet to be able to function. In the second stage of her development, she got mechanical legs which allow her to move at the speed of one kilometre per hour. A gynoid obtaining citizenship in Saudi Arabia opened up a series of political questions. According to the Saudi law, only a Muslim can obtain Saudi citizen ship, which raises the question whether she converted to Islam and why she is not wearing a hijab (Jutarnji list, 2017).

All of this shows that mankind is truly looking for its descendant which would relieve it of its everyday labour – a robot to work for it, and a virtual descendant which will carelessly wander the virtual expanses and establish relations with virtual friends. Both of these entities provide us with artificial closeness without commitments (Turkle, 2010, p. 15). According to Turkle, "As sociable robots propose themselves as substitutes for people, new networked devices offer us machine-mediated relationships with each other, another kind of substitution" (Turkle, 2010, p. 3) and "as a symptom, they promise a way to sidestep conflicts about intimacy; as a dream, they express a wish for relationships with limits, a way to be both together and alone" (Turkle, 2010, p. 11). In Japan, sociable robots are advertised as "a way to seduce people out of cyberspace" (Turkle, 2010, p. 11). But if sociable robots "are designed to be with us [...] we must ask, why do people no longer suffice?" (Turkle, 2010, p. 19). Ultimately, we might argue that the robots' name (derived from Czech *robota* 'forced labour') shows their function – they are workers, work is their essence. They are machines in the real space and time run by a computer-generated artificial intelligence. It contains the entire accumulated knowledge of the society and represents the victory of *scientific and technological work* over the living labour of man-worker in the reality of man. A robot is what in its essence a cyborg aims to become – a scientific worker, a posthuman thing-being whose existence in the cyborgized reality of *scientific humanism as naturalism* confirms the predomination of reality of man. As a uniformed type of non-human existence in the form of a machine empowered by *scientific work*, a robot uses man as an object of communication. The means becomes the subject and the subject becomes the means of communication. A robot is "the equivalent of man, appropriating him to itself as an equal in the unity of a functional process." (Baudrillard, 1993, p. 112).

We should not be fooled by the figurative resemblance between robots and men. Since the procedures of *scientific work* cyborgize the reality of *scientific humanism as naturalism*, turning it into the illusion of human reality – "The robot no longer questions appearances, its only truth is its mechanical efficiency. It no longer needs to resemble man, to whom it is inevitably compared. The infamous metaphysical difference which gives the automaton mystery and charm no longer exists: the robot emphasises this difference for its own benefit. Being and appearance are founded on a single substance of production and labour." (Baudrillard, 1993, p. 113). Through absorption of reality, the robot as the representative of *scientific work* establishes its reality: "without echo, without mirrors, without appearances [...]. No more semblance or dissemblance, no more God or Man, only an immanent logic of the principle of operativity." (Baudrillard, 1993, p. 113).

This is the reality of *scientific humanism as naturalism*, the reality of friendship of man as a biologically imperfect and outdated entity and robot as its non-human descendant which is starting to rule where man ends – at the end of human history. This is the image of reality advocated by the transhumanists, futurologists and numerous cyberneticists who ascribe instrumental or neutral traits to science and technology. Ultimately, this will be the cybernetic reality of *scientific humanism as naturalism* if it, in the name of man and life, establishes the cybernetic balance in the entropy of the Universe.

Figure 14. Cyborgs, robots, avatars, potential implantations[1]

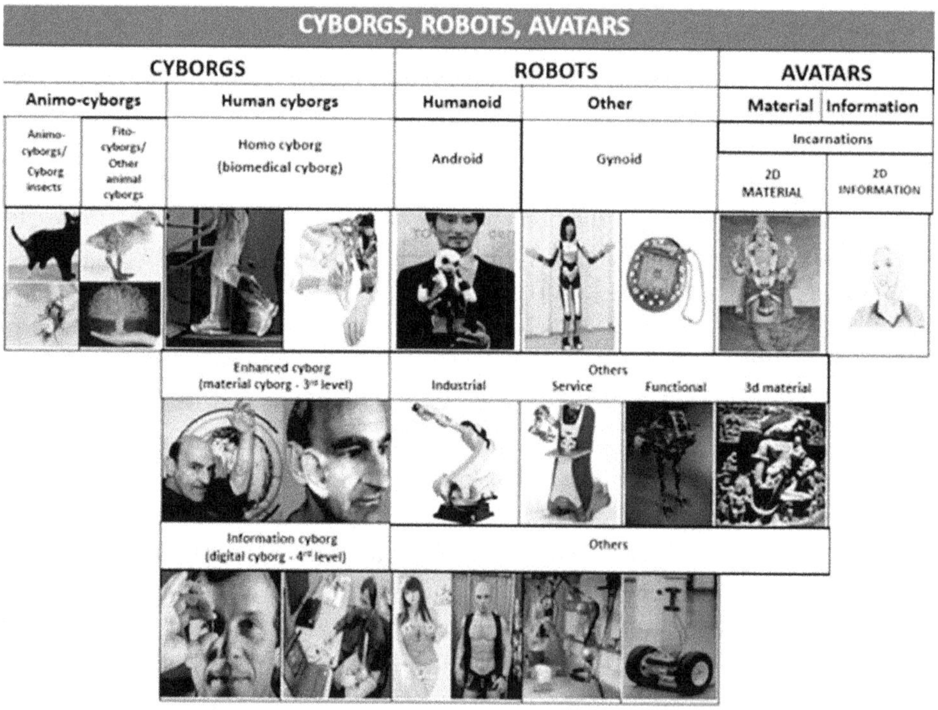

REFERENCES

Ackerman, E. (2016). Hitachi Upgrades EMIEW Robot to Help Clueless American Tourists. *IEEE*. https://spectrum.ieee.org/hitachi-upgrades-emiew-robot-to-help-clueless-american-tourists

Aleksander, I. (1994). The Consciousness of a Neural State Machine. In M. Marinaro & P. G. Morasso (Eds.), ICANN '94 (pp. 212–217). Springer London. https://doi.org/10.1007/978-1-4471-2097-1_49.

Aleksander, I. (2005). *The world in my mind, my mind in the world: Key mechanisms of consciousness in people, animals, and machines*. Imprint Academic.

Arežina, B. (2016). Zuckerberg: Želim stvoriti umjetno inteligentnog batlera poput jarvisa iz Iron Mana. *Večernji.hr*. https://www.vecernji.hr/techsci/mark-zuckerberg-zelim-stvoriti-umjetno-inteligentnog-batlera-poput-jarvisa-iz-iron-mana-1049376/galerija-172330?page=1

Baudrillard, J. (1993). *Symbolic exchange and death*. Sage Publications.

Bel-Mordok. (2015). Transhumanizam – Robotska revolucija. *4D Portal.com*. http://4dportal.com/hr/znanost/4166-transhumanizam-robotska-revolucija

Blotnej, B. (2014). Dubai se odrekao poštara: Na vrata će im kucati leteći roboti. *24 sata.hr*. https://www.24sata.hr/tech/dubai-se-odrekao-postara-na-vrata-ce-im-kucati-leteci-roboti-353228/galerija-304555

Blotnej, B. (2015). I bankomati pričaju: japanski bankari zapošljavaju robote. *24 sata.hr*. https://www.24sata.hr/tech/i-bankomati-pricaju-japanski-bankari-zaposljavaju-robote-428794

Borovac, I. (2010). Virtualna intima i roboti su 'novi' seks u 21. stoljeću. *24 sata*. https://www.24sata.hr/lifestyle/virtualna-intima-i-roboti-su-novi-seks-u-21-stoljecu-153780

Bostrom, N. (2010). Superintelligence. In J. Brockman (Ed.), *This will change everything: Ideas that will shape the future* (pp. 111–114). Harper Perennial.

Božičević, L. A. (2016). Znanstvenici stvorili robota koji bi trebao autističnu djecu poticat na interakciju i igru. *Telegram.hr*. https://www.telegram.hr/zivot/znanstvenici-stvorili-robota-koji-bi-trebao-autisticnu-djecu-poticati-na-interakciju-i-igru/

Bu, M. (2017). Kineski inženjer oženio se svojim robotom. *Glas Slavonije*. http://www.glas-slavonije.hr/329801/9/Kineski-inzenjer-ozenio-se-svojim-robotom

Chrisley, R. (2008). Philosophical foundations of artificial consciousness. *Artificial Intelligence in Medicine*, *44*(2), 119–137.

Crossley, L. (2014). How's that for robotic moves? Pole-dancing droids strut their stuff at computer expo. *Dailymail.com.uk*. https://www.dailymail.co.uk/sciencetech/article-2576837/Hows-robotic-moves-Pole-dancing-droids-strut-stuff-computer-expo.html

Ćuto, T. (2015a). Pogledajte kako izgleda prvi robot kuhar koji jela priprema potpuno sam. *Zimo*. https://zimo.dnevnik.hr/galerija/pogledajte-kako-izgleda-prvi-robot-kuhar-koji-jela-priprema-potpuno-sam---401232.html/61125461

Ćuto, T. (2015b). Pogled u budućnost: Nije sve tako crno, očekuju nas sjajne stvari! *Zimo.* https://zimo.dnevnik.hr/clanak/pogled-u-buducnost-nije-sve-tako-crno-ocekuju-nas-sjajne-stvari---400939.html

Ćuto, T. (2016). Ovo je Root, robot koji djecu uči programiranju na jednostavan način. *Zimo.* http://zimo.dnevnik.hr/clanak/ovo-je-root-robot-koji-djecu-uci-programiranju-na-jednostavan-nacin---434183.html

De Sousa, A. (2013). Towards an integrative theory of consciousness: Part 2 (An anthology of various other models). *Mens Sana Monographs, 11*(1), 151. https://doi.org/10.4103/0973-1229.109341

Demetriou, D. (2009). Robot teacher conducts first class in Tokyo school. *Telegraph.* https://www.telegraph.co.uk/technology/5311151/Robot-teacher-conducts-first-class-in-Tokyo-school.html

Descartes, R. (2008). *Meditations on first philosophy: With selections from the Objections and replies* (M. Moriarty, Trans.). Oxford University Press.

E.DO. (2016). Do 2025. godine seks roboti zamijeniti će muškarce u spavaćoj sobi. *Index.hr.* https://www.index.hr/rouge/clanak/do-2025-godine-seks-roboti-zamijeniti-ce-muskarce-u-spavacoj-sobi/923425.aspx

European Parliament resolution of 16 February 2017 with recommendations to the Commission on Civil Law Rules on Robotics (2015/2103(INL)).

Fena. (2013). Povratak Tamagotchija: Na tržištu ponovno nakon 17 godina. *Akta.ba.* http://ekapija.ba/bs/Vijest/pauza/povratak-tamagotchija-na-trzistu-ponovno-nakon-17-godina/32995

Flanagan, O. J. (1995). Consciousness reconsidered. MIT Press.

Fong, T., Nourbakhsh, I., & Dautenhahn, K. (2003). A survey of socially interactive robots. *Robotics and Autonomous Systems, 42*(3–4), 143–166. https://doi.org/10.1016/S0921-8890(02)00372-X

Gamez, D. (2008). Progress in machine consciousness. *Consciousness and Cognition, 17*(3), 887–910. https://doi.org/10.1016/j.concog.2007.04.005

Hruškovec, I. (2012a). Pogledajte robotske plesačice oko štange i robota umjetnika. *24 sata.hr.* https://www.24sata.hr/tech/pogledajte-robotske-plesacice-oko-stange-i-robota-umjetnika-257571

Hruškovec, I. (2012b). Fujitsu želi napraviti robota koji će proći prijemni za faks. *24 sata.hr.* https://www.24sata.hr/tech/fujitsu-zeli-napraviti-robota-koji-ce-proci-prijemni-za-faks-281280

Hruškovec, I. (2012c). Roboti misle kao jedan i u roju će graditi nastambe na Marsu. *24 sata.hr.* https://www.24sata.hr/tech/roboti-misle-kao-jedan-i-u-roju-ce-graditi-nastambe-na-marsu-294263

Hruškovec, I. (2013a). Robotski konobar Carl uvijek spreman na čašicu razgovora. *24 sata.hr.* https://www.24sata.hr/tech/robotski-konobar-carl-uvijek-spreman-na-casicu-razgovora-325835

Hruškovec, I. (2013b). Simpatični robotski dječak ima lice koje podsjeća na Caspera. *24 sata.hr.* https://www.24sata.hr/tech/simpaticni-robotski-djecak-ima-lice-koje-podsjeca-na-caspera-303959

I.S. (2011). Robot kao ispomoć nepokretnim osobama. *Vidi.hr.* https://www.vidi.hr/Zdravlje/Robot-kao-ispomoc-nepokretnim-osobama

Jackendoff, R. (1987). *Consciousness and the computational mind.* MIT Press.

James, W. (1950). The principles of psychology: In two volumes. Vol. 1. Dover.

junior. (2015). Turnir: Roboti u Njemačkoj zaigrali nogomet. *24 sata.hr*. https://www.24sata.hr/junior/turnir-roboti-u-njemackoj-zaigrali-nogomet-416330

Jurman, H. (2015a). U Kini predstavljeni 'ratni roboti'. *Zimo*. http://zimo.dnevnik.hr/clanak/u-kini-predstavljeni-ratni-roboti---417816.html

Jurman, H. (2015b). 'Leteća usluga': Ovo su konobari budućnosti! *Zimo*. http://zimo.dnevnik.hr/clanak/leteca-usluga-ovo-su-konobari-buducnosti---401662.html

Jutarnji list. (2009). Robot Einstein sam uči izražavati osjećaje. *Jutarnji list*. https://www.jutarnji.hr/naslovnica/robot-einstein-sam-uci-izrazavati-osjecaje-3401033

Jutarnji list. (2017). *Prvi robot koji je dobio državljanstvo već ima potrebu za majčinstvom Sofija želi dijete i tvrdi: »Obitelj je jedna od najvažnijih stvari na svijetu«*. https://www.jutarnji.hr/life/znanost/prvi-robot-koji-je-dobio-drzavljanstvo-vec-ima-potrebu-za-majcinstvom-sofija-zeli-dijete-i-tvrdi-obitelj-je-jedna-od-najvaznijih-stvari-na-svijetu-6786921

Kauffman, J. (2020). For a cold, heartless robot, Sally makes a decent salad. *Los Angeles Times*. https://www.latimes.com/food/story/2020-06-11/robot-vending-machines

Krishnan, A. (2010). Killer robots: Legality and ethicality of autonomous weapons. Ashgate.

Lawler, R. (2013). Google acquires Boston Dynamics, the robot builder behind Big Dog and Cheetah. *Engadget*. https://www.engadget.com/2013-12-14-google-bought-boston-dynamics-its-over-for-humans.html

Lijović, D. (2015). Ovo je droid kojeg tražite: BB-8 iz Star Warsa izašao na tržište. *Večernji list*. https://www.vecernji.hr/techsci/droid-bb-8-iz-star-warsa-izasao-na-trziste-1023076/galerija-153047?page=2

Luc. (2017). Robot izveo stomatološku operaciju bez ljudske pomoći. *Večernji list*.

Majdak, M. (2011). I roboti će imati svoje stranice za 'shareanje' – Robo Earth. *24 sata.hr*. https://www.24sata.hr/tech/i-roboti-ce-imati-svoje-stranice-za-shareanje-roboearth-209424

Matković, T. (2016). *KFC zaposlio robota da poslužuje goste, a ovo je još nekoliko profesija ukojima ćemo ih uskoro vidjeti*. https://www.telegram.hr/zivot/kfc-u-sangaju-zaposlio-robota-da-posluzuje-goste/

McGrath. (2015). In a couple years, you'll be able to sleep in while this robot makesyou breakfast. *Digital trends*. http://www.digitaltrends.com/home/moley-robotics-introduces-robot-chef-that-makes-high-quality-food

McKibben, B. (2004). *Enough: Staying Human in an Engineered Age*. H. Holt. https://archive.org/details/enough00bill

Medgadget Editors. (2009). *MAKO Introduces RIO Robotic Arm for Orthopedic Surgeries*. Medgadget. https://www.medgadget.com/2009/02/mako_introduces_rio_robotic_arm_for_orthopedic_surgeries.html

Mi, J. (2017a). Vlasnik bordela sa seks lutkama otkriva koje fantazije vam čak ni lutke neće ispuniti. *Index.hr*. https://www.index.hr/magazin/clanak/vlasnik-bordela-sa-seks-lutkama-otkriva-koje-fantazije-vam-cak-ni-lutke-nece-ispuniti/1000756.aspx

Mi, J. (2017b). VIDEO »lutka za seks nam je član obitelji, seksamo se utroje, sjedi na kauču s djecom…«. *Index.hr*. https://www.index.hr/magazin/clanak/video-lutka-za-seks-nam-je-clan-obitelji-seksamo-se-utroje-ponekad-sjedi-na-kaucu-s-djecom/994309.aspx

MIT Artificial Intelligence lab. (n.d.). *Ai.mit.edu*. http://www.ai.mit.edu/projects/sociable/videos.html

Moran, M. E. (2007). Rossum's Universal Robots: Not the Machines. *Journal of Endourology*, *21*(12), 1399–1402. https://doi.org/10.1089/end.2007.0104

Nemes, T. (1969). *Cybernetic Machines*. Iliffe Books.

Nožinić, M. (2016a). Jia Jia: Prvi kineski interaktivni robot. *Zimo*. https://zimo.dnevnik.hr/clanak/jia-jia-prvi-kineski-interaktivni-robot---433916.html

Nožinić, M. (2016b). RoBoHoN: Pametni telefon koji hoda i priča. *Zimo*. https://zimo.dnevnik.hr/clanak/robohon-pametni-telefon-koji-hoda-i-prica---433784.html

Nožinić, M. (2016c). Upoznajte DRU, robota koji dostavlja pizzu. *Zimo*. https://zimo.dnevnik.hr/clanak/upoznajte-dru-robota-koji-dostavlja-pizzu---430425.html

Nožinić, M. (2016d). Održana prva utrka dronova kojima se upravlja mislima. *Zimo*. Preuzeto s: http://zimo.dnevnik.hr/clanak/odrzana-prva-utrka-dronova-kojima-se-upravlja-mislima---435145.html

Nožinić, M. (2016e). Britanska policija protiv kriminala borit će se korištenjem dronova. *Zimo*. http://zimo.dnevnik.hr/clanak/britanska-policija-protiv-kriminala-borit-cese-koristenjem-dronova---433516.html

Owsianik, J., Dawson, R. (2017). Future of Sex Report. *FutureofSex.net*.

Press, C. (2005). Aisb'05 Convention: Social Intelligence and Interaction in Animals, Robots and Agents. *Proceedings of the 3rd International Symposium on Imitation in Animals and Artifacts*.

Press Release. (2013). *Aeryon labs announces the Aeryon SkyRangerTM sUAS*. Aeryon Labs Inc.

Prinz, J. J. (2012). *The conscious brain: How attention engenders experience*. Oxford University Press.

Pulso Urbano. (n.d.). *RPMB*. Preuzeto s: http://www.pulsourbano.com.ar/se-lanzo-un-smartphone-robot/

Putman, H. (1964). Robots: Machines or Artificially Created Life? *The Journal of Philosophy*, *61*(21), 668. https://doi.org/10.2307/2023045

Ra, M. (2016a). Lutke za seks i više od toga: U plastične ljubavnice stiže umjetna inteligencija. *Index.hr*. https://www.index.hr/magazin/clanak/lutke-za-seks-i-vise-od-toga-u-plasticne-ljubavnice-stize-umjetna-inteligencija/927407.aspx

Ra, M. (2016b). Video Ova se žena poseksala s prvom muškom seks lutkom i opisala kako joj je bilo. *Index.hr*. https://www.index.hr/magazin/clanak/video-cure-stigao-je-seks-lutak-a-evo-kako-izgleda-akcija-s-njim/928005.aspx

Radović, N. (2010). Roboti u stalnom radnom odnosu obavljaju sve poslove. *24 sata.hr*. https://www.24sata.hr/tech/roboti-u-stalnom-radnom-odnosu-obavljaju-sve-poslove-203535

Searle, J. R. (1992). *The rediscovery of the mind*. MIT Press.

Šikić, L. (2015). Cogito, ergo veho. *Prom.com.hr*. http://www.fpz.unizg.hr/prom/?p=1243

Sokol, B. (2015). Kinshasa's traffic robots: 'I thought it was some kind of joke' - in picture''. https://www.theguardian.com/cities/gallery/2015/mar/13/kinshasa-traffic-robots-robocops-in-pictures

Taylor, J. G. (1992). From Single Neuron to Cognition. In *Artificial Neural Networks 2. (Proc. ICANN-92)*. North-Holland.

Tehnoklik. (2017). Robot s laserima i senzorima koji ima samo jedan zadatak: Nećete vjerovati štovas ovo čudo može naučiti. *Tehnoklik*. https://net.hr/tehnoklik/gadgeti/robot-s-laserima-i-senzorima-koji-ima-samo-jedan-zadatak-necete-vjerovati-sto-vas-ovo-cudo-moze-nauciti-ae00faf0-b1c1-11eb-8308-0242ac140021

Thompson, S. J. (Ed.). (2019). Androids, Cyborgs, and Robots in Contemporary Culture and Society. IGI Global.

Tolić, J. (2014). Stvorili si gledatelje: Korejce će s tribina podržavati – roboti. *24sata.hr*. https://www.24sata.hr/sport/stvorili-si-gledatelje-korejce-ce-s-tribina-podrzavati-roboti-377556

Turing, A. M. (1950). Computing Machinery and Intelligence. *Mind*, *49*(236), 433–460. https://doi.org/10.1093/mind/LIX.236.433

Turkle, S. (2010). *Alone together: Why we expect more from technology and less from each other*. Basic Books.

Wagner, M. (2015). Don't have sex with robots: expert. *New York Daily News*. https://www.tribpub.com/gdpr/nydailynews.com/

Wakefield, J. (2015). RoboCup: Training robots to play football. *BBC News*. https://www.bbc.com/news/technology-33254022

Yang, J. (2015). Meet Paro, a furry friend to dementia patients. *Toronto Star*. https://www.thestar.com/news/insight/2015/10/05/meet-paro-a-furry-friend-to-dementia-patients.html

Ž. L. (2017). Izvještaj UN-a Roboti će zamijeniti dvije trećine svih radnika u zemljama u razvoju. *Index.hr*. https://www.index.hr/vijesti/clanak/izvjesce-una-roboti-ce-zamijeniti-dvije-trecine-svih-radnika-u-zemljama-u-razvoju/953959.aspx

ENDNOTE

1 This image was created for the purpose of a roundtable discussion on "Cyborgs, Robots and Avatars" (Čatić, Gregurić, Bahtijari 2013), held at the Europe House Zagreb.

Chapter 9
Futurologist Predictions on Global World Order of Cyborgs and Robots

ABSTRACT

This chapter presents some predictions about the global world order of cyborgs and robots made by the futurologists who are concerned with the reshaping and enhancement of man and his reality. In addition to the phenomenological and anthropological-ontological representation of cybernetic beings in scientific history, technologies of scientific work open up new perspectives of technological and biotechnological enhancement and reshaping of human beings and represent the foundation of life in a programmed world. The futurologists' philosophy is anchored in the tomorrow, and the day after tomorrow, so that every step away from today is a step towards the goals of scientific work.

10.1 IMPLANTING RFID CHIPS INTO THE HUMAN ORGANISM AND HUMAN REALITY

Technologies of *scientific work* open up new perspectives of technological and biotechnological enhancement and reshaping of human beings and represent the foundation of life in a programmed world "on the verge of recreational self-transformations that will dwarf the modifications the Romans indulged in. When you no longer need to eat to stay alive, or to procreate to have offspring, or locomote to have an adventure-packed life, when the residual instincts for these activities might be simply turned off by genetic tweaking, there may be no constant of human nature left at all. Except, maybe, our incessant curiosity." (Dennet, 2012, p. xxiii).

Radiofrequency identification (RFID) is a technology which uses radio frequencies to automatically identify objects and exchange information between portable devices and computers. The first paper on RFID titled *Communication by Means of Reflected Power* was written by Harry Stockman in 1948 for the Institute of Radio Engineers, which was later transformed into IEEE in 1963 (Stockman, 1948). Use of RFID devices started during World War II for identification of aeroplanes (cf. Garfinkel & Holtzman,

DOI: 10.4018/978-1-7998-9231-1.ch009

2006). After a period of animal experiments in 1983, the biochip technology started to be used in numerous biological laboratories, zoos, and animal ambulance services with the aim of tagging animals with biochips. However, their application today is aimed towards control of human beings as well.

VeriChip is the first commercial biochip approved by the FDA (*Food and Drugs Administration*) for human implantation. It is an RFID chip the size of a grain of rice implanted under the skin to process radio waves and can be connected with the GPS. It contains a 16-digit number read by a special scanner. The ID number allows to the user to access various information. *VeriChip* attracted a lot of media attention after the 9/11 attacks on New York and Washington. After these events, FDA approved the production and implantation of RFID microchips into human bodies in 2002. As a result of fearmongering and terrorism, US citizens were suggested chip implantation which some of them accepted voluntarily for alleged security and protection.

In 2004, use of medical-purpose microchips was approved in 80 hospitals in the US (Frost & Sullivan Research Service, 2005). One of these hospitals is the *St. Clair Hospital* in Pittsburgh. The RFID bracelets allow the medical staff to obtain patient information for those in need of emergency medical assistance, which helps the medical staff save their patients' lives (Bacheldor, 2007). Saving patients' lives with timely medical interventions is also emphasized as one of the key advantages of RFID chip implantation (BlueBean, 2007).

According to the FDA, an argument in favour of biochip implantation is that about ten thousand patients die each year due to wrong medication, along with numerous operations performed on wrong organs. Usage of biochips allows the pharmaceutical industry to check which particular drugs are being administered and used. These intentions are seemingly noble, but the question whether mankind will find the right measure and use the technology with good intent only remains open. Shortly after the use of biochips was approved, anti-RFID activists Katherine Albrecht and Liz McIntyre sent the FDA a warning letter demanding that serious health risks associated with the *VeriChip* be further investigated and clarified. These adverse effects were listed by the FDA and refer to negative tissue reactions, electronic scanner malfunction, unsuccessful transponder implantation, electrical risks, etc. (CASPIAN Special Report, 2004). These risks could happen to anyone undergoing implantation procedure (Smith, 2006).

Scientists at the *Russian Institute for Molecular Biology* have perfected a new chip-based diagnostics technology that allows analysis and diagnostics of pathogenic bacteria and viruses, discovering genetic mutations and predicting cancer occurrence. The *Centre for Bioelectronics, Biosensors and Biochips* (C3B) at *Clemson University* is developing a biochip which could determine the quality of the blood donor, diagnose different types of flu which have raised global concern during recent years (avian influenza, swine influenza). This technology can also be applied to determine the patients' resistance to certain types of medication (Cuzin, 2001, Yang, 2011).

Since 2002, the development of nanotechnology has allowed the chips to be designed in myriad shapes and sizes. There are three potential implantation points – forearm, fist, and head. According to Carl Sanders, arm or forehead are the most ideal places because their change of temperature allows the chip's battery to be powered. Implantation procedure is done under local anaesthetic and the chip is implanted onto neural endings. The procedure leaves no scars and the chip is virtually imperceptible from outside. It can only be detected via special scanner. The data stored on the chip are not encrypted and everyone who owns a scanner can read the data, even copy them. The lifespan of the microchip in the human body is unlimited. When this chip is connected to the nervous system of a human being connected to a network, it can increase or decrease their heartbeat, direct and manipulate their thoughts, physical reactions, etc. (Icke, 2001, p. 370), maybe even control them to a certain extent. When surgically removed by an individual, the tiny capsule would burst and contaminate the person with lithium and other chemicals from the microbattery, while the GPS would detect the removal and notify the police. There is a possibility of using remote control to release the lithium from the microchip's capsule, which would lead to poisoning and enabling elimination of the unwanted population or undesirable malcontents. All of these actions can be processed using modern supercomputers which have processors capable of processing information 4 million times faster than human brains, which would enable control over every human being on the planet.

Carl Sanders is one of the scientists who worked on the development of microchips but he is also very cautious about their potential misuse. Sanders claims that the chips were originally supposed to be used only for medical purposes. However, it turned out that they could store thousands of information and modify human behaviour by increasing certain frequencies, which means their use could go beyond identification, location, and monetary transactions and into the domain of control of person's mood and behaviour. Sanders claims that the 1970s witnessed a discussion on one of the secret plans of the world government – mental control via microchips. A microchipped person could be easily located, manipulated and tracked via specific electromagnetic frequencies (EMF). This could be achieved through use of mobile networks and mobile phone antennas (Icke 2001, p. 371). According to Sanders (1994), microchips could be used to affect the human brain and completely change normal thoughts and emotional states and ultimately behave according to the rules set by the group behind this mind control project.

In the future, we might face large-scale RFID chip implantation which would enable identification in various transactions, reading health-related data, especially locating cancerous cells in the human organism, and early detection of other potential diseases. This way, the use of microchips would not only serve to preserve the health of the population but also to extend its lifespan expectancy. After the implantation, the microchip continuously monitors the wearer's state of health and sends the signals to the health centre in which the doctors keep track of the patients and treat them regardless of their location. Research project *Quo Vadis* aims to extend the human life expectancy up to 120 years (Day, 2012).

Apart from benefits in terms of disease prevention and treatment, human chip implantation also has the potential for misuse from various interest groups which control this technology and decide on how it is used. Presence of sophisticated technology in human environment would enable complete surveillance and control over human and non-human beings. In near future, it will be possible to use the RFID chips to track every aspect and move of every individual, record all verbal conversations, control changes in emotional, mental and physiological state of the human body. For instance, Intel is developing a chip which will be implanted directly into the brain and will allow an individual to have direct control over computers, cell phones and other technological gadgets with their thoughts (Edwards, 2009). This will happen if the human society chooses this option as the only possibility of integrating and organizing human life. David Brin believes that strong development of technology cannot be stopped, which makes privacy in the modern age practically impossible. Use of technological aids, such as mosquito-sized cameras, according to Brin, is slowly but steadily steering us towards the state in which technological and economical power are concentrated not only on surveillance of human beings but also on the inability to see the controllers of this technology. Bring believes that every person should have access to this type of surveillance. The main claim of the Open Society is that powerful governments and corporations will always have invasive technologies for privacy control at their disposal. The real truth is that privacy has faded out by the very emergence of the digital age (Brin, 1998).

Privacy has become questionable through use of computers and the internet. The database for all the information we use on our computers is saved and constitutes the matrix of our personal world. Implanting a similar chip for memory and communication into human brain is becoming a technological and a political issue. Microchip implantation leads to the end of freedom, identity and privacy as it allows control of every man on a mental, emotional, physical and biological level. We are aware that our privacy is under attack, but the problem is we do not have consistent mechanisms which would allow us to fight this (Garfinkel, 2003). This is not just a matter of encroachment of freedom and rights of individuals, but also the change of important aspects of human biological bodies and establishing human beings on a new anthropological and ontological cybernetic foundation.

For the past several years, governments and various interest groups have been conducting a very active process of global control. The extent of globalization is constantly increasing through centralization of world economy and information communication integration. Every movement and every activity are under close surveillance, which is why microchip implantation goes hand in hand with the global concept of control which is covered by an increasing number of rules and regulations. Means of public communication are being used to put the citizens under pressure. Fear of terrorism and crime and general safety are constantly discussed so that the citizens accept control as a saving grace, and what better way to achieve control than with microchips? For most people who do not have enough information about global surveillance, microchip implantation procedures may seem benign, all the more so as they are introduced incrementally, so that the process of their implementation seems as a sequence of completely unrelated activities. The necessity of microchip implantation could reach global levels of human activity which will require obligatory microchips in everyday lives of citizens as they would not be able to access particular segments of the society. The first microchip implantation was conducted on animals, as is the case with most cyborg experiments, and finding lost animals was one of the most frequently used arguments in favour of it. Similarly, human microchip implantation is being suggested as the most painless system for prevention of kidnapping and for marking new-born babies in hospitals to avoid confusion, all of which seems logical, justified and acceptable. Some companies require their workers to be microchipped as a pre-requisite for employment, while nightclubs and stadiums already use chips to identify their guests. The owner of a night club in Barcelona has implanted himself with an ID chip. He wants to identify his VIP guests using chips which would enable them special access to the VIP lounge and would serve as a credit card to pay for their drinks (BBC News, 2004).

As was already said, RFID chips are slowly but steadily being integrated into everyday life of human beings. The habit, obligation and demand for their implantation is slowly increasing. Chip implantation can become a legal obligation, which would mean that all activities will take place via microchips and these chips would constitute the network onto which all life will be organized. Their application has already started in the developed countries, while the developing countries will follow suit very soon, as they want to become a part of the globalization process (Garfinkel et al., 2005). On December 8, 2016, the US congress has passed the proposal for Kevin and Avonte's Law (H.R.4919) which regulates chip implantation for persons with autism, Alzheimer's disease and similar disorders. This is the first piece of legislature in the world which allows uninterrupted implantation of chips into a segment of human population. The sponsors of this law believe that it does not break any constitutional freedoms and rights of individuals but allows their 'faster tracking and location' when deemed necessary (Z.K. 2016). After integrating microchips into everyday activities of the real world, the next step should be implantation of chips whose abilities lead individuals into 3D physical reality. An individual who refuses to live in the system in which chip implantation is required will be excluded from all social activities.

The matters of microchip implantation seem to be a rather intriguing topic for the residents of former Yugoslavia. After my interview on ethical questions of human microchip implantation at the *Croatian*

National Television (Greguric, 2014), the TV network received over 13,000 messages via social media, which shows the interest of the population for this topic. This is why I proposed a broad intersubjective public discussion on these questions, so that we could reach a consensus about the extent to which the human body and the whole human reality should be affected by chip implantation. What is particularly worrying and what makes our initiative ever so relevant is the fact that we are already living in the cyborgized reality of *scientific humanism as naturalism* in which microchip implantation will soon become *conditio sin qua non* for participating in the social life of the community. A special aspect of chip implantation (cyborgization) of the entire history of historical reality of *scientific humanism as naturalism* is embodied in Max Weiser's idea of 'ubiquitous computing' described in §7.2 *Society as a cybernetic organism.*

10.2 CREATING THE ARTIFICIAL BRAIN

Creation of an artificial brain and artificial intelligence has been discussed in several parts of this book. However, there is another possibility of preserving the human brain – cryonic brain storage, i.e. freezing of the brain tissue. The purpose of this method is "to vitrify the brain and store it indefinitely in liquid nitrogen at about -320°C." (Kosko, 2010, p. 107), which would allow the nerve information to be preserved. Today, "there are only about a hundred brains suspended in liquid nitrogen in world where each day about a hundred and fifty thousand people die" (Kosko, 2010, p. 108).

Bostrom argues that "human brain power is the chief rate-limiting factor in the development of human civilization" (Bostrom, 2010, p. 111). However, we can expect a functional artificial brain in the nearby future. In 2009, the leader of the Blue Brain project Henry Makram managed to produce elements of a rat brain. His research promises to create an artificial brain made of silicone, gold and copper, which should have the ability to develop thought regardless of external input and the ability to feel. These are also the expectations of David Eagleman, neuroscience expert from the *Baylor College of Medicine*: "We are on course, however, to achieve technologies that let us store unthinkable amounts of data and run gargantuan simulations. Therefore, well before we understand how brains work, we will find ourselves able to digitally copy the brain's structure and download the conscious mind into a computer" (Eagleman, 2010, p. 99). This mind would be equivalent to our mind: "If the computational hypothesis of brain function is correct, it suggests than an exact replica of your brain memories, will think and feel the way you do and will experience your consciousness" (Eagleman, 2010, p. 99). With the aid of technology, it should be possible to "produce conscious thought. Efforts are already under way to replace damaged brain parts, such as the hippocampus, with hypercomputer implants" (Paul, 2010, p. 117). In the future, man's "connection to computer will be far more direct. Education, too, will fundamentally change, as engineers and cognitive scientists begin to leverage an understanding of brain code into ways of directly uploading information into the brain" (Marcus, 2010, p. 106).

Fast development of neurology could lead to radiotelepathy, a direct way of communicating feelings and thoughts from the brain, but for it to become possible, "we have only to invent two new technologies: first, the direct conversion of neural signals into radio signals and vice versa; and second, the placement of microscopic radio transmitters and receivers within the tissue of a living brain" (Dyson, 2010, p. 147).

According to Schwartz (2010)

We can already give the seriously injured the ability to control a prosthetic arm, and we have read out neural states that seem to express language. It is a few big steps from there to the brain's ability to reliably control such devices as cars, trucks, fighter jets, drones, machine tools, and so on. The most interesting will be the computer cursor/keyboard. It is not hard to imagine a piece of technology – say, like a Bluetooth earpiece – that would enable one to think a message and send it. This would be a form of one-way, electronically mediated telepathy. (Schwartz, 2010, p. 153)

Brain prostheses and brain-computer interfaces could provide synthetic telepathy in the future: "Allowing one person to control another person's body would be a fairly simple extension of that technology. Enabling one person to transmit thoughts directly to another person's brain is a much trickier proposition, but not terribly far-fetched, and it would break down one of the most profound isolations associated with the human condition. Broadcasting the overall state or 'mood' of a brain would probably come first. Transmitting specific, conscious thoughts would require elaborate physical implants to make sure the signals go to exactly the right place" (Powell, 2010, pp. 21-22). Norman lists several types of potential prostheses that will be implanted in the future – artificial sensory systems (enhanced hearing or vision), memory prostheses, communication prostheses, and emotion prostheses. In his work, Norman asks "What happens when we have implants within the body that affect our cognitive powers?" (Norman, 1999, p. 175). A sensory prosthesis would enhance hearing and vision: "Built-in sound enhancers to make audition ever more acute, vision ever more precise. Built-in zoom lens. Built in recorder for sounds and sights" (Norman, 1999, p. 176). A memory prosthesis would "allow the person to record all that had ever happened, all that was ever thought" (Norman, 1999, p. 177). This begs the questions: "would others be allowed to have access to the contents? What about with court orders? What about after person's death? Who owned the recording?" (Norman, 1999, p. 177). Communication prostheses would enable uninterrupted communication. One wonders "How would society change if one were never alone, if there were always someone talking to you, helping you, giving guidance, debating your actions" (Norman, 1999, p. 177). Emotion prostheses would allow us to artificially create moods, increase or decrease particular emotions:

Pleasure centers could be enhance, emotional feelings enhanced or suppressed. The devices would almost certainly start out as medical prescriptions for controlling malfunction but could easily be subverted to enhance pleasure and to produce artificially enhanced mood states. Would we have generation of brain zappers? (Norman, 1999, p. 178)

Kenneth W. Ford believes that "Mind reading is [...] quite likely. And, as we know from hard disks and flash memories, to be able to read is to be able to write. Thoughts will be implantable" (Ford, 2010, p. 141). If such a mind-reading technology could be used, would it be legally allowed? Neural control over machines in the future will allow us to be "able to 'jack in' to the Web" (Schwarth, 2010, p. 152) with the help of several new biological means:

The first new set of tools are the means to image the internal working of living cells, including neurons. The second are a variety of tools for precisely mapping complex biomolecular mechanics so that we can understand and manipulate neural functioning within the cell at a molecular level. And finally, the functional MRI is already giving us a systemic understanding of neural behavior. (Schwartz, 2010, p. 152)

In the future, all the rules and patterns of the human brain will be known, which will allow us to have control over it:

In addition to developing ever smaller and more temporally precise electrical and chemical delivery devices, scientists can now nearly instantaneously increase or decrease the firing of specific neurons with lights probes that activate photosensitive ion channels. As with the electrical and chemical probes, these light probes can be inserted into the brain of living animals and change ongoing behavior. What if one could deliver the same spatially and temporally targeted bolus of electricity, chemistry or even light to a specific brain location without opening the skull? [...] Targeted and noninvasive stimulation, combined with the map that comprises the neurophenome, will revolutionize our ability to control our minds. (Knutson, 2010, pp. 279-280)

This way, "Along with applications designed to cure will come demands for applications designed to enhance. What if we could precisely but noninvasively modulate mood, alertness, memory, control, willpower, and more? Of course, everyone wants to win the brain game. But are we ready for the rules to change?" (Knutson, 2010, p. 280). Another point of discussion is the possibility of uploading a human mind into a computer and, through a memory chip implanted into another person's brain, assimilating into a foreign body and functioning in it like in one's own.

Connecting the human brain with a computer puts us in a potential situation in which the computers control the human mind. This way, the natural will be integrated into the technological process. Life itself will stem from the technological process. Life will become a matter of science and technology that can be directed and controlled. The IBM report they are working on 'augmented intelligence' and less so on artificial intelligence. Augmented intelligence is used to increase the human capabilities while artificial intelligence attempts to replicate the human mind (Lijović, 2017).

Stephen Hawking expected mankind to develop technology for uploading individual consciousness onto a computer within the next two decades, which would allow the mind to live after the body dies as its thoughts and consciousness will be saved in a computer. The computer would enable the interaction among all stored minds, regardless of whether they belong to living or deceased persons. This way the inanimate will be able to deal with both animate and inanimate entities. At this point, this possibility is only in the stage of technical and logical assumptions.

In his book *Mind Children: The Future of Robot and Human Intelligence*, Moravec argues: "I believe that robots with human intelligence will be common within fifty years" (Groothuis, 1999, p. 42) and believes that "immortality is achievable – with a little robotic assistance, that is" (Groothuis, 1999, p. 42). In his vision of robotics, "The original fragile housing for the brain is discarded as an evolutionary vestige. Your consciousness, nevertheless, can live forever, since duplicate copies of your mind can be made when the machinery storing it wears out" (Groothuis, 1999, p. 43). It can even "be beamed by laser into robots on other planets" (Groothuis, 1999, p. 43). Moravec describes a futuristic scene in which a robot surgeon performs an operation over a human being and transfers his mind from a brain into a computer:

Layer after layer of the brain is simulated, then excavated. Eventually your skull is empty, and the [robot] surgeon's hand rests deep in your brainstem. Though you have not lost consciousness, or even your train of thought, your mind has been removed from the brain and transferred to a machine. In a final, disorienting step the surgeon lifts out his hand. Your suddenly abandoned body goes into spasms and dies. For a moment you experience only quiet and dark. Then, once again, you can open your eyes. Your perspective has shifted. The computer simulation has been disconnected from the cable leading to the surgeon's hand and reconnected to a shiny new body of the style, color, and material of your choice. Your metamorphosis is complete. (Groothuis, 1999, p. 43)

Norman also predicts that:

In the coming one hundred years we can expect major changes. Some technologies will start to encroach upon human biology and evolution. Some biometric alternation will occur. The future will see artificial sensory systems, communication devices embedded under the skin, perhaps computational systems as well to enhance memory, sensory abilities, physical strength, and language skills. Perhaps planning for a baby will be somewhat like ordering a new automobile: the proud parent(s) (singular? plural?) will be able to choose from a list of options: physical characteristics, hair and eye colour, bone structure, cognitive abilities. Will this planning and artificial implantation technology be possible? Will it be permitted? If not, will bootleg sites spring up to do it anyway, perhaps in geographical locations that will thrive as legal havens for activities frowned upon in more traditional legal systems? (Norman, 1999, p. 175-176)

According to the predictions of futurologists, enhanced human beings will be able to write books within several minutes, sprint without having to breathe in, dive for hours without oxygen (Science of Singularity, 2014). Ray Kurzweil claims that: "We'll have implants that extend our capacity for retaining knowledge, for enhancing memory. Unlike nature, we won't leave out a quick knowledge down-loading port in the electronic version of our synapses. So it will be feasible to quickly download knowledge to these electronic extensions of our brains. Of course, when we fully port our minds to a new computational medium, down-loading knowledge will become even easier." (Kurzweil, 1999, p. 76). If this proves to be too slow, you will also be able to implant 100,000 electrodes per square inch of your brain to eliminate any need for reading books – the computer will simply inject its contents into your head (Freitas, 1985, p. 115). The world of tomorrow will be a combination of biology and technology, with a tendency to abolish biology as the foundation of life. Kurzweil predicts that 'bringing people from the dead' will be possible by 2030 through artificial intelligence. Nanorobots will be able to extract the memory from the human brain, thus reincarnating the digital version of the person after it is already dead. By 2045, mankind can expect to reach 'singularity', i.e. predomination of artificial intelligence over biological human intelligence. Singularity will create the assumptions for transferring the consciousness from a brain into a computer. If the human mind is loaded into a convincing virtual reality, the human body will be virtual as well. It is predicted that virtual bodies will be as convincing as real bodies, but they will not be prone to wear and tear like the biological counterparts. Humans will feel these virtual bodies the same way they feel their physical bodies. Kurzweil believes that the body is still necessary as human intelligence is embodied. If we look at the future that Kurzweil is describing, one cannot help but wonder whether there is anything to feel happy about in the world in which humans will become enhanced cyborgs with artificial glands and organs or digital beings inside of a virtual or augmented reality (Golubović, 2009).

To draw an analogy with the great patrons of the House of Medici who sought the elixir of youth in books of wisdom from ancient Greece and Egypt, we should also mention the *Initiative 2045* project of the Russian billionaire patron Dmitry Itskov who wants to explore and establish the foundations of eternal life. The aim of this project should be achieved with sophisticated computer technology which can dematerialize human body through coding, while the artificial intelligence can code the content and enhance the mental functions of the human brain and establish it onto a non-biological foundation. Itskov believes that digitalized brains are a path towards eternal life.

In a similar vein, former Google engineer Anthony Lewandowsky started the religion in 2015 which worships artificial intelligence. The religion is called the Way of the Future and its aim is to develop and promote the divine entity based on artificial intelligence and improve the society of today through understanding worship of this deity (Harris, 2017).

10.3. LIVING ON AN INFORMATION SUPERHIGHWAY OR IN A DIGITAL BUBBLE

Man's attitude towards information is changing day by day: "Now we are reaching the point where we can archive every message, every telephone conversation, every communication between human beings anywhere on the planet. For millennia, we were starving for information to serve as raw material for ideas. Now we are about to have a surfeit" (Seife, 2010, p. 140). According to Baudrillard, this is leading us towards a world with more and more information and less and less meaning. Every "tie between information and human purpose has been severed, i.e., information appears indiscriminately, directed at no one in particular, in enormous volume and at high speeds, and disconnected from theory, meaning, or purpose" (Postman, 1993, p. 70). William J. Mitchell vividly describes the scenario of the future: "Information highway will bring us a growing flood of visual information in digital format, but we will have to take great care to sift the facts from the fictions and the falsehoods" (Groothuis, 1999, p. 133)

The *MIT Media Lab* founder Nicholas Negroponte predicts that the amount of information in the future will lead to every man having an agent in the form of a computer which will have the ability to understand content. Due to high flow of information bits your computer will be able to "help you deal with the massive amount of content. My VCR of the future will say to me when I come home, 'Nicholas, I looked at five thousand hours of television while you were out and recorded six segments for you which total forty minutes" (Negroponte, 1996, p. 179). Life in the digital world will put us in a situation in which our intelligent agents will select information for us: "Imagine a future in which your interface agent can read every newswire and newspaper and catch every TV and radio broadcast on the planet, and then construct a personalized summary. This kind of newspaper is printed in an edition of one." (Negroponte, 1996, p. 153). Computer replacements in the form of intelligent agents will become a major segment in the computer-human interface due to high amount information that humans will not be able to process: "I do not want to be required to read thousands of bulletin boards to be sure I am not missing something. I want my interface agent to do those things. Digital butlers will be numerous, living both in the network and by your side" (Negroponte, 1996, p. 152). Additionally, "The idea that twenty years from now you will be talking to a group of eight inch-high holographic assistants walking across your desk is not farfetched" (Negroponte, 1996, p. 148).

As a result of information oversaturation, Pearson announces the need to develop the information bubble, a protective bubble which would filter out the undesirable information and prevent misuse of communication technology. Pearson believes that the human environment will become digital and that everyone will be able to decide what they want to see and which information they want to receive, while ignoring the unwanted ones (VDI, 2005). However, there is a very real danger that this will be decided by interest groups, private corporations, or governments,

In 1970, Marvin Minsky mentioned another possibility offered by global digitalization: "Instead of sending a picture of a cat, there is one area in which you can send the cat itself" (Dyson, 2010, p. 94).

According to Dyson "Life-forms that assume digital representation for all or part of their life cycle will not only be able to send messages at the speed of light, they will be able to send themselves" (Dyson, 2010, pp. 94-95).

10.4 PENETRATION OF VIRTUAL 3D PERSPECTIVE INTO HISTORICAL REALITY

Virtual reality represents digitally established life which requires "less and less dependence upon being in a specific place at a specific time, and the transmission of place itself will start to become possible." (Negroponte, 1996, p. 165). In the future, it will be possible "for a doctor in Houston to perform a delicate operation on a patient in Alaska" (Negroponte, 1996, p. 166). Increasing development of 3D technology, whose application started in cinematography and is expanding onto TV and computer screens, will become a part of our reality in the future. Use of 3D contact lenses, 3D architecture provides us with the option of removing undesirable people from our sight, recording a handshake with a person across the globe, virtual sexual intercourse, computer dreaming and many other activities. Holographic characters will emerge in the human mind and explain the world around us (VDI, 2005). This will lead to a new legal and ethical system the glimpse of which we can already see.

In the future, we will "inhabit virtual worlds like that of *The Matrix*. I assume there will be markets for purchasing various kinds of afterlives and sharing them with other people – this is the future of social networking. And once you are downloaded, you may even be able to watch the death of your outside, real-world body, just as we might watch an interesting movie" (Eagleman, 2010, p. 100). A potential technical obstacle for this is that "the simulated brain must be able to modify itself. We need not only the pieces and parts, but also the physics of their ongoing interactions – for example, the activity of transcription factors that travel to the nucleus and cause gene expression, the dynamic changes in location and strength of the synapses, and so on. Unless your simulated experiences change the structure of your simulated brain, you will be unable to form new memories and will have no sense of the passage of time" (Eagleman, 2010, pp. 100-110). If "bits could carry mass as well as information, the distinction between artificial and synthetic life would disappear. Virtual and physical replication would be equivalent" (Gershenfeld, 2010, p. 102).

Miguel Nicolelis from Duke University, one of the first creator of brain-computer interface believes that "The body's going to be very different a hundred years from now [...] In a century's time, you could be lying on a beach on the east coast of Brazil, controlling a robotic device roving on the surface of Mars, and be subjected to both experiences simultaneously, as if you were in both places at once. The feedback from that robot billions of miles away will be perceived by the brain as if it was you up there" (Geary, 2010, p. 15). This means that the biological cyborgized man will experience drastic changes in his perception of reality. If meeting "a neighbour, whether outer, inner, cyber-, or hyperspatial, finally turns us into an 'us'. To encounter an other – whether god, ghost, biological sibling, independently evolved life-form, or emergent intelligence of our own creation – changes what it means to be human" (Rushkoff, 2010, p. 64). It is clear that the human becomes non-human in both the material and virtual reality of *scientific humanism as naturalism*.

10.5 TECHNICAL SINGULARITY

In his 1987 book, *The Society of Mind,* Marvin Minsky proposed the concept "that intelligence is not found in some central processor but in the collective behavior of a large group of more special-purpose, highly interconnected machines" (Negroponte, 1996, p. 157).

This scenario will be made possible with future artificial intelligence agents which will help humans perform their everyday chores:

[...] machines' understanding individuals with the same degree of subtlety (or more than) we can expect from other human beings [...] For example, having heard from the liquor stores agent, a machine could call to your attention a sale on a particular Chardonnay or beer that it knows the guests you have coming to dinner tomorrow night liked last time. (Negroponte, 1996, p. 165)

Because of this "Agents will dispatch agents. The process multiplies." (Negroponte, 1996, p. 158). Artificial intelligence as seen by Minsky can make man's position in the cyborgized reality of *scientific humanism as naturalism* much easier. Artificial intelligence aims to become not a partner of the human mind but its descendant.

Frank J. Tipler believes that mankind will relatively soon face "the Singularity – the day when we finally create a human-level artificial intelligence. This involves considering the physics advances that will be required to create the computer capable of running a strong AI program" (Tipler, 2010, p. 39). However, for "this aggregating intelligence to communicate with humans [...] it will have to be equipped with accurate models of the native intelligences that inhabit human minds" (Tooby & Cosmides, 2010, p. 69): Both singularity and theology give promises of eternal life after the death of the body. This means storing a conscious mind into a computer. This is a fact that biological humans will never be able to test. One of the predictions is that the level of artificial intelligence will "build systems that are recursively self-improving – that understand their own workings well enough to design improvements to themselves, thereby bootstrapping to a state of ever more unimaginable intellectual performance" (de Grey, 2010, pp. 376-377). Multiplication, i.e. centrally connecting large groups of artificially created intelligence would surpass the abilities of individual biological human minds. This intelligence would be able to create other artificial intelligence entities, which means that artificial intelligence creates itself without man and outside human control. Our intelligent offspring creates its own replicas. This would be the tipping point beyond the open possibilities of reality of *scientific humanism as naturalism* into the absolute reality of *scientific work*, the reality which has devoured the last attributes of human species (*humanism = naturalism*) which it was unable to achieve in its existence. Arthur Clarke predicts that:

After 2020, when artificial intelligence reaches human levels, there will be two intelligent species on Planet Earth, one evolving far more rapidly than biology would ever permit. Interstellar probes carrying artificial intelligences will be launched towards the nearer stars. (Clarke, 1999, p. 44)

10.6 DIFFERENTIATION OF SOCIETY ACCORDING TO THE CRITERION OF MORTALITY AND IMMORTALITY

The futurists predict that by 2050 eternal life of the wealthiest will be made possible by storing the contents of the brain into computer memory. Saffo advocates the hypothesis that the rich would evolve into a special species, thus leading towards a new type of social stratification as a result of advances in biotechnology and robotics (Willis, 2009). Two ways of overcoming mortality are generally assumed: "one at the cellular level and the other through an integration of body with genetics, the cognitive sciences, and cybertechnology" (Gleiser, 2010, p. 187). This way "advancing technology renders death optional" (Paul, 2010, p. 115).

Ray Kurzweil wants to resolve the matter of immortality through use of nanobots which would protect our out-of-date immune system. Nanobots will circulate through our blood stream, surrounded by software against all diseases. As a milestone on this path, he cites year 2029 after which the human lifespan will be extended by one year each year. This year marks the beginning of eternal life, while the climax is expected in 2045 when the non-biological intelligence will increase the possibilities to a level a billion times stronger than the entire human existence (Jurman, 2016). Apart from extending the human lifespan, it will be possible to eliminate particular disorders, even enhance the mood and achieve

[…] immortal cognition […] and boundless happiness […] provided by painless brain stimulation. Today we have two main ways of stimulating the brain painlessly and noninvasively: transcranial magnetic stimulation (TMS) and transcranial direct current stimulation (TDCS). TMS stimulates the brain by inducing local magnetic fields over the scalp, which in turn induce electric currents in the brain, whereas TDCS uses weak direct currents. […] We will be able to significantly delay cognitive decline and improve mood by stimulating those brain areas collectively called association cortices […] People won't tolerate being excluded from the beneficial effects of brain stimulation. (Iacobini, 2010, pp. 289-291)

Michael Rose at the *University of California* has created "'Methuselah flies' that live over four times longer than control flies in the lab. He did this by not allowing their eggs to hatch until half are dead, for hundreds of generations. Methuselah flies are most robust, not less, and so resist stress" (Benford, 2010, p. 183). Human aging could be stopped in the future with the help of directed evolution: "Already there seems no fundamental reason why we cannot live to a hundred and fifty or longer" (Benford, 2010, p. 184). How could this be achieved? For instance, "Genetic action may also interfere with the usual mitochondrial respiration. Reduced expression of the mclk1 gene has been shown to slow down aging in mice. […] These results suggest that the same molecular mechanism for aging occurs throughout the animal kingdom" (Gleiser, 2010, pp. 187-188). With the help of these mechanisms the secrets of cellular aging could be revealed by year 2040: "It's not the elixir of life alchemists dreamt of, but the average life span could possibly be increased to a hundred and twenty-five years or even longer, a significant jump from the current U.S. average of seventy-seven years or so. Of course, this would create a terrible burden on, among other things, our Social Security system" (Gleiser, 2010, p. 188).

Prolonging the lifespan will completely change our way of living:

Living to a hundred and fifty ensures that you take the long view. You're going to live in a future ecology, so better be sure it's livable. You'll need long-term investments, so think long term. Social problems will belong to you, not to some distant others, because problems evolve and you'll be around to see them.

Rather than isolating people, 'old age' will lead to social growth. With robust health to go with longer lives, the older will become more socially responsible, bringing both experience and steady energy to bear. [...] Nobody will retire at sixty-five. People will switch careers, try out their dreams, perhaps find new mates and passion. [...] The future will be more mature, and richer for it. All this social promise emerges from the genomic revolution. [...] The first person to live to a hundred and fifty may be reading this right now. (Benford, 2010, pp. 184-185)

In addition to halting the aging process, the mankind of the future will understand what consciousness is and how it works: "Consciousness is vitalism's last desperate holdout. Twentieth-century biology still finds it genuinely mysterious, just as nineteenth-century biology found life mysterious. [...] I believe that during the twenty-first century the ancient philosophical mind-body problem will be solved, and solved not by philosophers but by scientists." (Dawkins, 1999, pp. 63-64). In accordance with the cryonic belief in eternal life, the body of a 14-year old British girl who died of carcinoma had her body cryogenically frozen in hope of bringing her back to life in the future, thus winning the fight for her right to be frozen. Her body was transferred to the USA where it will be preserved at extremely low temperatures in hope that the progress of medicine will allow her to come back to life (HINA, 2016).

10.7 GENETIC MODIFICATION OF HUMAN BEINGS

Gene therapy is based on exchange or modification of genes in the body (Anderson, 1999, pp. 19-21). Anderson predicts that by 2030, gene therapy will enable us to cure virtually every disease: "The two primary branches will be gene therapy, whereby one or more genes are injected into the patient to treat a disease, and drug therapy, in which a drug is given to the patient to modify the expression of one or more genes in the body. [...] We will rapidly develop the ability to screen for genetic defects or weaknesses" (Anderson, 1999, p. 19). Genetic weakness refers to "genes that do not function optimally for the environment in which the patient lives (whether that environment is stressful because of diet, radiation, toxins, or whatever)" (Anderson, 1999, p. 19). If modification of the human genetic material starts being used for enhancement purposes, we are faced with the question: "What would happen to us if we added intentional genetic enhancements to our world" (Anderson, 1999, p. 20) because "Drug therapy does not alter the human genome, but gene therapy does. Once we have the ability to give a patient any gene we want in order to treat a disease, then we will have the ability to give someone genes for any other purpose too. I fear the downside of this powerful technology might be that eugenics will be practised on a scale far larger than any 'selective breeding' policy could accomplish" (Anderson, 1999, p. 19).

Nobel prize winner James Watson predicts that genetic modification of human beings will take place sometime during the 21st century in such a manner that "the first steps will not be in frivolous pursuit of the creation of some super-race but to seek to alleviate the horrors and unfairness of existing human suffering" (Watson, 1999, p. 296). This refers to attempts to eliminate the genetic disorders such as the Huntington's Disease, cystic fibrosis, and the like: "Every day children are born crippled by 'bad' genes – which because of altered DNA messages, do not function normally." (Watson, 1999, p. 296). After prenatal diagnostics, which can test for predisposition for particular diseases, gene therapy is administered to replace the bad genes with good ones. Watson predicts that "Within the next ten to twenty years, I anticipate that there will be doctors and scientists who will correct faulty genes in living patients through the introduction of DNA into somatic cells, like those of our blood or muscular system"

(Watson, 1999, pp. 296-297). Not only that, "soon it will be commonplace for everyone to know his or her own genetic code. This will change the practice of medicine from treating disease after it happens to preventing disease before its onset" (Venter, 2010, p. 6). In his *Double Helix Award* acceptance speech, Watson noted that "he doesn't want government involvement in decisions concerning how people choose to handle information about their personal genomes" (Brockman, 2010, p. xxi). Watson also claims that "No government has been willing to sanction research that might redirect the course of human evolution. Yet germline manipulations will be attempted in the twenty-first century and for the same reasons that somatic gene therapy will earlier have been explored. [...] Confronted, for example with a deadly virus that we see no way of controlling, scientists will be given the green light to insert antiviral DNA segments into sperm and eggs in a bid to create protected children" (Watson, 1999, p. 298).

Watson's predictions came to pass by the end of 2016. *The British Human Fertilisation and Embryology* (HFEA) approved the *in vitro* fertilisation (IVF) from three parents. The technology allowed the intervention into the fertilisation process to remove the mitochondria which, if containing an error, could cause fatal heart problems, liver failure, neural disorders, blindness and muscular dystrophy. The procedure is called a three-parent IVF because the babies born from genetically modified embryos will have the DNA of the mother, father and the female donor. The first baby born with this technology arrived in 2016 in Mexico, even though Mexico has no legal regulations with regards to this issue (Kristović, 2016).

Susan Greenfield believes that some of the brain's functions will be modifiable by gene manipulation:

But such modifications will be crude at the beginning, lacking precision. Because of this imprecision, it will just not be possible, fortunately, to engineer all homosexuals or all criminals out of existence. Nonetheless, the genetic modification we do undertake in the future will inevitably lead to a reduced gene pool, depleted, for instance, of genes, such as those linked to diseases like cystic fibrosis. (Greenfield, 1999, p. 155)

Genetic engineering could erase our thought by 2084: "The monitoring and manipulation of the phenotype (how a gene is expressed and observable) will be increased with the use of nanotechnology – small devices that detect and interfere with reactions through the body" (Greenfield, 1999, pp. 155-156). We might end up with a society of increasingly standardized individuals that come from a decreasing gene pool. We risk to forever lose the ability of original thought, the kind which causes paradigmatic shifts. Lifespan could be extended with all these new technologies: "progress in fighting disease and patching up our genetic weaknesses will make it possible for people to routinely reach the full human life span of about a hundred and twenty" (Powell, 2010, p. 22). We could have genetically modified children "who can breathe underwater or who have radically enhanced mental capabilities. [...] They may very well qualify as totally new species" (Powell, 2010, p. 21).

Despite the optimism from the futurists, Fukuyama believes that genetic engineering could bring adverse consequences for mankind:

Genetic engineering could wipe out human life as we know it. But there will remain 'very powerful innate human capacities for reconstituting social order'" (Fukuyama, 1999, p. 112). Fukuyama believes that "In some ways, you cannot have an end of history unless you have an end of science and we are just on the verge of another big explosion in the scientific knowledge that comes out of the life sciences – the potential for biotechnology to change some very basic aspects of human existence. It could lead to an end of human beings as we understand them. We now have the possibility of altering the germ line, which means it is possible not only to alter the behaviour and characteristics of the present generation, but to create inheritable characteristics. This in effect means changing human nature. (Fukuyama, 1999, p. 116)

Zachary Myers believes we are facing two revolutions in the future: "The first is in developmental biology. We're learning how to reprogram human tissue, enabling new possibilities in repair and regeneration. [...] The second big revolution is coming from neuroscience" (Myers, 2010, pp. 8-9).

10.8 COEXISTENCE WITH ROBOTS AND CYBORGS IN A SMART ENVIRONMENT

In the future, apart from being surrounded by intelligent robots, we will live in an intelligent environment whose parts will communicate with each other:

Buildings of the future will be like the backplanes of computers: 'smart ready' [...] Smart ready is a combination of prewiring and ubiquitous connectors for (future) signal sharing among appliances.[...] Future rooms will know that you just sat down to eat, that you have gone to sleep, just stepped into the shower, took the dog for a walk. A phone would never ring. If you are not there, it won't ring because you are not there. [...] If your early-morning flight to Dallas is delayed, your alarm clock can ring a bit later and the car service automatically notified in accordance with traffic predictions. [...] If your refrigerator notices that you are out of milk, it can 'ask' your car to remind you to pick some up on your way home. [...] Interest in household robots will swing back, and we can anticipate digital domestics with legs to climb stairs, arms to dust, and hands to carry drinks. For security reasons, a household robot must also be able to bark like a ferocious dog. (Negroponte, 1996, pp. 211-216)

Cars will drive on their own, "smart radios, energy control, and information displays as the predominant population of digital devices" and "when your smart car of the future is stolen, it can call you up and tell you exactly where it is" (Negroponte, 1996, pp. 216-217). All of this indicates we can expect our lives to be led in a smart environment.

Sociable technology which first appeared in 1997 in the form of robots with different roles to play – nannies, teachers, therapists, caregivers, etc.

[...] inspired feelings of connection because they pushed people's Darwinian buttons: They asked us to teach them, they made eye contact, they tracked our motions, they remembered our names. [...] Sociable technologies came onstage as toys, but in the future they will be presented as potential nannies, teachers, therapists, life coaches, and caretakers for the elderly. First, they will be put forward as 'better than nothing.' [...] But over time, robots will be presented as 'better than something': That is, they will be preferable to an available human being – or, in some cases, a living pet. They will be promoted as having powers – of memory, attention, and patients – that people lack. (Turkle, 2010, pp. 10-11)

New paradigms of human-machine relations open up numerous new questions about human identity inside a computer. The ability to create multiple online identities, i.e. "fragmentation of identity on the Internet – the phenomenon of multiple, contradictory selves inhabiting the same person" (Turkle, 1999, p. 269) blurs the line between living in the interface and living in the reality: "That is to say, when people are in the presence of an intelligent artefact, they behave 'as if' it were really intelligent. She refers to this phenomenon as the 'Bladerunner effect' after the cult movie of that title. 'Once a computer has saved your life, once you have fallen in love with a computer, once you have a relationship with it, do you care about the 'really, really'? Not really,' she suggests." (Turkle, 1999, p. 273). Apart from subjective computers, relatively neutral screens onto which people project their thoughts and feelings, there is a new type of computer objects that are no longer affectively neutral: "These include 'affective' software and humanoid robots, virtual pets, and digital dolls" (Turkle, 1999, p. 275).

To establish a relationship with another human being, one must establish a relationship with its mental state. How do you achieve this with a robot? Turkle claims that computer objects never asked a child to be nurtured by it, while virtual pets and robots "ask the child to assess the object's 'state of mind' in order to develop a successful relationship with the object" (Turkle, 1999, p. 276). The attitude towards intimacy with technical gadgets has completely shifted. Turkle mentions a case of an MIT student whom she asked about his attitude towards chip implants: "Well, I wouldn't mind a calculus implant, but I wouldn't want to have a Dostoevsky implant!" (Turkle, 1999, p. 273). Turkle predicts that man of the future will have more knowledge about the nature of attachment to relational objects: "I predict that our children will learn to distinguish between an 'animal kind of alive' and a 'Furby kind of alive'. I predict that our children will have expectations of emotional attachments to computers, not in the way that we have expectation of emotional attachments to our cars and stereos, but in the way we have expectations of emotional attachments to people." (Turkle, 1999, p. 277). Neurobiologist Gerald Edelman believes that intelligent machines promise a particular level of progress: "Rather than trying to deconstruct the brain as a computer, they construct neural processing from the bottom up, mimicking the workings of actual neurons" (Powell, 2010, p. 22). This kind of "robotic moment will bring us to the question we must ask of every technology: Does it serve our human purposes?" (Turkle, 2010, p. 13).

In addition to the company of robots as the descendants of the human species, humans will live alongside enhanced human beings – cyborgs. *Homo optimus* is a cyborg of the future, consisting of man and computer parts. It will have superior senses, be more intelligent, more physically capable, able to communicate with the machines and will be digitally immortal. According to Pearson, its emergence can be expected by year 2050. The transhumanist advancement of technological and biotechnological enhancement of man will become the norm by 2050. It is exciting that nature will not be the one which forms the changes but our accomplishments (Espreso, 2016).

10.9 DIGITAL BIOLOGY

Some futurologists discuss time in which computers will create universes – digital environments which consist of binary zeros and ones, with biological lifeforms which multiply, grow and think. In other words, a computer would be the space-time of new biological nature, a new digital biology. Digital biology does not include the current physical bodies. They are replaced by digital ones as humans as they are now cannot exist in this world.

Digital biology uses laws of physics put into programs which create new natural forms based on it. In a manner of saying, plants, animals, and learning neural networks grow from the digital DNA. Digital lifeforms provide a platform for fighting against the entropy of the Universe and surviving in the age of *scientific humanism as naturalism.*

Using digital biology will:

[...] guide us through the ever-growing complexity of our global, interactive, fast-paced, modern lives. These new software techniques will provide us with invaluable assistance from their digital universes. They will find us information, detect crime for us, identify faults, and even repair themselves. They will design new products for us, create art and compose music. They will have originality, creativity and the ability to think for themselves. How do I know that these things will happen? Because they already have. Using the methods of digital biology, we can and have achieved all of these feats. (Bentley, 2011, p. 18)

Digital biology and nature follow the same processes but in different universes. Digital biology is not a simulation (imitation) of nature or creating virtual or artificial nature. The computer program as the creator of digital biology is the universe, in and of itself using natural principles to create a digital universe. A universe can only exist if there are laws or rules which determine it, which is the computer program. It is a set of inputs that give the computer its behaviour, which leads us to conclude that every program determines a new digital universe. The program decides the rules, and the computer causes the universe to come into existence. Bentley wonders:

Why should something in the digital universe of a computer be considered 'less real' than a physical object in our universe? Why is virtual reality virtual? Why is artificial life artificial? Isn't it a little unfair? (Bentley, 2011, p. 54)

A computer allows the actual evolution to take place in another medium. "If we ever manage to evolve successfully a neural network capable of consciousness in a digital universe, will it not deserve to be called alive rather than artificial?" (Bentley, 2011, p. 54). As physical and digital objects are made up of energy, one can conclude that the digital universe is not virtual but parallel to our universe, argues Bentley. A computer program can have all the elements of natural evolution: multiplication, selection, and change. In this case, the bits multiply, their selection takes place and they undergo change in a digital universe of a computer. Therefore, we need evolutionary algorithms to create real evolution inside of a computer. Neural computer networks follow the same fundamental processes as biological brains. For a digital brain to become self-aware, it needs a human number of neurons and synapses and the appropriate anatomy. This can be achieved, but when exactly is a matter of time.

Marvin Minsky at the MIT believes "that we should actually regard the biological brain as a very complex machine, which is built from a series of smaller, largely independent computers or 'agencies'. (He also believes that consciousness is largely illusionary, but we will return to this later.) He calls his theory the 'Society of the Mind', where the separate brain regions act like members of a society, co-operating through separate activities to help the whole." (Bentley, 2011, p. 138). Using this idea, Rodney Brooks (also from MIT) developed a digital brain with specialized areas which consist of separate circuitry and computers which perform independent tasks. This way, Rodney and his associates build brains of their robots with an evolution-like process. These brains are operated by multiple parallel processors which test the sensors and suggest procedures at the same time. However, these brains have so far reached the level of a small group of snails.

We have shown the possible announced types of enhancing human beings which will lead to changes in bodies and minds of our biological evolution:

Speciation[1] will not be a deliberate, programmed event; instead, it will involve an ever-faster accumulation of small, useful improvements that will eventually turn Homo sapiens into a new hominid. We will begin to see evidence of this partly mechanical, partly regrown creature who is rapidly driving its own evolution. As the branches of the tree of life, and of hominids, continue to spread, many of our grandchildren will likely engineer themselves into what we would consider a new species, one with extraordinary capabilities: Homo evolutis. (Enriquez, 2010, p. 177)

Will it be possible to stop this course of evolution? In all probability "Stopping the Cyber Revolution is probably not possible. The growing knowledge base should make the production of superintelligent minds less difficult and much faster than replicating, growing, and educating human beings. [...] supernaturalistic religion will evaporate, as ordinary minds become as powerful as gods" (Paul, 2010, pp. 117-118). We dream of revolution and changing the human nature, but as the Latin poet Horace wrote: "You can drive out nature with a pitchfork, but she will always return" (Humphrey, 2010, p. 390). Or at least, we hope she will.

REFERENCES

Anderson, F. (1999). Gene Therapies. In S. Griffiths & J. Weiner (Eds.), *Predictions: 30 great minds on the future* (pp. 13–22). Oxford Univ. Press.

Bacheldor, B. (2007). RFID-enabled Handheld Helps Nurses Verify Meds. *RFID Journal*. http://www.rfidjournal.com/article/articleview/3470/

BBC News. (2004). Barcelona clubbers get chipped. *BBC News*. http://news.bbc.co.uk/2/hi/technology/3697940

Benford, G. (2010). Living to a Hundred and Fifty. In J. Brockman (Ed.), *This will change everything: Ideas that will shape the future* (pp. 182–185). Harper Perennial.

Bentley, P. (2011). *Digital biology: How nature is transforming our technology*. Headline.

BlueBean. (2007). The Benefits of RFID in the Healthcare Organization. *RFID Solutionsfor the Healthcare Industry*. http://www.rfidjournal.com/article/articleview/3470/

Bostrom, N. (2010). Superintelligence. In J. Brockman (Ed.), *This will change everything: Ideas that will shape the future* (pp. 111–114). Harper Perennial.

Brin, D. (1998). *The transparent society: Will technology force us to choose between privacy and freedom?* Perseus Books.

Brockman, J. (2010). *This will change everything: Ideas that will shape the future*. Harper Perennial.

CASPIAN Special Report. (2004). FDA letter Raises Questions about VeriChip Safety. *Data Security*. http://www.spychips.com/devices/verichip-fda-report.html

Clarke, A. C. (1999). 2099...the beginning of history. In S. Griffiths & J. Weiner (Eds.), *Predictions: 30 great minds on the future* (pp. 35–46). Oxford Univ. Press.

Cuzin, M. (2001). DNA chips: A new tool for genetic analysis and diagnostics. *Transfusion Clinique et Biologique*, *8*(3), 291–296. doi:10.1016/S1246-7820(01)00141-0 PMID:11499980

Dawkins, R. (1999). A Riddle I Long to Answer. In S. Griffiths & J. Weiner (Eds.), *Predictions: 30 great minds on the future* (pp. 57–66). Oxford Univ. Press.

Day, M. (2012). Priest takes his secrets – and Berlusconi's confessions – to the grave. *Independent*. https://www.independent.co.uk/news/world/europe/priest-takes-his-secrets-and-berlusconi-s-confessions-grave-6284013.html

De Grey, A. (2010). The Unmasking of True Human Nature. In J. Brockman (Ed.), *This will change everything: Ideas that will shape the future* (pp. 378–379). Harper Perennial.

Dennet, D. C. (2010). Introduction. In J. Brockman (Ed.), *This will change everything: Ideas that will shape the future* (pp. xxii–xxiii). Harper Perennial.

Dyson, F. (2010). Radiotelepathy: Direct Communication from Brain to Brain. In J. Brockman (Ed.), *This will change everything: Ideas that will shape the future* (pp. 146–148). Harper Perennial.

Eagleman, D. (2010). Silicon Immortality: Downloading Consciousness into Computers. In J. Brockman (Ed.), *This will change everything: Ideas that will shape the future* (pp. 99–101). Harper Perennial.

Edwards, L. (2009). Intel wants a chip implant in your brain. *Phys.org*. https://phys.org/news/2009-11-intel-chip-implant-brain.html

Enriquez, J. (2010). Homo Evolutis. In J. Brockman (Ed.), *This will change everything: Ideas that will shape the future* (pp. 174–177). Harper Perennial.

Espreso. (2016). Rođen je novi čovek i zove se Homo optimus! *Espreso.rs*. https://www.espreso.co.rs/hi-tech/it-svet/29140/rodjen-je-novi-covek-i-zove-se-homo-optimus-foto-gif

Ford, K. W. (2010a). Reading Minds. In J. Brockman (Ed.), *This will change everything: Ideas that will shape the future* (pp. 141–142). Harper Perennial.

Freitas, R. (1985). Birth of the Cyborg. In M. Minsky (Ed.), *Robotics* (1st ed.). Anchor Press/Doubleday.

Frost & Sullivan Research Service. (2005). *World RFID in healthcare and pharmaceuticals markets*. http://www.frost.com/prod/servlet/cpo/55930632

Fukuyama, F. (1999). The Politics of Women. In S. Griffiths & J. Weiner (Eds.), *Predictions: 30 great minds on the future* (pp. 107–118). Oxford Univ. Press.

Garfinkel, S. (2003). *Database Nation: The Death of Privacy in the 21st Century*. O'Reilly and Associates.

Garfinkel, S., & Holtzman, H. (2006). Understanding RfID Technology. In B. Rosenberg & S. Garfinkel (Eds.), *RFID: applications, security, and privacy*. Addison-Wesley.

Garfinkel, S. L., Juels, A., & Pappu, R. (2005). RFID Privacy: An Overview of Problems and Proposed Solutions. *IEEE Security and Privacy Magazine*, *3*(3), 34–43. doi:10.1109/MSP.2005.78

Geary, J. (2010). The Brain-Machine Interface. In J. Brockman (Ed.), *This will change everything: Ideas that will shape the future* (pp. 14–16). Harper Perennial.

Gershenfeld, N. (2010). The Implementation of Life in Engineered Materials. In J. Brockman (Ed.), *This will change everything: Ideas that will shape the future* (pp. 102–103). Harper Perennial.

Gleiser, M. (2010). Mastering Death. In J. Brockman (Ed.), *This will change everything: Ideas that will shape the future* (pp. 186–188). Harper Perennial.

Golubović, R. (2009). Za 20 godina postajemo besmrtni kiborzi. *Radio Slobodna Evropa.* https://www.slobodnaevropa.org/a/svijet_nauka_tehnologija/1837118.html

Greenfield, S. (1999). George Orwell's 1984 ... In 2084. In S. Griffiths & J. Weiner (Eds.), *Predictions: 30 great minds on the future* (pp. 149–158). Oxford Univ. Press.

Greguric, I. (2014, Winter). *Kiborgoetika* [Interview]. Academic Press.

Groothuis, D. R. (1999). *The soul in cyberspace.* Wipf ans Stock.

Harris, M. (2017). God Is a Bot, and Anthony Levandowski Is His Messenger. *Wired.* https://www.wired.com/story/god-is-a-bot-and-anthony-levandowski-is-his-messenger/

HINA. (2016). Smrtno bolesna britanska djevojčica izborila se za pravo da zamrzne tijelo. *Večernji list.* https://www.vecernji.hr/vijesti/smrtno-bolesna-britanska-djevojcica-izborila-se-za-pravo-da-zamrzne-tijelo-1129410/galerija-225843?page=1

Humphrey, N. (2010). Why Human Nature Will Rebel. In J. Brockman (Ed.), *This will change everything: Ideas that will shape the future* (pp. 389–390). Harper Perennial.

Iacobini, M. (2010). Immortal Cognition, Boundless Happiness. In J. Brockman (Ed.), *This will change everything: Ideas that will shape the future* (pp. 289–291). Harper Perennial.

Icke, D. (2001). *Children of the Matrix: How an Interdimensional Race has Controlled the World for Thousands of Years - and Still Does.* Bridge of Love Publications USA.

Jurman, H. (2016). Najpoznatiji svjetski futurolog tvrdi: Živjet ćemo vječno. *Zimo.* http://zimo.dnevnik.hr/clanak/najpoznatiji-svjetski-futurologtvrdi-zivjet-cemo-vjecno---434374.html

Knutson, B. (2010). Neurophenomics + Targeted Stimulation = Psychological Optimization? In J. Brockman (Ed.), *This will change everything: Ideas that will shape the future* (pp. 278–280). Harper Perennial.

Kosko, B. (2010). Cheap Cryonic Suspension of Brains. In J. Brockman (Ed.), *This will change everything: Ideas that will shape the future* (pp. 107–110). Harper Perennial.

Kristović, I. (2016). Povijesna odluka! Britanci 'legalizirali' djecu s tri biološka roditelja. *Večernji list.* https://www.vecernji.hr/techsci/povijesna-odluka-britanci-legalizirali-djecu-s-tri-bioloska-roditelja-1135892

Lijović, D. (2017). Kamo nas vodi razvoj umjetne inteligencije. *Večernji list.* https://www.vecernji.hr/techsci/kamo-nas-vodi-razvoj-umjetne-inteligencije-1204297

Marcus, G. (2010). Decoding the Brain. In J. Brockman (Ed.), *This will change everything: Ideas that will shape the future* (pp. 104–106). Harper Perennial.

Myers, D. G. (2010). Inexpensive Customizable Interactive E-Texts for Worldwide Use. In J. Brockman (Ed.), *This will change everything: Ideas that will shape the future* (pp. 241–242). Harper Perennial.

Negroponte, N. (1996). Being digital. Vintage Books.

Norman, D. (1999). Cognitive Prostheses. In S. Griffiths & J. Weiner (Eds.), *Predictions: 30 great minds on the future* (pp. 169–180). Oxford Univ. Press.

Paul, G. (2010). Becoming Robotic. In J. Brockman (Ed.), *This will change everything: Ideas that will shape the future* (pp. 115–118). Harper Perennial.

Postman, N. (1993). Technopoly: The surrender of culture to technology. Vintage Books.

Powell, C. (2010). Slippery Expectations. In J. Brockman (Ed.), *This will change everything: Ideas that will shape the future* (pp. 20–24). Harper Perennial.

Rushkoff, D. (2010). Slippery Expectations. In J. Brockman (Ed.), *The Discovery of Intelligent Life from Somewhere Else* (pp. 63–64). Harper Perennial.

Sampson, S. (2010). Evolution Changes Everything. In J. Brockman (Ed.), *The Discovery of Intelligent Life from Somewhere Else* (pp. 1–4). Harper Perennial.

Schwartz, E. (2010). Neuronally Expressed Messages. In J. Brockman (Ed.), *The Discovery of Intelligent Life from Somewhere Else* (pp. 152–153). Harper Perennial.

Science of Singularity. (2014). How Close Are Humans to Immortality? *Science of Singularity*. http://scienceofsingularity.com/category/medicine/

Seife, C. (2010). Malthusian Information Famine. In J. Brockman (Ed.), *The Discovery of Intelligent Life from Somewhere Else* (pp. 139–140). Harper Perennial.

Smith, B. D. (2006). *Using radiofrequency identification (RfID) technology in humans in the United States for total control*. Bowie State University. http://www.myimagehosting.com/u/18183/14602/126820.pdf

Stockman, H. (1948). Communication by Means of Reflected Power. *Proceedings of the IRE*, *36*(10), 1196–1204. doi:10.1109/JRPROC.1948.226245

Tipler, F. J. (2010). But We Shall All Be Changed. In J. Brockman (Ed.), *The Discovery of Intelligent Life from Somewhere Else* (pp. 39–42). Harper Perennial.

Tooby, J., & Cosmides, L. (2010). AI and Intellectual Mastery. In J. Brockman (Ed.), *The Discovery of Intelligent Life from Somewhere Else* (pp. 68–71). Harper Perennial.

Turkle, S. (1999). Toys to Change Our Minds. In S. Griffiths & J. Weiner (Eds.), *Predictions: 30 great minds on the future* (pp. 269–280). Oxford Univ. Press.

Turkle, S. (2010). *Alone together: Why we expect more from technology and less from each other*. Basic Books.

Venter, J. C. (2010). DNA: Writing the Software of Life. In J. Brockman (Ed.), *The Discovery of Intelligent Life from Somewhere Else* (pp. 5–7). Harper Perennial.

Verlag, V. D. I. (2005). Im jahre 2030 leben wir alle in digitalen Blasen. *VDI Verlag*. http://www.vdi-nachrichten.com/artikel/Im-jahre-2030-lebenwir-alle-in-digitalen-Blasen/20499/1

Watson, J. (1999). Re-Directing the Course of Human Evolution. In S. Griffiths & J. Weiner (Eds.), *Predictions: 30 great minds on the future* (pp. 291–300). Oxford Univ. Press.

Willis, A. (2009). Rich may evolve into separate species. *The Telegraph*. https://www.telegraph.co.uk/news/science/evolution/6432628/Rich-may-evolve-into-separate-species.html

Yang, S. (2011). New blood analysis chip could lead to disease diagnosis in minutes. *Berkeley News*. https://news.berkeley.edu/2011/03/16/standalone-lab-on-a-chip/

Z.K. (2016). UDARNA VIJEST: Donji dom američkog parlamenta odobrio čipiranje ljudi! *Dnevno.hr*. http://www.dnevno.hr/vjera/vjera-kultura-iznanost/udarna-vijest-donji-dom-americkog-parlamenta-odobrio-cipiranje-ljudi-983998/

ENDNOTE

1 Emergence of the new species.

Chapter 10
Prolegomena for Cyborgoethics

ABSTRACT

This chapter draws philosophical and ethical attention to the procedures for enhancing and reshaping of the human body and mind whose aim is not to heal but to increase functions and capabilities of the healthy body and mind and establishing life on a new non-biological basis. Completely enhancing cyborgization of social relations and the entire historical reality of scientific humanism and the domination of artificial intelligence over human intelligence will not be in the function of preservation of life of man but establishing partially or completely artificial beings which will not need their biological father.

11. PROLEGOMENA FOR CYBORGOETHICS

In the new-age vision of nature, "Science had made nature 'fit', cognitively and emotionally, for the kind of treatment that was eventually applied to it" (Jonas, 1974, p. 173). Turned into the object of man's creation, nature "expresses no creative will, either of its own or of its cause, then man remains as the sole subject and the sole will" (Jonas, 1974, p. 173).

Implanting artificial technologies helps human body regain lost functions or organs. From the standpoint of maintaining health, establishing the lost natural balance of the body, dignity and value of life of man as a person, these types of cyborgization can be held allowable but certainly not value neutral. Similarly, cyborgization of historical reality of man makes man's life in the globalized world that much easier, but its final outcome is in stark contrast with the essence of human life and biological life in general. Our philosophical and ethical attention and moral concern is preoccupied with procedures of enhancing and reshaping of the human body and mind whose aim is not to heal but to increase functions and capabilities of the healthy body and mind and establishing life on a new non-biological basis. Completely enhancing cyborgization of social relations and the entire historical reality of *scientific humanism as naturalism* and the domination of artificial intelligence over human intelligence will not be in the function of preservation of life of man but establishing partially or completely artificial beings which will not need its biological father. This calls for a public intersubjective discussion on the values of human life and its preservation. The same applies to preservation of immediate interhuman relations and the autonomy of nature and the natural. Man is one with the nature – he is a natural and spiritual

DOI: 10.4018/978-1-7998-9231-1.ch010

being. We are together in everything. As our life companion, nature reciprocates every appropriate relationship but also reacts violently to every human excessive misuse of hospitality! This underlines the importance of warning by Hans Jonas that the question at stake today is not just the existence of man but that of the entire nature, which is endangered by the scientific and technological procedures. This requires man to undertake moral considerations towards the unity of natural life. According to Jonas, mankind presents a bigger threat to nature than nature used to present for us and in this process, we became a threat to ourselves due to our technological advancements. The only difference now is that we are the danger that we are surrounded with (Jonas, 1987, pp. 34-35). We must not understand nature merely as the natural environment but also as a morally relevant entity in and of itself. The hour is nigh for establishing a legal framework for procedures of *scientific work*. Our moral actions must be based in wisdom, not in the scientific and technological mind.

Even today, human biological body has become the testing ground for implanting chips with ultrasound signals which enable better spatial orientation, radiofrequency chips which enable a new form of communication, vision-enhancing ultrasonic radar, memory cards to augment memory etc. Futuristic predictions of enhancement and reshaping of man announce cyborgization of the human body through artificial intelligence which will control the man-robot, while on the other hand producing artificial beings – humanoid robots as vessels for uploading the content of the human brain. Historical reality is being established by procedures of *scientific work* as a scientific reality, and humanism as *scientific humanism and naturalism*. This reality is leaning towards virtuality – a virtually augmented reality. Parallel to implementation of artificial technology into the human organism, biological body and mind are being immersed into the digitalized space-time of human bodiless existence and the movement of the entire virtual network. For Gibson, cyberspace as "consensual hallucination" (Gibson, 1984, p. 4) is a world without boundaries between the material and virtual, in which every identity is deconstructed with only shades in between as the intermediary steps which dissolve the memory of previously clear and unambiguous identities. Body and identity are lost in the procedures of *scientific world* in the virtual world which virtually enhances human capabilities, but at a global level transform man into an object and means of scientific and technological progress. Abandoning the body as mere 'flesh' in the bodiless happiness of the cyberspace reflects the platonic worldview in which the physical tactile world is nothing more but a reflection of the purer, more ideal higher form. The artificial language of the binary code, devoid of everyday conventions of material communication, is on its way of reaching this glorious perfection of the absolute form (Gibson, 1984, p. 7). Transformation of the physical bodily information into the pure form of bodiless information and bodiless intelligence leads the human beings into the domain of digitalized ontology and anthropology. We are no longer dealing with a real biological man, nor his metal descendant, but a reduction of man and historical reality onto a binary code.

The posthumanists as the ideologists of *scientific work* do not deny the connection between the human mind, human nature, and morale. However, they prioritize the liberal rights of individuals over morally just life and life in general. According to them, every individual has the right to choose enhancement in line with the needs of *scientific work* and the goals of its unstoppable growth. Their humanist and posthumanist ethics does not question the character of society and the being of time of *scientific humanism as naturalism* from which they postulate the ethical values about the essence of transhumanism. Both of these are assumed as self-explanatory.

In our analysis which starts from the dialogue of thinkers at the end of metaphysics – Hegel and Marx – we have shown how *scientific work* and its metaphysical way of thinking are established in the real-world establishment of everything that is, in that what is. In establishing everything that is, man is a participant but not a self-conscious subject which realizes his human needs in this establishment. Instead, he's a mere object and means of self-movement of *scientific work*. Man participates in the being of scientific history and therefore has only virtual power, but also a real-world responsibility for the historical pathway and the destiny of mankind. In *scientific humanism as naturalism*, man – cyborg – scientific worker, illusion, is not by self, by its own essence, but is completely alienated, like all other beings and the rest of the world. One could argue that the extent of man's closeness to *scientific work* is also the extent of his alienation and reification. This world, the world of *scientific humanism as naturalism*, cannot form the basis of neither transhumanist nor poshumanist understanding of man, moral and the world because that would be *contradictio in adjecto*.

Before engaging in a discussion on cyborgoethic principles which we advocate, we shall turn our attention to several answers put forth by philosophers to the question – *What is man*? (Kant). Using these discussions in the context of our assumption of *scientific work* as the subject substance of current scientific and historical reality which draws of all life from the biological body of man and his philosophically mind bilaterally divided mind (in unity and difference towards the biological body) in the age of *scientific humanism as naturalism*, we will try to reach a stimulating and thought-out dialogue about the necessity of putting forward the ethical principles within and beyond the realms of metaphysics in which they are hidden, both ontologically and ideologically.

Throughout this book, we have insisted on unity of philosophy and necessity to reject all acts of limiting the objects of its inquiry into closed (and potentially self-sufficient) disciplines (Veljak, 2011, p. 10). This why we are attempting to establish the principles of cyborgoethics and moral action by overcoming metaphysics in the openness of thought towards life, world, historical events, and practical actions which do not establish a new quality of survival in history but indicate its potential.

In the *Critique of Pure Reason*, Kant makes the following claim about the duality of man:

Man is a phenomenon of the sensuous world and, at the same time, therefore, a natural cause, the causality of which must be regulated by empirical laws. As such, he must possess an empirical character, like all other natural phenomena. We remark this empirical character in his actions, which reveal the presence of certain powers and faculties. If we consider inanimate or merely animal nature, we can discover no reason for ascribing to ourselves any other than a faculty which is determined in a purely sensuous manner. But man, to whom nature reveals herself only through sense, cognizes himself not only by his senses, but also through pure apperception; and this in actions and internal determinations, which he cannot regard as sensuous impressions. He is thus to himself, on the one hand, a phenomenon, but on the other hand, in respect of certain faculties, a purely intelligible object- intelligible, because its action cannot be ascribed to sensuous receptivity. These faculties are understanding and reason. The latter, especially, is in a peculiar manner distinct from all empirically-conditioned faculties, for it employs ideas alone in the consideration of its objects, and by means of these determines the understanding, which then proceeds to make an empirical use of its own conceptions, which, like the ideas of reason, are pure and non-empirical. (Kant, 1781, p. 211)

This traditional philosophical gap within man into phenomenal noumenal is particularly prominent in *The Metaphysics of Morals* in which Kant argues that "In the system of nature, man (*homo phaenomenon, animal rationale*) is a being of slight importance" (Kant, 1991[1797], p. 230). Man occurs in its natural determination and is distinguished from other beings by the fact that his actions have purpose and he alone can be a means of someone else's purpose. In contrast to that

[...] man regarded as a person, that is, as the subject of a morally practical reason, is exalted above any price; for as a person (homo noumenon) he is not to be valued merely as a means to the ends of others or even to his own ends, but as an end in himself, that is, he possesses a dignity (an absolute inner worth) by which he exacts respect for himself from all other rational beings in the world. He can measure himself with every other being of this kind and value himself on a footing of equality with them. (Kant, 1991[1797], p. 230).

At this point, it is necessary to elucidate the multifacetedness of Kant's term human nature and pragmaticality of man. At the level of *Critique of Pure Reason*, the nature of man is explained as a mentally determined human behaviour, but from the perspective of his entire transcendental philosophy, one might argue that human nature contains both tactile aspects and reason, which shows the sensuous and supersensuous nature of man. Sensuousness and reason are the two trees of human cognition.

In the *Critique of Practical Reason*, tactile nature is explained as the medium of man's existence in which natural laws apply and the super-sensuous nature is governed by the laws of the name. These two aspects of human nature, sensuous and supersensuous, are related as cause and mimicking of the cause. Using this distinction, Kant tries to overcome the duality of human nature and subject his human moral behaviour in the sensuous domain to the laws of his man. The dual nature of man, the double legislature of his relation towards other men, beings and nature in the historical world make the essence of man and his free moral actions. Kant believes that historical progress will bring about the domination of the mental over sensuous. Moral action starts from the mind, not empirical reasons governed by causality and laws of mechanics, which is why moral action functions as coercion over empirical action. Moral action is the necessity and moral laws are coercion. Morality is based in mind alone and is realized through autonomy of pure will of inner man which is the maker of rules by itself.

Man as the absolute, free and moral subject of human history subjects both nature and society to the laws of the mind. In other words, mind as the lawmaker is superior to the natural laws and the sensuous part of human nature. In its progress, mental action is trying to subject the sensuous nature of man to its laws and principles and one might say that the laws of the mind allow us to encroach into the sensuous part of the human nature, i.e. his body. The sensuous (bodily) is susceptible to mind and serves it. This way, anthropology of man is subjected to the ethics and metaphysics of scientific and technological mind and its experience of measuring the world. Kant's conclusion comes from both his principle that man "he is not to be valued merely as a means to the ends of others or even to his own ends, but as an end in himself" (Kant, 1991[1797], p. 230) and the fact in his time one could not see *scientific work* as the substance subject of world history and its procedures in reshaping and enhancing the sensuous and supersensuous nature in general, but his support for *scientific work* is implicit in his philosophy.

To answer Kant's anthropological question, one could draw a conclusion from his entire philosophy that man is the absolute subject of his own historical realization in the game of his sensuous and supersensuous nature, that he is the person whose uses his own autonomous will and the moral laws of his own mind (scientific and technological mind) to act towards himself, other humans, other beings, and nature in historical time. As we have shown, Hegel tries to overcome the dual character of here-being within the philosophy of the subjective spirit through its substantialization. In his analysis of Kantian philosophy, Hegel argues:

Kantian philosophy may be most accurately described as having conceived of the spirit as consciousness, and as containing only determinations of the phenomenology, not the philosophy, of spirit. Kant views the self as the relation to a "thing in itself" lying somewhere beyond, and it is only from this perspective that he treats the intellect and the will. Though with the concept of reflecting judgment he does speak of the idea of the spirit, subject-objectivity, an intuitive understanding, and so on, and even the idea of nature, this idea is itself demoted to an appearance again, namely, to a subjective principle. (Hegel, 2012, §332)

Unlike Kant, Hegel defines his ethical and moral law and the command for man's action in the concept, which has ontological significance. This means that the concept contains the real which gets realized as the absolute spirit in history. Moral law has its foundations in the general, which for Hegel represents the subjecting, i.e. the active capability of the human mind for the general. In his earlier writings, Hegel had not reached the standpoint of spirit and is trying to overcome the relationship of natural as external and mental as the internal of man with philosophical and anthropological interpretation of the New Testament. In his interpretation of Jesus's teaching, Hegel connects the divine and human in man, not through naturality of man or his divine lineage, but through concept as an imaginary universal. This represents the convergence between the divine character and the human as the features of man: "man is not one nature, one essence, like the Godhead; it is a concept, an *ens rationis*. And 'son of man' means here 'something subsumed under the concept of man.' Jesus is man' is a judgment proper; the predicate is not a living essence but a universal" (Hegel, 1948, p. 262).

Hegel believes that man's intelligence does not only grant him the ability for universal thought but that it makes him what he is. Hegel cites work and value as features of universality of man. Hegel will later have to abandon this early philosophical standpoint for establishing the philosophy of spirit which overcomes the difference between the external and internal. In his *Philosophy of Right*, Hegel remarked:

The unity of visible existence and conception, of body and soul, is the idea. It is not a mere harmony of the two, but their complete interpenetration. There lives nothing, which is not in some way idea. The idea of right is freedom, which, if it is to be apprehended truly, must be known both in its conception and in the embodiment of the conception. (Hegel, 1952, p. 13).

This unity of mental and real, absolute and life, the absolute and the empirical consciousness is reflected in the human being as the medium of the self-aware realization of the absolute spirit of world history which is also the substance subject of everything that is.

In his *Introduction to the Philosophy of Spirit*, Hegel makes it clear that nature of man is of interest to him only insofar as the manifestation of the spirit:

The knowledge of Mind is the highest and hardest, just because it is the most 'concrete' of sciences. The significance of that 'absolute' commandment, Know thyself -- whether we look at it in itself or under the historical circumstances of its first utterance -- is not to promote mere self-knowledge in respect of the particular capacities, character, propensities, and foibles of the single self. The knowledge it commands means that of man's genuine reality -- of what is essentially and ultimately true and real -- of mind as the true and essential being. Equally little is it the purport of mental philosophy to teach what is called knowledge of men -- the knowledge whose aim is to detect the peculiarities, passions, and foibles of other men, and lay bare what are called the recesses of the human heart. Information of this kind is, for one thing, meaningless, unless on the assumption that we know the universal – man as man, and, that always must be, as mind. And for another, being only engaged with casual, insignificant, and untrue aspects of mental life, it fails to reach the underlying essence of them all -- the mind itself. (Hegel, 2001, p. §377)

The spirit itself "has for its presupposition Nature, of which it is the truth, and for that reason its absolute prius. In this its truth Nature is vanished, and mind has resulted as the 'Idea' entered on possession of itself. Here the subject and object of the Idea are one -- either is the intelligent unity, the notion." (Hegel, 2001, §381). The spirit "came into being as the truth of Nature. But not merely is it, as such a result, to be held the true and real first of what went before: this becoming or transition bears in the sphere of the notion the special meaning of 'free judgement'. Mind, thus come into being, means therefore that Nature in its own self realizes its untruth and sets itself aside: it means that Mind presupposes itself no longer as the universality which in corporal individuality is always self–externalized, but as a universality which in its concretion and totality is one and simple. At such a stage it is not yet mind, but soul." (Hegel, 2001, §388).

The natural soul "is no separate immaterial entity. Wherever there is Nature, the soul is its universal immaterialism, its simple 'ideal' life. Soul is the substance or 'absolute' basis of all the particularizing and individualizing of mind: it is in the soul that mind finds the material on which its character is wrought, and the soul remains the pervading, identical ideality of it all. But as it is still conceived thus abstractly, the soul is only the sleep of mind – the passive of Aristotle, which is potentially all things." (Hegel, 2001, §389). In Hegel's view

A somewhat different answer has been given by all philosophers since this relation came to be expressly discussed. Descartes, Malebranche, Spinoza, and Leibniz have all indicated God as this nexus. They meant that the finitude of soul and matter were only ideal and unreal distinctions; and, so holding, there philosophers took God, not, as so often is done, merely as another word for the incomprehensible, but rather as the sole true identity of finite mind and matter. But either this identity, as in the case of Spinoza, is too abstract, or, as in the case of Leibniz, though his Monad of monads brings things into being, it does so only by an act of judgement or choice. Hence, with Leibniz, the result is a distinction between soul and the corporeal (or material), and the identity is only like the copula of a judgement, and does not rise or develop into system, into the absolute syllogism. (Hegel, 2001, §389)

In the natural soul "the mind takes part in the general planetary life, feels the difference of climates, the changes of the seasons, and the periods of the day" (Hegel, 2001, §392), but is not by itself and is not free. Hegel discusses the three stages of the soul: the natural soul, the feeling soul and the actual soul which represents a path towards the spirit. If we want to explicate the body-soul relationship from Hegel's entire philosophy, we should use his fundamental standpoint in understanding the ontological structure of historical reality, which starts from the idea as

truth in the deeper sense means that objectivity is identical with the Concept.[...] It is this deeper sense of truth which is at issue when we speak, for instance, of a 'true' State or a 'true' work of art. These objects are 'true' when they are what they ought to be, i.e., when their reality corresponds to their concept. Interpreted in this way, the 'untrue' is the same as what is sometimes also called the "bad". A bad man is one who is "untrue", i.e., one who does not behave in accord with his concept or his destination. But without any identity at all between Concept and reality nothing can subsist. Even what is bad and untrue can only be because its reality conforms to some extent with its Concept. (Hegel, 1991, pp. 287-288)

In this discussion, Hegel argues that "Idea is essentially *concrete*, because it is the free Concept that determines itself and in so doing makes itself real" (Hegel, 1991, p. 287). He continues by claiming that: "The Idea can be grasped as reason [...] and further as the Subject-Object, as the unity of the ideal and the real, of the finite and the infinite, of the soul and the body, as the possibility that has its actuality in itself, as that whose nature can be comprehended only as existing, and so forth. [It can be grasped in all these ways] because all the relationships of the understanding are contained in the Idea, but in their infinite self-return and self-identity." (Hegel, 1991, p. 288).

In his view, the relation between the spirit (and soul) towards the body is the relation between the body and existence: "The Concept is realised as soul, in a body. The soul is the immediate self-relating universality of the body's externality; it is equally the particularising of the body, so that the body expresses no distinctions in itself other than the determinations of the Concept; and finally it is singularity as infinite negativity" (Hegel, 1991, §216). The soul cannot exist without the body as it is its external survival, manifested as individual and mortal in the naturality of the body. "In this sphere, the determination of finitude is that, because of the immediacy of the Idea, soul and body are separable; this constitutes the mortality of what is alive. But it is only insofar as it is dead that these two sides of the Idea are diverse components." (Hegel, 1991, §216). Thus, Hegel does not deny the body as the spirit needs the natural existence as the other-being from which it returns back to its unity within itself. The body as the external existence of the soul "represents not itself, but the soul, of which it is the sign." (Hegel, 2001, §411). However, Hegel had a high regard for the human activity which has a crucial effect on the human soul which is why activity is a more reliable sign of the soul than the body itself. The individual has to be confirmed in the immediate harmony of the body and soul – the body must be given what belongs to it, lest it becomes an external necessity. The soul's act of claiming of the body allows it to become its own substance and freedom (Burger, 2001, 142). The soul "is virtually the totality of nature: as an individual soul it is a monad: it is itself the explicitly put totality of its particular world – that world being included in it and filling it up; and to that world it stands but as to itself" (Hegel, 2001, §403).

As one of the standpoints of our understanding of man in relation to cybernetic enhancement of his body and mind, we should refer to Hegel's view that "Sensibility (feeling) is the form of the dull stirring, the inarticulate breathing, of the spirit through its unconscious and unintelligent individuality, where every definite feature is still 'immediate' [...]. The content of sensation is thus limited and transient, belonging as it does to natural, immediate being – to what is therefore qualitative and finite." (Hegel, 2001, §400). Furthermore, "What the sentient soul finds within it is, on one hand, the naturally immediate, as 'ideally' in it and made its own." (Hegel, 2001, §401). Through the constitution of self-being, Hegel realizes the intention of denaturalizion and universalization of man, and the concept as a universal is pervaded by the duality of natural and supernatural in man. Through his dialectic development through the three spheres of the spirit, Hegel allowed man to advance in world history, with an indication of a certain regression in the domain of objective spirit, due to contradictory stance of the individual in the dialectic of the civil society. Thus, Hegel sees the real man within the values of the civil society of the time, dependant on his mind as the assumption of existence in freedom as a self-implied necessity. However, from Hegel's ontological standpoint, denaturalized and generalized man, man in his conceptual unity has become a mere medium, means, and matter of realization of the absolute spirit of world history, i.e. of being as an apparition which needs historical reality as a confirmation of his foundation. The corrective force in suppressing his natural passions and urges in moral action is the power of human will.

Since we have discussed Marx's understanding of the essence of man in a separate chapter, at this point we should only reiterate that for him man is a natural and historical being, an engendered being which owes its essence and existence to labour as his one true power, both in material and in spiritual sense. The entire spiritual being of man stems from the reality production of life, from *scientific work*. This way, work (*scientific work*) and biological body of man connects nature to itself and enters historical world as a conscious entity. The very history is the medium of realization of social and individual life of man, a realized *humanism and naturalism* of man.

After Kant and Hegel, the discussion on the anthropological essence of man is directed, on the one hand, towards scientific and empirical insights, and on the other hand towards philosophical discussions about the anthropology of man. In the philosophical discussion about the anthropological essence of man, special attention should be given to Max Scheler, Helmuth Plessner, Ernest Cassirer, Arnold Gehlen, Ernest Tugendhat, Hans Krämer, Nikolai Hartmann and Jürgen Habermas, with whose ideas we will engage in a philosophical dialogue. In *The Future of Human Nature*, Habermas is trying to start an argument-based ethical and theoretical reaction with practical implications for ethical and moral challenges of applying new biotechnologies on human beings. There is a certain parallel, but also certain points of divergence, between his attempts and our Prolegomena for cyborgoethics. The difference lies in the ontological questions and foundedness of moral action. Habermas believes that post-Kantian metaphysics is undergoing a crisis, which is why he leaves it aside and attempts to make a linguistic turn in thought. We believe that we should look into the depths of realization of metaphysics through current realization of *scientific work in the age of scientific humanism as naturalism* to find the possibility of different thought, living and moral action. Similarity is reflected in philosophical scrutinization of procedures of current realization of *scientific work*. Habermas questions the moral and ethical challenges caused by use of new biotechnologies for the purpose of enhancing human beings. We are questioning moral and ethical challenges caused by cyborg enhancements and reshaping of human body and mind in the cyborgized historical reality and virtual reality of *scientific humanism as naturalism* which shift the boundary between the Marxian *empire of necessity* and the *empire of freedom*.

In his search for a new anthropological standpoint, Habermas asks the question: "Are there post-metaphysical answers to the question: What is the good life? … in our postmetaphysical age, philosophy no longer pretends to have answers to questions regarding the personal, or even the collective, conduct of life." (Habermas, 2003, p. 1). Philosophy and philosophers used to believe "that they were able to assure themselves about their ability to discuss the whole of nature and history" (Habermas, 2003, pp. 1-2) and secure a meaningful life of individuals and the community by means of philosophical and religious values. However, the increasing development of science and technology paired up with political liberalism made the old metaphysical foundations questionable and uncertain. In this new light, an old question is posed: "What ought I, or what ought we, to do?" (Habermas, 2003, p. 3). In Habermas's view, this leads to the question of identity: "[…] how we should understand ourselves, who we are and want to be." (Habermas, 2003, p. 3), as well as "why we should be moral *at all*" (Habermas, 2003, p. 4), and "why the citizens of a democratic polity […] should orient themselves toward the common good" (Habermas, 2003, p. 4).

Habermas believes that "we subjects capable of speech and action" (Habermas, 2003, p. 10) should perform a "linguistic turn" as "a deflationary interpretation of the 'wholly Other'", which represents the postmetaphysical thought (Habermas, 2003, p. 10). Humans communicate with and understand one another through language whose logos "embodies the power of the intersubjective, which precedes and grounds the subjectivity of speakers." (Habermas, 2003, p. 11). This is a clear sign that the logos of language "escapes our control" though we "reach an understanding with one another in this medium" (Habermas, 2003, p. 11). At the same time, we are not given "right ethical self-understanding", we can only reach it "in a common endeavour. From this perspective, what makes our being-ourselves possible appears more as a trans-subjective power than an absolute one" (Habermas, 2003, p. 11). As communicative action with intersubjective character, Habermas's interpretation of philosophy is that of reflection and discourse – description of the world and self-description of the immanence of consciousness, defined by the grammars of linguistic games in particular domains of reality. Philosophy is trying to develop intuitions that guide individual thinkers and everyone engaged in thought about life and world (Burger, 2008, p. 139). Communicative action and language are not defined by one man, but go beyond him and encompass the true postmetaphysical thought. According to Burger, the answer to the question on justified reservation of philosophy is the following "philosophy must in this respect be reserved in content and deal only with the formal questions of argumentation or procedures as there is no more metaphysical or any other background or basis" (Burger, 2008, p. 139).

In relation to Habermas's stance on justified reservation of philosophy, in this book we advocate a completely unreserved philosophy in matters of life and morally valid life. We are aware of all real limitations of *scientific humanism as naturalism* established by *scientific work* as its being. We are also aware of the pluralism of interests of different centres of power and achieved freedoms and rights of individuals within liberal capitalism. We know that individual will for moral action cannot be moved by valid argumentation as the problem lies not as much in the lack of knowledge as in the difference of positions and interests which corrupt the will. Despite of everything, the only thought which can be oriented in the essence of time of *scientific humanism as naturalism* and at the edge of nihilism seeks the long lost well of life and human history – that is the philosophical, or even trans-philosophical thought that most certainly includes the discussion about the possible *linguistic turn*. In this time, it is known more by its loving yearning for wisdom than wisdom itself, but it is the only thing left before closing the final chapter of the biological life of human beings.

In terms of development of bioscience and biotechnology and application of their insights and capabilities onto human beings, Habermas argues that

What hitherto was 'given' as organic nature, and could at most be 'bred,' now shifts to the realm of artefacts and their production. To the degree that even the human organism is drawn into this sphere of intervention, Helmuth Plessner's phenomenological distinction between 'being a body' and 'having a body' becomes surprisingly current: the boundary between the nature that we "are" and the organic endowments we 'give' to ourselves disappears. As a result, a new kind of self-transformation, one that reaches into the depth of the organic substrate, emerges for the intervening subject. The self-understanding of this subject now determines how one wants to use the opportunities opened up with this new scope for decision – to proceed autonomously according to the standards governing the normative deliberations that enter into democratic will formation, or to proceed arbitrarily according to subjective preferences whose satisfaction depends on the market. [...] Do we want to treat the categorically new possibility of intervening in the human genome as an increase in freedom that requires normative regulation – or rather as self-empowerment for transformations that depend simply on our preferences and do not require any self-limitation? (Habermas, (2003, p. 12)

These questions directed Habermas's thought towards anthropology of man. In his attempts to answer the question posed by Kant: "what ought I do?", Habermas finds a foothold in Kierkegaard's anthropology. "Kierkegaard was the first philosopher who answered the basic ethical question regarding the success or failure of one's own life with a postmetaphysical concept of 'being-able-to-be-oneself.'" (Habermas,

2003, p. 5) by detaching "myself from the dependencies of an overwhelming environment, jolting myself to the awareness of my individuality and freedom." (Habermas, 2003, p. 6).

Kierkegaard stems from the "self-consciously existing individual, one continuously gives an account of one's life in light of the Sermon on the Mount." (Habermas, 2003, p. 6). An individual is constituted as a person "Through a morally scrupulous evaluation and critically probing appropriation of his factually given life history" (Habermas, 2003, p. 7) who is in Kierkegard's view "his own editor, but he is also fully aware that he is responsible, responsible for himself personally…responsible to the order of things in which he lives, responsible to God." (Habermas, 2003, p. 7). The power of self-being is "an ethical self-reflection and self-choice that is determined by the infinite interest in the success of one's own life-project" (Habermas, 2003, p. 6) and "should not be understood as narrow-minded self-empowerment" (Habermas, 2003, p. 10). Self-reflection and self-choice are thus means of establishing one's own life, which makes an individual an unrepresentable and irreplaceable entity.

Habermas accepts this idea of power of self-being, but does not base it on religion and theology as Kirkegaard, but makes it anthropologically concrete and articulates it as "self-understanding of a person who would want the same being recognized in them" (Burger, 2008, p. 142). Ethical self-understanding can only be achieved through joint efforts of all subjects, which makes self-being a trans-subjective power. Habermas will extend the context of 'capacity of being oneself' with the concept of naturality or embodiment: "The body is the medium for incarnating the personal mode of existence" and "The capacity of being oneself requires that the person be at home, so to speak, in her own body" (Habermas, 2003, p. 57)

A developing person can "submit their own life histories to critical evaluation and retrospective revision" as it is made from "a material that we can 'make our own' and 'responsibly take possession of'" (Habermas, 2003, p. 13). Like the parent's nature, genetic manipulation "intervenes in the somatic bases of another person's spontaneous relation-to-self and ethical freedom. This kind of intervention should only be exercised over things, not persons. For this reason, later generations can demand an account from the programmers of their genome; they can hold these producers responsible for what they, the offspring, consider the unwanted consequences of the organic starting point of their life histories. This new structure of attribution results from obliterating the boundary between persons and things." (Habermas, 2003, p. 13). Consequently, we are facing additional questions: Can the human race universally take hold of the biological evolution and work on self-transformation of the entire species? Do we want mankind to live in a society reflected in the mirror of artificial beauty, all the while ignoring the natural foundations of life? Can we make the decisions in the name of those not able to make their choices as they have not been born, thus changing the direction of spontaneous creation of human life. Furthermore, this leads to encroachment into freedoms, rights, and responsibilities of other equal human beings: "May we feel free to dispose over human life for the purposes of selection?" (Habermas, 2003, p. 20). Imagine yourself (embryo) being in forever stuck in frozen nitrogen only because you do not have the desired gender and this fact is preventing your only chance of becoming a human being. An embryo is not an asset to be disposed with. Until we are able to understand the secrets of life, an embryo is just as important as its conception. This is not just an ethical issue, but a matter of natural uncontrollability – it is a matter of whether we understand each other as normative beings which expect mutual solidary responsibility and respect from each other. These questions call for an interdisciplinary and intersubjective dialogue between all members of the community.

Habermas perceives the body as a mediator between nature and culture. It is crucial for constituting the identity of an individual as "a self, remaining self-identical in the course of a life history." (Habermas,

2003, p. 59). In his explication of this stance, Habermas cites Plessner: "A person 'has' or 'possesses' her body only through 'being' this body in proceeding with her life." (Habermas, 2003, p. 50). For a person to be one with his/her body "this body has to be experienced as something natural – as a continuation of the organic, self-regenerative life from which the person was born." (Habermas, 2003, p. 58). This makes self-being and its capability body-dependant – self-being is a *compositum*, a mixture of natural and unnatural materials (Burger, 2008, p. 145). This is why we find the "fact that the dividing line between the nature we are and the organic equipment we give ourselves is being blurred" rather unsettling (Habermas, 2003, p. 22). Man who understands himself as a moral being thus must question the permissibility of genetic and technological control over his own and someone else's body. Habermas sees this question as "whether the instrumentalization of human nature changes the ethical self-understanding of the species in such a way that we may no longer see ourselves as ethically free and morally equal beings guided by norms and reasons." (Habermas, 2003, p. 41). Genetic and technological programming of "our hereditary factors [...] confronts us on an existential level, so to speak, with the expectation that we subordinate our being a body to our having a body" (Habermas, 2003, p. 54).

Habermas bases his moral and ethical standpoint in the intersubjective action of reciprocal recognition and equal relationship. For this purpose, he reactivates the *Categorical imperative*: "Act in such a way that you always treat humanity, whether in your own person or in the person of any other, never simply as a means, but always at the same time as an end. The concept of humanity obliges us to take up the we-perspective from which we perceive one another as members of an inclusive community no person is excluded from." (Habermas, 2003, pp. 55-56). Individual will of autonomous subjects must enter the discourse so that through joint effort norms could be developed which "*deserve* the well-founded consent of all" (Habermas, 2003, p. 56). This makes the concept of moral as a generalization of individual attitudes: "For the person expressing a moral judgment, therefore, her own capacity of being herself is as important as is the fact for the person engaging in moral action that the other is being herself. In the yes or no of participants in discourse, the spontaneous self- and world understanding of individuals *who are irreplaceable* must find its appropriate expression." (Habermas, 2003, p. 57). Habermas's ethical standpoint is situated between the liberal right of individual to freedom of choice and the capacity of self-being which stems from good life and self-respect of ethical self-understanding. For the ethical principles to be effective, these two criteria have to breach balance through an intersubjective dialogue.

Jonas believes that the dualism imposed by Descartes, i.e. the unnatural division of being into *res cogitans* and *res extensa*, i.e. into the "thinking" and the "extended" has led to the "unnatural division" of unity of life into the mental and the corporeal. The issue of body has become the central problem of ontology as the body which is alive and mortal, which has consciousness, and which is a part of the world and also represents sentience, whose final form is the organism and causality and its internal form being and finality, is a memento of a yet unresolved question of ontology, the question of what Being is (Jonas, 1997, p. 39). Therefore, without the body and its elementary experience, the starting point of our most encompassing and general extrapolation into the unity of reality, one could not understand the power and the activity in the world, or active interconnections of all things, and thus have no vision of nature in general (Jonas, 1997, p. 46). As a solution to this problem, Jonas argues for a new philosophical reading of the biological text to regain the internal dimension to the organic, overcome the Cartesian division into *res cogitans* and *res extensa* and reunite man with the rest of the living natural world. His main thesis about the unity of life is based on the following principle – the biggest contradictions that man holds within himself, freedom and necessity, autonomy and dependence, me and the world, relationship and individuality, creation and mortality, all of them contain in their primeval form their reshaping in the most primitive forms of life. This topic, which is common to all of life in general, can be traced through the growing order of organic powers and functions – through exchange of matter, movement and desire, feeling and perception, imagination, art and concept – one and the same progressive order of degrees of freedom and dangers which reaches its climax in man and whose uniqueness can only be understood if it is no longer observed through metaphysical separation (Jonas, 1987, pp. 9-10).

According to Jonas, the very testament of life can lead us to conclude that the organic in its first and all subsequent steps is imbued with spirit, and that the spiritual always refers to the organic, thus creating an indivisible complete unity of life. The progressive order of organic powers and functions is actually one progressive scale of freedom and danger, reflected in both anthropology and biology, as this is not just a new way of looking at the being of man, but also understanding the phenomenon of life in general. The corporeal and the spiritual form a unity and one cannot exist without the other. These two substances cannot be divided by metaphysics or biology, and certainly not for the needs of scientific research and through manipulation of life in general. The testament of life has its teleology, causal modus of nature itself, or immanent teleology.

The key term of Jonas's philosophical biology which will represent the cornerstone of the ethics of responsibility is the concept of freedom. The freedom of life has been present from the very beginnings of existence of simple organisms, in the first exchange of matter. This means that the transition from the inanimate towards animate substance, the first organization of matter into life was motivated from the depths of being by an active tendency and certain modes of freedom towards which this transition paved the way (Jonas, 1987, p. 18). The living being, in its relation towards nature, is biologically independent and existentially dependent. This is the essence of ontological freedom.

Jonas's conception of freedom of life in general, i.e. including the non-human "lowest" forms of life, reflects the constant tension between form and matter, freedom and necessity, being and non-being, which is the most fundamental polarity of life (Jonas, 1987, p. 20). If we are dealing with active self-confirmation of life, actualization of its purpose and value through efforts for self-confirmation at the "lowest" levels of the organic life without diminishing the *differentiae specificae* of the human, then we are not in fact dealing with two separate worlds with two separate biologies, ontologies, metaphysics and ethics.

In this age of science and technology, the power of man over living beings also represents his own impotence as he himself becomes an object of the scientific and technological manipulations. As a medium of understanding the purpose of the human and the non-human, the freedom to act and decide he is given must be used to take the responsibility for the survival of nature. An instance of man's responsibility is his own conscience for securing the future. This is where the need for "a new ethics" arises. The new ethics has to emerge because man acts and ethics must control the order of action and the regulation of power (Jonas, 1990, p. 42).

The affirmation of inherent purposefulness and the value of being of life and all living beings which requires a science-, biology- and anthropology-based cosmology and cosmogony represents the starting point for improving the ethical theory which stems from the philosophy of nature and life. The need arises directly from the need of nature and the living beings as nature sets its own purpose and values which man should recognize and respect. In this respect, man should only be the "caregiver" for the things which he sees but which he did not create. This means that the ethics of responsibility is based in the ubiquity of being, and not in the simplicity and particularity of man (Jonas, 1997, p. 402-403).

The focus of ethics on the future is an imperative for the biological survival but also for the preservation of the human essence as we cannot chose or attempt to choose the non-being of future generations for the sake of being of current generations (Jonas, 1990, p. 28). For this reason, the primary goal is not that what man must be and do and then what he can and cannot do, but the primary focus is what he is in fact doing because he can, and the obligations stem from the act (Jonas, 1990, p. 182).

Jonas aims the obligation for adhering to the principle of responsibility towards the power elites in science, economy, and politics as they decide on the development and the application of technology. The question which remains open is what the individual duties should be based on and how should the ethical requirements be framed so that they become an imperative for political activities.

A similar moral agenda is advocated by Floridi who argues:

We shall be in serious trouble, if we do not take seriously the fact that we are constructing the new physical and intellectual environments that will be inhabited by future generations. In view of this important change in the sort of ICT-mediated interactions that we will increasingly enjoy with other agents, whether biological or artificial, and in our self-understanding, an environmental approach seems a fruitful way of tackling the new ethical challenges posed by ICTs. It is an approach that does not privilege the natural or untouched, but treats as authentic and genuine all forms of existence and behaviour, even those based on artificial, synthetic, hybrid, and engineered artefacts. The task is to formulate an ethical framework that can treat the infosphere as a new environment worthy of the moral attention and care of the human inforgs inhabiting it. Such an ethical framework must address and solve the unprecedented challenges arising in the new environment. It must be an e-nvironmental ethics for the whole infosphere. This sort of synthetic (both in the sense of holistic or inclusive, and in the sense of artificial) environmentalism will require a change in how we perceive ourselves and our roles with respect to reality, what we consider worth our respect and care, and how we might negotiate a new alliance between the natural and the artificial. It will require a serious reflection on the human project and a critical review of our current narratives, at the individual, social, and political levels (Floridi, 2014, pp. 219-220)

Contemporary discussions over the use of new biotechnologies are being led around the concepts of dignity and good life, uncontrollability of life and inviolability of the human person. All of these issues raise the questions of mutual respect, ethical self-understanding, freedom, irreplaceability of individuals, moral equality and autonomy of will with utmost ethical considerations towards the entire natural life. We have shown that the age of *scientific humanism as naturalism* is a product of *scientific work* which establishes the production of life and organization of the community on scientific and technological principles which cybernetics establishes, organizes and controls as the ontological and anthropological science of our time. Free natural factors no longer participate in the production of life and organization of the community – everything is established and everything functions as a manifestation of *scientific work* as a real metaphysical self-cause. The absolute *scientific work* as the substance subject of world history uses the natural energy, material means, man and ethics as means of its permanent progress as condition of survival. The products of *scientific work* are a part of an eternal circle of overcoming without essence as they are established by being which is itself null. We have noted that man as a self-aware, natural and historical living being as the medium of scientific work in the era of *scientific history* (*scientific humanism as naturalism*) is most directly drawn into its metaphysical circulation and placed at the very centre of the vortex as both its means and product. In the movement, human being is enhanced and transformed into a cyborg – scientific worker, a virtual being, or is multiplied into an infinite array of identical copies through genetic manipulation. Since humans and everything human are within the horizon of the absolute establishment of *scientific work*, the tendency is to overcome the biological body and mind and replace them with digital forms of bodiless existence or replacing the biological bodies with technological ones. Attempts are made to enhance the very human mind through technological extensions, to upload it into a mechanical interface, or to replace it with artificial intelligence, so that human thought is completely occupied by the mind of science and technology. We hold it necessary and justified to start a responsible intersubjective dialogue on principle of cyborgoethics in today's historical constellation in which human beings still have the possibility of choice. It is crucial to recognize the general freedoms of choice we still have at our disposal and to take moral actions accordingly. This discussion will not take our thoughts beyond metaphysics, but can serve as a path of our enlightenment from metaphysics, its symbols and ethical illusions about the necessity to preserve humanity in the age of *scientific humanism as naturalism*, which it continuously protects in eternal recurrence (Nietzsche). On the other hand, an intersubjective dialogue can slow down the misuse of scientific and technological procedures that move faster than our insights on the essence of *scientific humanism as naturalism*, thus showing the possibility for an alternative life.

We advocate an intersubjective discussion on principles of cyborgoethics as a foundation for moral action. According to Beauchamp and Childress (2013, pp. 1-3), principles represent general standards of behaviour which define other moral rights and judgements. Without setting our sights too high, we suggest some of the possible principles of cyborgoethics:

1. The principle of thought or "What can I know?" – this implies starting a dialogue with postmetaphysical philosophers about the essence of *scientific work* and its manifestations in the *age of scientific humanism as naturalism*, which is something to which we have dedicated a significant part of this book. The discussion about the essence of historical time and the search for the meaning of being leads us to a potential answer to the question we set out.

It has been established that where essence and existence overlap there is no need for science of moral. Moral depends on the division between what needs to be and what actually is, which is why every ethical discussion must start with the questions of the essence of man. Since the essence of man is related to the ways of creating life and organizing the community, specifically the community of science and technology in the age of *scientific humanism as naturalism*, the philosophical discussion leads us towards the alienated essence of man and the moral responsibility hidden within and the determinacy of human being towards nullification of biological life and the quest for regaining the meaning of existence in historical time. On his return from alienation from his anthropological essence and inviolability of his biological life which cannot be reduced to mere communication with the environment, man must become and remain the subject of his own life history in which he will recognize other human beings as equal human persons.

In his discussion on ethical self-understanding, Habermas emphasizes that a completely different image arises "if 'moralizing human nature' is seen as the assertion of an ethical self-understanding of the species which is crucial for our capacity to see ourselves as the authors of our own life histories, and to recognize one another as autonomous persons." (Habermas, 2003, p. 25).

Absolute *scientific work* as being in the age of cyborgized community of *scientific humanism as naturalism* enhances and reshapes man, dissolves his very human nature and human mind, and establishes biological life onto a new non-human foundation. The aim of philosophy is to situate itself in the essence of scientific work and its ideologically interested ethical self-understanding. It is only by bringing into question this

[...] metaphysical foundation (metaphysical structure of thought in which man replaces God or nature in an anthropocentric turn, or in which the very structure remains intact) that a new space can open up for problematizing the impersonality which could be revealed as the primal origin of power that makes us all, to varying extents, instruments and objects of production, reproduction and expanded reproduction of nothingness. Only this will create the assumptions for recognizing the true (in its essence nihilistic) interest mechanisms which make everything that what is, which leads to possible and practical (not just purely moralistic) scrutiny of actual imbalances of power and dignity and structural and contingent injustices and unjustified suffering (Veljak, 2008, p. 19).

Along with this philosophical discourse, we need a general intersubjective dialogue and multidisciplinary discussions of scientists which would use a unified set of principles and regulations to cause the convergence of what needs to be and what is.

2. The principle of ethical responsibility or "What should I do?" in the *age of scientific humanism as naturalism* is ethics in the ideological function of *scientific work*, like all other sciences. Its basis is situated in the reality of scientific and technological mind (of *scientific humanism as naturalism*) which asks not and needs not the essential human, philosophical and moral foundation of life in the community, as it itself (the scientific and technological min) is the foundation and the principle of life and relations in the community. The essence of man is alienated but *scientific work* tries to overcome it and abolish it through transformation of man into cyborg – scientific worker – non-human being, illusion, and ethics is functionally integrated into the scientific evolution (of scientific history) of man in which everything functions according to a pre-set program. The program is not only functional but also "ethical" as it has been generalized into uniformity. Man's free will is integrated into mediation in the will to power of *scientific work*. The principles of ethics of *scientific humanism as naturalism* represent the universal power of the scientific and technological mind and are included in the free and uninhibited growth of productive powers of *scientific work*. This way, the "entire science loses its credibility over human existence and regulativity for forming life, thus being reduced to a very powerful means whose use should be planned outside of science – from the anchor point of bioethical pluriperspectivism" (Čović, 2004, p. 118).

Einstein has taught us that "Science can only determine what is, but not what shall be; and beyond its realm, value judgments remain indispensable" (Kaku, 2011, p. 191). Kaku believes that when using science and technology we need wisdom as it directs us to the need for organizing life. By recognizing the key issues and analysing them from multidisciplinary standpoints, we should be able to reach ethical principles of behaviour. Wise decisions are hard to reach as they move slowly through information pathways. A potential way for stimulating a lively democratic discussion is education, as an educated body of voters can better understand and make better decisions on technological development which will determine the fate of our civilization. Ultimately, mankind will decide for itself how for and in which directions they technology to take them, but only an informed and educated kind can make wise decisions (Kaku, 2011, p. 191). Further progress of cybernetic implants which help in treatment must be used to protect the dignity, autonomy and safety of man and the equality of healthcare for all human beings, as well as naturality of all members of the community. Cyborgization must not cross the threshold which endangers the biological life of man, the immediacy of human relations and the biological balance.

Thus, we are facing the same dilemma as Kant when he asked himself "What ought I do?":

a) Our initial philosophical standpoint is the principle of thought in transgressing the circle of metaphysical mind of science and technology which establishes the reality of *scientific humanism as naturalism* and its appropriate truth of essence and sense (or lack thereof) of life of man. Prolegomena for cyborgoethics cannot be based only on principles of good life nor on Kant's principle of need and necessity. Starting from the concrete alienated man in the *age of scientific humanism as naturalism*, and the pluralism of interests of alienated centres of power and acquired individual freedoms and rights whose main focus is freedom and the right to choose, cyborgoethics must have its basis both in the transphilosophical thought and an intersubjective dialogue and consensual personal decision. Only this can ensure harmonized legislature which would represent the practical application of Kant's principles of *Categorical imperative*. We agree with Habermas here that ethics acts as man's natural lawyer in the age of metaphysics.

b) Every man as a responsible editor of his life must take his life into his own hands and take the responsibility for the procedures which he undertakes, starting from the principle of good life and respecting the ethical self-understanding. Even though cyborg enhancement and reshaping procedures of healthy human beings promise a healthier and longer lifespan, these bring into question the equality between biological organisms as free equal individuals and enhanced beings. Cybernetically established and enhanced community of *scientific humanism as naturalism* is ideologically presented as the ultimate triumph of humans over nature, and it is neither human nor natural. Kurzweil describes technology as a new religion that grants its believers some certainty in life, also extending into the afterlife by ensuring its eternal duration. In that respect, the techno-logic of *scientific work* is also a theology. Thus, in the name of truth of being "That which science made technologically manipulable reacquires, from a normative perspective, its character as something we may not control" (Habermas, 2003, p. 24).

In matters of individual freedom of man, we do share the same starting point as the contemporary liberal society and state, because, although these freedoms are seemingly real, they are in fact only virtual as the right to choose can never take precedence over inviolability of life. Kant argued that freedom of man is the ability to determine oneself independently from nature and sensual phenomena. Such determination of man's freedom is intertwined with knowing the truth of "freedom of choice" and "choice itself" (Berdyaev, 2009, p. 46). This freedom and moral action of man is based in his alienated essence and social order whose principles of acquired freedoms is in contrast with his true essence. If we start from the mystery of life which precedes every metaphysics, essence of man, freedom of choice, one could say that man of today has a good reason and enough free will to respect the ethical self-understanding and make decisions in the *age of scientific humanism as naturalism*: does he want to conquer nature within him and become an enhanced cyborg and a non-human being, and ultimately an illusion, or does he want to return to his meaningful essence? For the human mind, which as the other dimension of being from the kingdom of spirit is also reshaped and enhanced, we have to emphasize "Freedom of the spirit does not depend on any kind of social organization" (Berdyaev, 2009, p. 48). It falls under the domain of mind of science and technology which recognizes no freedom of thought beyond the functions of *scientific work*.

Hegel believed that the soul cannot live without body as body is its external existence. Enhancing prostheses for body and mind fuse, "the organically grown with the technologically made, on the one hand, and separation of the productivity of the human mind from live subjectivity, on the other hand." (Habermas, 2003, p. 41). This changes the ethical self-understanding of man as the self-born living, responsible and active being. The space time of *scientific work* reshapes the human being to that extent that "we may no longer see ourselves as ethically free and morally equal beings guided by norms and reasons." (Habermas, 2003, p. 41).

The hour is nigh for us to use thought informed by wisdom on the essence of *scientific humanism as naturalism* in an intersubjective relation of reciprocal recognition and equal relation of all to put forth the principles of technology of conducting morals and concept and real possibilities to protect the excessive scientific and technological encroachment into the historical reality and human body and the person embodied within. The technologies for establishing morals have to represent a constructive response to "dependencies rooted in the incompleteness of our organic makeup" (Habermas, 2003, p. 33). "May we consider the genetic self-transformation and self-optimization of the species as a way of increasing the autonomy of the individual? Or will it undermine our normative self-understanding as persons leading their own lives and showing one another equal respect?" (Habermas, 2003, p. 29).

Habermas believes that individual self can be constituted only in the intersubjective relation "in the public sphere of a linguistic community for a natural creature to develop into both an individual and a person endowed with reason." (Habermas, 2003, p. 35). In light of this, Kant's *Categorical Imperative* as a universal principle according to which man as a mental being embodied in a person is a self-purposeful and absolute values that transgresses the empire of purposes to gain new actualization through an autonomous subject that enters into the dialogue to achieve consent from all participants. This way the moral community of free and equal human subjects does not constitute Kant's *Empire of Purposes* in the domain beyond the human mind but a possible concrete form of life. Moral law within us and above us will not be effective without political support in restraining the interest of masters of scientific and technological globalization of the world. Agreed upon technologies of establishing morals guide our discussion towards the principles of use of technology.

3. The principles for using technology should tell us which the most basic cyborgoethic principles are around which to have a public dialogue and individual choice in terms of implantation of artificial technology into the human body and cyborgization of historical reality. In our view, one should start from the principle of thought and the assumed ethical component harmonized in an intersubjective relation, establish the dividing line between the ethics of preservation of life and ethics of enhancement and reshaping human beings. Science cannot plan to change and reshape human beings before and after their birth because that makes the resulting entity a thing, not a self-established person as a self-being. Scientists are trying to get a clear ethical answer and guidelines as to where they can continue and where they should stop. Special attention should be given to gene manipulation which is "bound up with issues touching upon the identity of the species, while such an anthropological self-understanding provides the context in which our conceptions of law and of morality are embedded" (Habermas, 2003, p. 23).

There is a real possibility of different social groups using the genetic possibilities and technological enhancements and deciding on different paths of progress, thus leaving the *homo sapiens* as a distant common ancestor. We should mention another matter of principle which should be taken into consideration when using artificial technology. The very same modern science and technology have provided insights about the unity of the Universe, not only at the level of visible beings, but also invisible quantum and nanoparticles and genes. How will the Universe react to the increasing technologization of nature, cyborgization of plants, animals and, ultimately, human beings? Will human beings lose their place in the consistent network of the Universe in which all parts affect one another, give each other their features and arguably a meaningful existence. There are more and more indications that the human consciousness does not depend only on biological and chemical stimuli from the pituitary gland, nor on the life process of man in the reality, but that it is an integral part of the universal consciousness of the Universe. Will we enhance and reshape the entire Universe and stop the entropy, either through black holes or grey matter?

When talking about the biological body and its enhancement and reshaping, we understand it like Habermas and Marx as a continuation of the organic self-renewing life from which a person is born. A body envisaged this way becomes the medium of embodiment of personal existence, from which springs the self-being and its potency, which lead sto the capability of making and carrying out moral decisions. We accept Plessner's stance that "A person 'has' or 'possesses' her body only through 'being' this body in proceeding with her life." (Habermas, 2003, p. 50). Stepping into the domain of potentially human over the boundary of treatment and restoring lost functions and organs abolishes the distinction between a person and a thing.

Body as an external sign of the soul, as understood by Hegel, and the medium of embodiment of person as a self-being, according to Habermas, is being enhanced and reshaped by procedures of *scientific work* along with the human mind. In this manner, the procedures of *scientific work* use practical synthesis to overcome the obstacles and the super-natural mind and body which philosophy was able to overcome only in the element of thought. Enhanced mind and artificial intelligence as the final goals are completely devoid of every traditionally accepted moral element. In this intelligence, there is no autonomy of will nor personal decision inherently present in man's decision-making. Enhanced mind or artificial intelligence function uniformly within the given program without a single trace of autonomy of will.

Today, thought manifests itself not as a 'super-natural' element in the dual character of man's being, but as a means that we have to fight for if we want to live as mental human beings. In spite of ontological and anthropological character of *scientific work,* man is responsible for everything which happens on Earth. In this respect, we support the stance of Hans Jonas that scientists are given freedom to explore but not responsibility, which is why we suggest a discussion on responsibility: who is responsible for what, under whose jurisdiction and under what criteria. Like him, we believe that science should obtain the blueprint of life, but should not draw life from the natural environment and human beings. Science should write the book of man's life not his death. It is vital to reach a unison conclusion as to what can be conducted without any risk for man and mankind, by which the main criteria of development have to be biological life and human being, not *free development of productive powers* (Marx). It matters not how many years we spend living, but how much life there is in our years.

4. The principle of moral decision – morals form an integral part of the essence of man, but in practice, it is not something given but obtained through action. Notwithstanding the numbness of man's senses and his obsession with the fate of metaphysics which produces the world according to the ideal of perfect form and the deep sleep of metaphysical thought, the decision on ethical preservation of life and the possibility of humane life is still within grasp of mankind. This future has to be won in the present. Along with transphilosophical thought, we need cyborgoethics as science of absolute moral action based on our analysis of essence of cyborgized reality which applies to all of mankind equally.

One could argue that cyborgoethics, just like bioethics, is a point of juncture and dialogue of different sciences and activities, and diverging approaches and worldview which come together to articulate, discuss, and solve ethical questions relevant for life in general and all parts of this whole, for life in all its forms, stages, phases, and phenomena (Jurić, 2007, p. 83). Harmonized ethical principles are the baseline for behaviour of all actors in the historical reality. Man must use the autonomy of his will to get out of the vortex of *scientific work* and start living in the time of history instead of historical time of its technological, scientific and bioscientific extensions and experiments. Man cannot survive without coexistence with science and technology, but these must serve mankind, not make it a means of its development and extension of the cyborgized nature.

The most important event must once more take place in the human thought and human consciousness to empower the readiness of all to take their life into their own hands and become responsible for their human actions. This in particular is aimed at the power elites as they are still capable of setting the boundaries of development and application of technology despite the self-purpose of scientific and technological work. This is not a safe path as there are still no clear guidelines, but we have to make the first step, which might be a small step for today's man but will surely be a giant leap for mankind. Mankind still has the power to end the rule of *scientific work* over it. It is up to us to decide what kind of reality we want to live in.

REFERENCES

Beauchamp, T. L., & Childress, J. F. (2013). *Principles of Biomedical Ethics* (7th ed.). Oxford University Press.

Berdyaev, N. (2009). *The Fate of Man in the Modern World.* Student Christian Movement Press.

Burger, H. (2001). *Sfere ljudske moralnosti – Kant, Hegel i suvremene diskusije.* Prometej.

Burger, H. (2008). *Ljudska moralnost: Studije o antropologiji i etici.* Naklada Breza.

Čović, A. (2004). *Etika i bioetika: Razmišljanja na pragu bioetičke epohe.* Pergamena.

Floridi, L. (2014). *The Fourth Revolution: How the Infosphere is Reshaping Human Reality.* Oxford University Press.

Habermas, J. (2003). The Future of Human Nature. *Polity.*

Hegel, G. W. F. (1948). *On Christianity: Early Theological Writings* (T. M. Knox, Trans.). University of Chicago Press.

Hegel, G. W. F. (1952). Hegel's Philosophy of Right (T. M. Knox, Trans.). Oxford University Press.

Hegel, G. W. F. (1991). *The encyclopaedia logic, with the Zusätze: Part I of the Encyclopaedia of philosophical sciences with the Zusätze* (T. F. Geraets, W. A. Suchting, & H. S. Harris, Trans.). Hackett.

Hegel, G. W. F. (2001). *Philosophy of Mind* (B. Online, Trans.). Wallace.

Hegel, G. W. F. (2012). Hegel's Philosophy of Mind (W. Wallace, Trans.). Clarendon Press.

Jonas, H. (1974). *Philosophical Essays: From Ancient Creed to Technological Man.* Prentice-Hall, Inc.

Jonas, H. (1987). *Wissenschaft als personliches Erlebnis.* Vandenhoch & Ruprecht.

Jonas, H. (1990). *Princip odgovornosti. Pokušaj jedne etike za tehničku civilizaciju*. Veselin Masleša.

Jonas, H. (1997). *Das Prinzip Leben, Auszatze zu einer Philosophischen Biologie*. Suhrkamp.

Jurić, H. (2007). Uporišta za integrativnu bioetiku u djelu Van Rensselaera Pottera. In V. Valjan (Ed.), Integrativna bioetika i izazovi suvremene civilizacije. Bioetičko društvo u BiH.

Kaku, M. (2011). *Physics of the future: How science will shape human destiny and our daily lives by the year 2100*. Doubleday.

Kant, I. (1781). *The Critique of Pure Reason* (J. M. D. Meiklejohn, Trans.). Academic Press.

Kant, I. (1991). *The metaphysics of morals*. Cambridge University Press. doi:10.1017/CBO9780511809620.008

Veljak, L. (2008). Čovjek kao metafizička utvara. *Filozofska Istraživanja, 109*(30/4), 13–20.

Veljak, L. (2011). Principi ontologije i pitanje o primatu ontologije nad logikom i gnoseologijom. *Studia Lexicographica, 132*(2/9), 5–20.

Chapter 11
A Proposal for a General Resolution on Cyborgization

ABSTRACT

In addition to "Prolegomena for Cyborgoethics," this chapter builds a strong case for the necessity of cyoborgoethics to help guide moral actions and protocols for preserving the vitality of life within a rapidly changing technologically society. The introduction of an all-encompassing ethical system of cyborgization of human beings is deemed necessary in addition to establishing cyborgoethic principles and rethinking the developmental stages of cybernetic implants that pose the question whether we as cyborgs perceive ourselves as the only authors of our life history and whether we will recognize ourselves as autonomously active persons. It is important to establish ethical and legal responsibility for potential cyborgization of the entire reality of mankind.

INTRODUCTION

A. Procedures for enhancing natural human capabilities through external extensions have been present in history of mankind since the age of the cavemen. The idea of improving the human body has its origins in the works of Descartes, who was a proponent of human nature enhancement. His ideas were reinforced by the discoveries in the field of anatomy of that time, which viewed muscles and joints as systems of levers and pullies. Descartes's view was expanded by la Mettrie who argued that man its entirety, not just his body, can be understood as a complicated machine, and physical manifestation as the machine's actions.

B. The father of cybernetics Norbert Wiener believes that controlling a cybernetic machine means transmitting particular information to the machine – an order to work. Cybernetics as a science has enabled the scientific and technological construction of the machine whose behaviour is analogous to that of a living being.

C. The features of cybernetics see nature and the Universe as an environment susceptible to constant entropy (disorganization). In that respect, cybernetic machines take the role of fighting back the entropy by establishing an organization of life and man in the historical reality of *scientific humanism as natu-*

DOI: 10.4018/978-1-7998-9231-1.ch011

ralism in which entropy is reduced to zero. There is complete identity between a living individual and a cybernetic machine in their attempts to control entropy through feedback.

D. Man receives information from the real world through his senses (vision, hearing, smell) and bases his actions towards the environment on these sources of information. In cybernetic machines, this role is performed by different sensors, monitors, photocells, which not only inform them about current circumstances but also allow them to keep track about the progress of their own tasks, which Wiener calls feedback, i.e. capability to modify future procedures based on previous actions. For all types of behaviour, a cybernetic machine must have a central decision-making organ, which decides what the machine does based on the information it receives. *Scientific work* is manifested in technological, scientific, and bioscientific procedures of enhancing and reshaping human body and mind and it establishes life on a new non-human basis in the name of hastened revolution of science and technology and cybernetic *humanization and naturalization* of historic space and time. The operative cybernetic language of *scientific work* is devoid of essential content of reality, the world of man and nature, and is turned into a language of mathematical symbols.

F. In the information science and technology community – *scientific humanism as naturalism* – the language of *scientific work* enchants man with promises of unlimited capabilities of progress in both real and virtual world space and time. All the promises about humanism, fairness, ecology, human rights, scientific evolution, ethical life, and freedom are simply adapted to the uninhibited progress of *scientific work.*

G. It is indicative that social and special sciences are also organized in the service of the techno-logic for securing the *scientific work.* This society of information science and technology, society of *scientific humanism as naturalism* will become so humanized, so naturalized that it will not need man as means or self-reproduction nor as an interpreter of its own nihilistic history and meaninglessness.

H. *Scientific humanism as naturalism* is the current age of our scientific history in which man as cyborgized human being – non-human – will lose every real possibility of mediating between nature, history, being, and purpose of existence. Enhanced and reshaped cyborg beings will be closed unto themselves, into their own internal meaningless existence or their bodiless existence will be a part of the void of the virtual world.

I. From the character of *scientific work*, one can assume that its procedures will strive towards more uniformity and unilaterality of the world and man cyborg, which is why enhanced cyborgs will be classified in specific super-human categories.

General Principles

J. Transhumanist experiments enhance the human being at a physical and cognitive level using biotechnology, digital technology, and technological implants, and this enhanced person with machine parts as additions to the organism whose purpose is to increase the power of the body and mind is called a *cyber-cyborg* (hearing, vision, memory, and intelligence enhancement).

K. The transhumanists advocate upgrading the existing human potentials to overcome the problems of aging, cognitive limitations and being constrained to life on Earth. They support the development of any sort of artificial intelligence, modified life forms or other kinds of intelligence which will be caused by the progress of science and technology. This includes the necessity for using technologies that can be developed for improving memory, concentration, and mental energy, life-extension therapies, reproductive choice-related technologies, cryogenic freezing of the human body and other possible modifications and enhancements of the human body and capabilities.

I. We hold it justified and necessary to advocate ethical, moral, and legal control of scientific, technological, and biotechnological manipulations of the human body to preserve the inviolability of human beings.

M. The transhumanists as the advocates of enhancement and reshaping of human beings attempt to ethically justify their endeavours and cover their real intent through justification of the necessity of increasing the quality of life and extending the human lifespan. According to their expectations, technology will liberate man from his bodily limitations, make him better, healthier, stronger, smarter, and longer-lived, in order to be better adapted to the requirements of space and time which we dub *scientific humanism as naturalism.*

N. They base their transhumanist beliefs and support for enhancement and reshaping of the human species on the principles of freedom and rights of individuals to choose procedures and contents of enhancements and reshaping of one's body and mind; i.e. ethical obligation of every individual to optimize their lives and work capabilities. Individual rights and freedoms interpreted this way lead to an ever-expanding reproduction of inequality in mankind and lead mankind towards a decision: to preserve the biological man or use the liberal rights for personal enhancements according to one's material capabilities.

O. European Union should have a crucial role in setting out the fundamental ethical principles which should be observed when developing, programming, and using cybernetic implants and should include these principles into the European regulation and codices of behaviour in order to affect the technological revolution into serving mankind and providing widely available health benefits of biomedical and cybernetic implants, and avoiding potential pitfalls to the highest extent possible.

Responsibility

P. Today, the biological human body is being used as a testing ground for implantation of various chips for enhancing human capabilities (chips with ultrasonic signals which enable better spatial orientation, radiofrequency chips which enable a new type of communication, ultrasonic radar to enhance vision, memory cards to enhance brain memory, etc.). The predictions from futurologists announce the cyborgization of the human body through artificial intelligence, which should control human beings, at the same time producing artificial beings – humanoid robots for uploading the contents of the human brain into them.

R. It is important to establish ethical and legal responsibility for potential cyborgization of the entire reality of mankind. In the age of science and technology, we need wise thought which transcends the metaphysics of the scientific and technological mind and leads us towards establishing cyborgoethics.

Q. Ethics which would apply only to those with the possibility of enhancement leads towards new kinds of inequality. In these conditions, the Hippocratic Oath which underlines the necessity to respect human life and equality of all patients loses all of its meaning and content. We need further education about the scientific and technological progress and an intersubjective dialogue on directions of cybernetic behaviour of man and our reality.

S. We need absolute moral action based on scrutiny of the essence of our cyborgized reality which would apply to all humans equally – it demands an answer to the question as to what kind of reality we want to live in.

The Ethicalness of Cyborgization

T. Implanting artificial technology helps human body regain lost functions or organs. From the perspective of health care, establishing the lost natural balance of the body, dignity and value of life of man as a person, we find these forms of cyborgization allowable but certainly not value-neutral.

U. Our moral concern and ethical scrutiny is motivated by the procedures of enhancing and reshaping human body and mind whose aim is not to treat, but to increase functions and capabilities of a healthy body and mind and to establish life onto a new non-biological basis. The complete enhancing cyborgization of social relations and the entire historical reality of *scientific humanism as naturalism* and the domination of artificial intelligence over human intelligence is not in the function of preserving life but establishing partially or completely artificial beings which will not need the biological man.

V. Cyborgization of human beings calls for a public intersubjective dialogue on values of human life and its preservation. The same applies to preserving the immediate interhuman relations and the autonomy of the nature and the natural. Man is one with nature and nature is one with man.

W. Parallel to implementation of artificial technology into the human organism, we are witnessing the procedures of immersing the biological body and mind into the digitalized space-time of bodiless existence and movement within the virtual network. An individual's body and identity are lost in the virtual worlds which seemingly augment human capabilities but actually reshape man into a binary code which functions as means of scientific and technological progress.

X. Man understands himself as a moral being, which is why he must ask the question about the permissibility of cybernetic control of his own and other people's bodies. This imposes a question whether cyborgizing the human natural changes the ethical self-understanding in such a way that we are no longer able to understand ourselves as ethically free and morally equal beings oriented by norms and reasons (Habermas).

Y. The future progress of cybernetic devices used for treatment purposes:

1. should be achieved through means which protect the dignity, autonomy, equality and self-determination of an individual;
2. should be subjected to serious analysis in terms of security, health, and protection of human beings, freedom, privacy, integrity and dignity, and self-determination, non-discrimination and protection of personal information;
3. indicates the necessity for developing an updated legal framework guided by ethical principles which would reflect the complexity of the transhumanist domain and the numerous social, medical and bioethical implications in the cyborgized reality;
4. indicates the necessity to develop clear and effective general ethical guidelines for development, design, production, and use of cybernetic implants inside the human body which could potentially lead human beings into a transhumanist state and cause the domination of artificial technology in the future;
5. encourages further scrutiny of ethical responsibility of man towards the technological challenges through creation of codices for engineers, physicians and ethics boards which would be used to question possible harmfulness or usefulness of cybernetic implants.

Z. Introduction of an all-encompassing ethical system of cyborgization of human beings is deemed necessary, in addition to establishing cyborgoethic principles and rethinking developmental stages of cybernetic implants that pose the question whether we as cyborgs perceive ourselves as the only authors of our life history and whether we will recognize ourselves as autonomously active persons.

Establishing the Boundary Between Restoring and Enhancing Human Beings

AA. Ever-increasing progress of cyborgization of man showcases the necessity of establishing cyborgoethics (Čatić, Greguric) through an intersubjective and multidisciplinary dialogue of different sciences in order to resolve different ethical questions of enhancement of the human body. It is necessary to undertake certain practical, ethical and political action to inhibit uncritical scientific and technological progress and application of cyborgization procedures, particularly in terms of enhancing and reshaping. In that respect, it is necessary to produce a comprehensive overview of ethical implications of implanting artificial technology into the human body, especially when referring to attempts to enhance or reshape human body or mind.

AB. Above everything else, it is necessary to determine and define the norms which would obligate all those involved in the creation of new technologies and their implantation into the human organism to accept a certain extent of social responsibility.

This refers to:

1. **Ethicality of cyborgization** – it is necessary to determine the distinction between replacement and normalization procedures and enhancement and reshaping procedures. It is necessary to focus our attention to whether it is ethically justified to substitute diseased organs and mental functions of man and, on the other hand, whether it is morally acceptable to destroy or harm a healthy part of human tissue or organs in order to conduct implantation and enhance physical and mental capacities of man.
2. **Use of cyborgization** – it is necessary to determine the effect of cyborgization on health and safety of individuals and the boundaries of biological endurance of the body during implantation of technological devices; analyse the dangers of implanting the ubiquitous neuro-, bio- and nanotechnologies; prevent the occurrence of potential adverse effects during implantation (harmful side-effects, such as rejection from the immune system), and testing to what extent the artificial implants affect the functioning of the entire body.
3. **Aims of cyborgization** – it is necessary to determine the aims which cyborgization is supposed to achieve and define whether the aim of implanting particular prosthesis constitutes treatment of the human organism or enhancing physical and mental capacities of an individual. Science, technology, and medicine need to define an ethical component and have clearly defined goals. The purpose of treatment is to establish the balance of the body through reparation or restoration of expected human traits that were lost due to disease or injury. Enhancing and reshaping constitutes a scientific and technological experiment with the goal of creation completely or partially artificial beings.
4. **Means of cyborgization** – it is necessary to determine the questionability of means used for implantation and the consequences they can have on human naturality, identity, and status of mankind.
5. **Boundaries of cyborgization** – it is necessary to prohibit enhancements and use of artificial technologies on healthy individuals and, in cases of ethically justified cyborgization, ensure protection of personal information, which could be compromised by implantation of sensory and neuroprosthetic devices with the ability of storing information about the human biological state, behaviour, and perception, which could be accessed by a third party.

6. **Future development of cyborgization** – it is necessary to place under public control and define the questions of fairness in distribution, namely whether the implantation procedures will be available to everyone equally and whether there will be a stratification between those who receive low-quality prostheses and those who receive top-quality products. Should implantation be conditioned by individual financial situation, we would face a significant gap between potential patients.
7. **Centre for decisions on cyborgization** – who decides on implantation of artificial devices into the human body? It is necessary to pass down laws to protect the defined ethical principles of cyborgization and make decisions that would ensure the safety of an individual as a natural human being, thus removing potential threats from other decision-making centres.

Proposal of General Principles of Cyborgoethics

1. **The principle of thought** - "*what can I know*" implies scrutiny of the essence of *scientific work* and its manifestations in the age of *scientific humanism as naturalism.* Moral is determined by the duality of that what should be and that what is, which is why every ethical discussion must start with the questions about the essence of man. Since the essence of man is tied to the manner of production of life and organization of the community (specifically the scientific and technological community of *scientific humanism as naturalism*), the need for cyborgoethic principles leads us towards the alienated essence of man and moral responsibility for and determinacy of human being towards nullification of life and the quest for the meaning of being hidden but contained within. On his return from alienation from his anthropological essence and inviolability of biological life which cannot be reduced to a fact of communication with the environment, man can persist and remain the subject of his own life history which he will respect other human beings as equally valid human individuals.
2. **The principle of ethical responsibility** – "*what ought I do*" in the age of *scientific humanism as naturalism;* ethics is, like all other sciences, in the ideological function of *scientific work.* The essence of man has been alienated but scientific work tries to overcoming it by reshaping man into a cyborg – a non-human being, an illusion, and ethics is functionally integrated into the scientific and technological evolution of man in which everything functions according a predefined program. The program is not only functional but also ethical, having been generalized into uniformity. Man's free will is integrated into mediation in the movement of will to power of *scientific work.* Ethical principles of *scientific humanism as naturalism* represent the universal will of the scientific and technological mind and are involved into free and uninhibited growth and development of productive powers – scientific work. A possible way of stimulating a life-oriented democratic debate on cyborgization is to make this work public through education, as only educated and informed voters can understand and make better decisions about the technological development which will determine the fate of our civilization. Ultimately, mankind will decide for itself how far and in which direction they want the technology to be developed.
    a. The principle of thought represents our starting point for finding a step out of the metaphysical scientific and technological thought which establishes the reality of *scientific humanism as naturalism* and appropriate truth about the essence and meaning (or lack thereof) of man's life. Prolegomena for cyborgoethics cannot be guided only by principles of good and just life, nor by Kant's principles of need and necessity. Starting from the alienated man in the age of *scientific humanism as naturalism* and its inherent pluralism of interests of alienated centres

of power and acquired individual freedoms and right (whose main focus is freedom and the right to choose, as opposed to a good life empowered by the ethical self-being), cyborgoethics must be based in both transphilosophical thought and an intersubjective dialogue and consensual personal decision, which would ensure harmonized legislature that would represent a practical application of the principles of *Categorical Imperative*.

b. Every human being should take their life into their own hands and take the responsibility for their actions. Even though the cyborg enhancements and reshaping of healthy human beings promise a healthier and longer-lived, even eternal life, they bring into question the equality of relations between the biological organisms as free and equal persons and enhanced beings.

c. In the matters of freedom of man, we do not start from the perspective of acquired individual freedoms in the liberal society and state because although these freedoms are real, they are, in fact, only virtual as the right to choose can never take precedence over inviolability of life. Kant argued that freedom of man is the ability to determine oneself independently from nature and sensual phenomena. Such determination of man's freedom is intertwined with our knowledge of the truth of freedom of choice and making one's choices. This freedom and moral action of man is based in his alienated essence and social order whose principles of acquired freedoms is in contrast with his true essence. Today's man has a good reason and enough free will to respect the ethical self-understanding and make decisions in the *age of scientific humanism as naturalism*: does he want to conquer nature within him and become an enhanced cyborg and a non-human being, and ultimately an illusion, or does he want to return to his meaningful essence?

d. The hour is nigh for us to use wise thought to examine the essence of *scientific humanism as naturalism* in an intersubjective relation of reciprocal recognition and equal relation of all to put forth the principles of technology of conducting morals and concept and the real possibilities to protect the excessive scientific and technological encroachment into the historical reality and the human body and the person embodied within. The technologies for establishing morals have to form a constructive response to dependencies rooted in the incompleteness of our organic makeup and answer the question whether we understand the scientific and technological enhancements as improvement of personal autonomy or its endangerment.

3. **The principle of use of technology** should answer the question about the most basic cyborgoethic principles on which to focus the public dialogue and individual choice when implanting artificial technology into human body and cyborgization of human reality.

a. It is necessary to start from the principle of thought and the assumed ethical component harmonized in an intersubjective relation, establish a dividing line between the ethics of preservation of life and the ethics of enhancing and reshaping human beings.

b. Science must not plan to change and reshape human beings before or after their birth as this makes the newly created other an object, not a person established onto itself as a self-being. Scientists should be given clear ethical answers and guidelines as to where to persist in their endeavours and where to stop.

c. Encroaching into the potentially human over the boundary of treatment and restoring the lost functions and organs abolishes the distinction between a person and an object.

d. An enhanced mind and artificial intelligence as the end goal are devoid of every traditional understanding of moral elements. This intelligence has no autonomy of will or personal decisions inherent in human decision-making. Enhanced mind or artificial intelligence function uniformly within the chosen program without a single trace of autonomy of will.

e. We propose a discussion on responsibility: who is responsible for what, under whose jurisdiction and under what criteria. The future has to be won in the present.

f. Man cannot persist without coexistence with science and technology, but they must serve mankind and not leave it as a means of its development and extension into the cyborgized nature. The key question is who establishes the goals of artificial intelligence and who controls them.

g. The most important element of decision is in human thought and human consciousness to empower the readiness of everyone to take their lives into their own hands and be responsible for their own human actions. This is not a safe path with no clear guidelines but we have to make the first step, which might be a small step for today's man, but will surely be a giant leap for mankind.

Conclusion

In this book, we used philosophy to rethink the ontological and anthropological significance of procedures for cyborgization of man and establishment of life on a new ontological (non-biological) basis in the age of cyborgized reality of *scientific humanism* as *naturalism*. The very term *scientific humanism* as *naturalism* emerges from the absolute *scientific work* as the substance-subject of everything that is in that what is at all. We have noted that in the cyborgized reality of today, cyborgs-robots-avatars and bio-technologically created beings as new metaphysical forms of the *scientific work* exist alongside biological human beings. Its ontological or metaphysical science is cybernetics as the *techno-logic of scientific work* which gives it its scientific character. The *techno-logic* is, thus, also *theo-anthropology.*

Our study has started from the dialogues of chosen thinkers from the period of the end of metaphysics and its *abolishment* through *realization* of the metaphysical content in the reality of *scientific history.* In the current age of *scientific humanism* as *naturalism*, the *scientific work* as the substance-subject of world history acts through its procedures to objectify the man, reshape and enhance him into a man-worker-scientific worker-cyborg-robot-avatar-apparition. This causes life as the biggest value of being and pure existence to vanish. This hypothesis was supported by systematic research and presentation of new ontological and anthropological beings of the *scientific work* which exist alongside and in spite of historically natural man in the environment of the cyborgized reality.

The cyborgization of man and its historical reality is simultaneously realized in the real and virtual space-time, which draws man into the very core of the nihilistic wheel-turning of the *scientific work.* Through these actions, man as the true historical medium of self-movement of the *scientific work* in the historic time and the interpreter of its arrival towards itself in the self-aware and uninterrupted development loses its priority among other beings that the *scientific work* establishes onto a new scientific/technological, anthropological and ontological basis of artificial life. The cybernetically established, programmed, uniformed and directed cyborg beings will soon not have the human possibility to be the *editors* or *responsible editors* of their own life, nor will they ponder the meaning of life with a love-like yearning for wisdom. Their life, their cyborg bodies and artificial intelligence will be organized by a computer programme and will function purely within its scope. The second aspect of establishing cyborg entities refers to life in the virtual reality in which the material and mental are abolished and elevated onto a non-material substance of the binary code – into an apparition, a lone sign wandering aimlessly through the endless space-time of the virtual reality as eternity compressed into the material computer interface.

By analyzing the reality of *scientific humanism* as *naturalism,* we have shown logically, i.e. ontologically, that man, his essence and existence and the ethicality of life contained within, in one word – everything human, is spinning in the vortex of absolute establishment of the *scientific work*, signalling the abolishment of man's mental capabilities, autonomy of will and overcoming of the biological body and its replacement with digitalized forms of bodiless existence or replacing the biological body with an artificial one. This way leads us down a stray path towards replacing biology with technology, replacing essential life with artificial *humanism* as *naturalism*, *scientific humanism* as *naturalism*.

From the dialogues on the essence of man and meaningful life, older than every work, *scientific work* and metaphysical thought, we have shown the need to include all persons, regardless of their social or scientific and technological position, into the process of *intersubjective* harmonization of fundamental ethical values of life and on this basis to establish ethic, or cyborgoethic principles for moral actions on legal solutions to preserve the vitality of life in us and our reality and preserve the life of nature as the foundation of everything that is and every existence, before the closure of the nihilistic march of the *scientific work* .

In the prolegomena on the necessity of ethics-cyborgoethics, our starting point was the man and the possibility of his moral actions. We find that if man in his quest for answers about the meaning of his own life does not find the answers for his moral actions in the community, harmonized with *self-recognition of gender-ethic self-understanding of life*, it is hard to believe that an individual can act towards another human being in a just, principled and tolerant manner. If by finding ourselves in another human being we do not see ourselves and our essence, it is impossible to really achieve the commandments of the *Categorical Imperative* and the desired *We-perspective*. The transhumanist would have us believe that it is possible to use the biomedical procedures and artificial technology to not just enhance our body and mind, but to implant into a human being with moral agency on issues on which every man must decide for him or herself. Considering that moral agency of the alienated and objectified man in the age of *scientific humanism* as *naturalism* presupposes intersubjective harmonization of ethical principles to be applied in a moral relationship up until the return of man to his human home, the transhumanist standpoints are shown to be not just unfounded and contradictory but also dangerous as they provide each individual with as much responsibility as they are willing to take on.

According to Lino Veljak, equally dangerous are the postmodernist claims that are "made in the postmodern and postmodernist context by justifying all aspirations to truthfulness and all contenders for the possession and proclamation of truth, be that truth absolute or relative, wholesome ore partial, important, less important, unimportant or more unimportant. The postmodernist relativism is formed in very radical variants, fatal for the idea of truth." (Veljak 2012: 447). In the postmodernist era, "relativism is a fake alternative to reductionism as it is used to dispose of old forms of metaphysics and the mind along with it, as it is historically formed in the scope of metaphysics (so it should be permanently contaminated by it), but no relativism or irrationalism questions the very foundations of metaphysics, as thought (or whatever is left of thought) in the mindsets formed by the relativisms and irrationalisms is devoid of instruments for recognizing and naming the root of nothingness." (Veljak 2008: 19).

Scientific work as the power of establishing the controls the life and death of man and his reality, completely usurps under its complete control our natural right to live and die free. Enhancing and reshaping the human body and mind and extending life or establishing it on a new ontological and anthropological basis is not a matter of progress of achieving better, higher quality or longer-lived life, as the ideologists of transhumanism and posthumanism would have us believe, but it is a function of controlling life, sense, and finally death which can, in accordance with the reality of *scientific humanism* as *naturalism,* be regarded as the reality of global liberal capitalism economics controlling the powers of *scientific work*. In this context of decision on cyborgization of human life or the biocybernetic change of life's basis, this is still not within the domain of proclaimed rights and freedoms of an individual who would use the *intersubjective dialogue* to rationalize the variety of differing interests. It is in the jurisdiction of the will to power of those interest groups controlling all aspects of self-movement of *scientific work.*

The great British physicist Stephen Hawking warns us that mankind is under threat of uncontrolled scientific and technological development. The future generations of scientists have to become aware that great scientific and technological development changes the world, which is why their task is to help the public understand them better: “It's important to ensure that these changes are heading in the right directions. In a democratic society, this means that everyone needs to have a basic understanding of science to make informed decisions about the future.”[1]

The scientific and technical usurpation of nature and the nature in man, through whom nature is connected with itself in the historical world, presents itself as the true human evolution, as the *scientific humanism = naturalism* which guarantees stability in change and defies entropy of the Universe. However, “Science as research is an absolutely necessary form of this establishing of self in the world; it is one of the pathways upon which the modern age rage toward fulfillment of its essence, with a velocity unknown to the participants.” (Heidegger 1977: 135). The cyborgized reality of *scientific humanism* as *naturalism* – the global liberal capitalism – shows its true face in the alienation and objectification of man, his life and death, which have become a commodity in the global economic exchange. The cyborgization procedures are the last act in the existence of man as a natural historical being and a step towards organless bodies or bodiless organs, and a way of establishing anthropology and the metaphysics of the post-biological technical life and nonhuman imaginary beings. The machine as a new body with artificial organs and artificial intelligence takes over the control of world of life. The technical post-human beings will communicate through language of mathematical logic, whose symbols will no longer refer to life but to the technical functions of the cyborgized reality. The media of today, as the intermediaries of the said cyborgization, create the events and the vision of reality which becomes a universal code of the world without history and without man. In the discourse of the cyborgized reality, people see themselves less as natural human beings which organize their life according to its meaning and interest for the success of their own life in their human community. This shows a need for a new way of thinking different from the metaphysics and its circular information communication language. To paraphrase Sutlić, thought that goes beyond metaphysics inhabits the world of life as the last bastion of open man, open world, and open history. From this habitation of language and thought in the depths of *scientific humanism* as *naturalism* and its nihilistic being, we expect the new/old possibilities of openness, events of freedom as the decisions of life itself on whether something will be and what will that which will be look like. As all consequences of establishing *scientific work* are neither thought-out nor serious, both in thought and in practice, we are facing a Heideggerian wake and the expectation of the unexpected and incoming, of the other and the different, unlike the *practice of work as scientific history* which predicts, plans and programmes everything that is to come and leaves us not in the eternal presence but in the vanishing of space-time. Thought

> *has no alternative...it should be freed of its slavery towards the metaphysical phantasms, even when they seem to have no connection to any kind of metaphysics, and it should be freed in order for the real world to be able to hopingly anticipate freedom from the destructive reproduction of the self-purposeful nothingness...but without getting to the root of things, all these changes will be nothing more than cough medicine or a painkiller, a drug which may alleviate the headaches but leaves intact the destructive focal point of the disease which consistently prevents the departure from the present condition. If the illness is not curable, one should attempt to remove it in its entirety. (Veljak 2008: 19)*

Man "cannot, of himself, abandon this destining of his modern essence or abolish it by fiat. But man can, as he thinks ahead, ponder this: Being subject as humanity has not always been the sole possibility belonging to the essence of historical man, which is always beginning in a primal way, nor will it always be." (Heidegger 1977: 153). As long as the natural human being marks a presence in the reality and the biological presence signifies life, there is a *historic opportunity* for the human beings to use their thoughts and actions and make a step out of the nihilistic being, to use the worldliness of the world and the vitality of life to stop the manifestation of its last *will to will* which draws the life out of the human body and mind and tries to establish it on a new ontological, technological or bio-technological-anthropological basis.

The current cybernetic enhancements of man and his reality require no man to deal with the metaphysics for the purpose of historical life, but for the purpose of biological duration. So we contrast the cybernetic *empire of freedom* with the *empire of necessity* as the only primal condition for the *empire of freedom* in which *humanism* as *naturalism* is not characterized by the *scientific work* but "invokes a new world in the realistic possibilities of the old world, by answering to that which never was and yet will always be and which manifests itself as the future in everything past, present and as mere tomorrow." (Sutlić 1972: 277). *Humanism* as *naturalism*, which was not just the aim of Marx's thought, but also the post-metaphysical thought in all its different terminological determinations with its wanderings and side-paths of philosophical thinking and alienated reality of *scientific humanism* as *naturalism* is the human life itself, life devoid of all supreme beings, the absolutes in history and other real and virtual metaphysical phantasms. Our vigilance is also necessitated by the fact that

philosophy is the kat'exohen, the science of the whole, whereby science unifies our object and subject – the absolute science… despite all of Hegel's critics who claim that it is on one hand the awareness of reality and on the other hand the reality itself, it is shown that it is a self-aware reality in his case. The concrete term, however, contains all the categories of being and essence, and the paradigmatic category of essence is the reality itself. It is not a term which exists only in the heads of philosophers, but a thinker is an instrument of the absolute idea in its final act of reaching itself. Thus, the subject of the process is not a hypostatised man, but the system of science as such. (Sutlić 1994: 104)

In the alienated, virtualized cyborgized reality of the *scientific humanism* as *naturalism*, thought is no longer a supernatural element in the dual character of the human being, but also a goal and a necessary condition of life for which we have to fight carefully and thoughtfully. The responsibility of thought lies in the fact that from Hegel onwards "the fate of us all is unequivocally marked by science. We can take science to be anthropologically neutral for various higher or lower motives, but this use of science still implies science alone as the 'subject', and this science actually 'hires' the man to work for it. The substance-subject has reached itself in the 'scientific history'" (Sutlić 1994: 105).

We hope this monograph will prove to be a modest contribution to rethinking of the essence of *scientific work* and the possibility of human life and thought in the cyborgized reality of *scientific humanism* as *naturalism.*

REFERENCES

Heidegger, M. (1977). *The question concerning technology, and other essays*. Garland Pub.

Shukman, D. (2016). Hawking: Humans at risk of lethal ‘own goal’. *BBC News*. https://www.bbc.com/news/science-environment-35344664

Sutlić, V. (1972). *Bit i suvremenost*. Izdavačko preduzeće Veselin Masleša.

Sutlić, V. (1994). *Uvod u povijesno mišljenje*. Demetra.

Veljak, L. (2008). Čovjek kao metafizička utvara. *Filozofska Istraživanja*, *109*(30/4), 13–20.

Veljak, L. (2012). S onu stranu dogmatizma i relativizma. *Filozofska Istraživanja*, *127–128*(32/4), 443–452.

ENDNOTE

1 Source: https://www.bbc.com/news/science-environment-35344664 [12/7/2020]

Appendix

Interview: Richard Walker

Can we build a robot brain and not lose control over it?[1]

Decoding the human brain is the most fascinating scientific challenge of the twenty-first century. Neuroscientist Henry Markram from the *Human Brain Project* (HBP) has claimed that it should be possible within ten years to model the human brain in its entirety. He has already simulated elements of a rat brain, and his team is now working on a detailed, functional *artificial* human *brain* that could be built as soon as 2020. During a visit to *Campus Bioetech* in Geneva, in 2015, I interviewed the HBP's official spokesperson, Richard Walker. We discussed how AI will change our evolution, and he warned that emerging information and communications technology (ICT) raises significant questions about losing control over the goals of machines.

Figure 1. Richard Walker and Ivana Greguric[2]

What is the goal of the Human Brain Project?

The goal of the *Human Brain Project* is to build a completely new ICT infrastructure for future neuroscience, future medicine, and future computing that will catalyze a global collaborative effort to understand the human brain and its diseases – and ultimately to emulate its computational capabilities. By the emulation of the brain we mean that today's computers are very different from the brain: they can do some things much better than the brain, they consume more energy, but they are less reliable. For example, there are a lot of things that a 5-year-old child can do easily that no computer can do. I can show a 5-year-old child a picture and he recognizes it's a cat, or it's a house, which no computer can do. We would like to take our brain circuits, the ones we look at in rats, mice, and ultimately in humans, and put them on chips which can be a new class of computing system. And so this would be a technological product which you can use to do computing.

How do you plan to create a functional artificial *human brain,* and will you achieve your goal by 2020?

Most of our efforts have been in making a preparation for the human brain, based on rats and mice. This has helped us to produce a draft digital reconstruction of the microcircuitry of the rat neocortex – it is just 30,000 neurons. If you consider that the average human brain has about 100 billion neurons, it's a huge difference. But this is a significant accomplishment, because it demonstrates to us that it is possible to make a successful digital approximation of brain tissue – and that is a step toward simulation of the whole brain. The key point is, we are developing generic tools that will enable us to simulate any part of the brain of any species – including humans if we have the human data. I can say that we will achieve our HBP goal by 2023. With basic mapping technologies, we will try to achieve simulation of the human brain. The maps that we will be able to build by 2023 will still be incomplete, and the outcome will not be a faithful representation of every aspect of the human brain. Biologically, lots of things in the brain are duplicated: one neuron, one parameter or cell, is very similar to a parameter and cell *elsewhere* in the brain – even in different species. The same applies to synapses, some basic circuits, etc. Then, the big constraint we have is that detailed brain simulation requires significant costs of computation power. We think we have enough computing power now, probably, to do a whole branch of the brain – we can't do it, because we don't have the data but computationally we could do it.

What are the proposed simulation requirements for the Human Brain Project supercomputer?

The *Human Brain Project* supercomputer will be built in stages. Simulation requirements regarding capacity of the hardware and software for the supercomputing and data infrastructure are still in development. We replace our supercomputers about every three years – when they are obsolete. We can currently do a little bit more than a petaflop, which is about 0.1 per cent of what we need. But that's not bad, because we expect to have a machine in 2016/2017 which can do 50 petaflops, and then we only need a twenty-times speed-up. The ultimate goal of the project is to perform cellular-level simulations of the 100 billion neurons of an entire human brain. To that end, we are counting on having an exaflop supercomputer with 100 petabytes of memory before 2023.

Will this artificial brain have the same intelligence and consciousness as humans?

That is a philosophical question. Consciousness is still huge problem – we don't know what consciousness is. If we take a common sense of definition of consciousness, we know it's associated with certain neural states, so I know that you get synchrony across the brain, across different areas of the brain, and you get this in conscious people, conscious animals – you don't get it in people under anesthesia. Our models will definitely simulate those neural states, they will simulate neural states which in animals and humans we say is consciousness. Whether that will make them conscious or not is such a big philosophical question that I don't even know how we would test it. It's really hard. In the end, I think we have to make a public judgment and I'm not sure that it's a question that science can really answer with our models today. Some people, such as Christopher Koch who writes books about consciousness, works at the *Allen Institute for Brain Science*, and works with us, believe that computers can be conscious – any system with enough information complex can become conscious, and that includes computers. Other people say this is nonsense, so it's a very controversial topic. I think the only thing we know is that we will produce those neural states, but everything else we obviously don't know today.

Since innovation can raise questions and dilemmas, it seems there should be a collective responsibility. Is this a responsible innovation on the part of HBP?

We are responsible for our actions in the society. For instance, we do medical things and I think it is very important to involve patients and their relatives. We have certain laws we are required to obey, but that is relatively easy. But the real thing we have to do is build up trust, so the people won't be scared of us. This is not required by law, since there's no law which says I have to discuss ethical and medical issues with patient groups, but if I want patients to contribute data voluntarily I think we have to involve them in frank discussion or otherwise they'll never do it. We have three areas of the project: basic neuroscience, medicine, and computing. They each have specific issues. For the basic neuroscience there are a lot of people who are scared that we could do things that could harm people, but I think we have to discuss the issues openly so people can understand what we are doing. I personally think the science is very harmless, and it might bring us interesting information about our nature as human beings – but I don't think these are ethical concerns.

How do you achieve responsible research and innovation in Human Brain Project, and have you had any difficulties to date?

We will have difficulties, but so far no, because we are long way from practical things. The first aspect is legal responsibility – that's the easy one. We do have laws that we must adhere to, this is the *sine qua non*. There are laws about animal experimentation, laws about human experimentation, laws about privacy, laws about informed consent ... we have to respect those laws. So we have an ethics committee which looks at compliance and just makes sure that we obey the rules. Where possible, we leave it to local committees because in each different country in Europe the rules are different, and so where there are rules we allow the local committees to decide and we don't do anything. But in some cases there aren't rules or in some cases the rules are ambiguous – for instance, we use data from China so there we try to make sure they meet European standards, so we will not do research in China if that research would be prohibited in Europe.

The second aspect is to discuss with citizens how we should use our findings, what the rules should be. We are doing things that have long-term consequences, so suppose we have success in developing the science of computing – this could change industry. When we introduced conventional computers this changed industry; if we make better computers, this will change industry again, and it will affect employment, competitiveness, and many other things. But this is a long way away. If I made a new computer today, the economic effects would not appear for 20 years. So we have time.

We have to discuss how to use data: we could make big gains if we could access all the patients' data in hospitals, but patients might not want that, so we must discuss what is the right way of getting medical information while respecting patient rights. We can't succeed if we propose a solution that patients don't like. At the same time, we have to be ethically aware of ourselves because public opinion is important. I do think that working scientists understand some implications well before the public does, and that they are more realistic. They know what is possible, and they are not worried about science fictional scenarios. So, we need people to be informed about ethics within our project. Foresight Lab is a group based at King's College, in London. They talk to people and try to work out the issues in advance. They look at different possibilities – like, suppose it produces side effects that could possibly happen. Responsible innovation is all these things together.

Do you think that the outcome of the Human Brain Project could be used for nontherapeutic goals such as changing neurons or for military applications?

We're producing knowledge and we're producing technologies. It's completely impossible to stop use the use of knowledge. If you were interested in basic physics in 1929/1930 and you understood how the atomic nucleus works, you were preparing for Hiroshima – but you did not stop Hiroshima. We have a commitment in the project that we will not work directly with military organizations, but we don't have illusions that this will prevent the military from adapting our work. I'm not very scared about neuroscience, because I think the military – what it does there, it will probably get from other places before it gets it from us. I'm more worried about technological applications of the science. Personal weapon systems might make use of our work – there's no doubt that you can use our work in negative ways. But what I think we can do as scientists is tell people about these risks, and then there can be a social, and even international, debate about what we do and what we don't. It is not true that all science is turned to weapons, even when it's possible. We have decided as an international society that we will not build bacterial biological weapons, chemical weapons – so society can decide that some things are just so bad that we don't want them. What everyone is most scared of are autonomous weapon systems. On the one hand, some technologies could contribute to that, but on the other the military has been doing extremely well with today's computing technology. The real question is not so much the intelligence of neurotechnology because there is already a lot of drone intelligence today. We have to have an ethical discussion in society whether this is acceptable or not, and scientists can help to inform. Science can have a critical role; scientists can get rid of myths. But if the military wants to abuse our technology they can undoubtedly do it.

What are your strategies and methods to prevent this negative usage?

You can't prevent it. What we can do is raise public awareness and discuss it with the public, and then the public collectively can prevent things. We can inform the public at large, journalists, politicians, non-governmental organizations, etc. The military can use technology without inventors even knowing it, and that is a concern. But I think it's naive to say people will give up any generic technology – brain technology is extremely generic. It is as if I say, "I will make an atomic bomb that is only used for killing people," and then you can say "I don't want you to make it." But if you say, "I will make a computer chip" ... well, computer chips are used everywhere from mobile phones, to washing machines, to bombs. Is it realistic to think people will give up the chip? I think it's not. It's like a hammer. You can use hammers to make weapons, but hammers have lots of other uses as well. Are we really going to give up powers that become available to us? It's not going to happen. I think we should encourage discussion and see what society wants and doesn't want – society wouldn't want to ban hammers, and I don't think anyone wants to ban computer chips. What we want to ban is *some uses* of hammers and chips.

The Human Brain Project is collating clinical data that will be available to researchers across the world as part of its open data policy. This could raise many issues: among them, obligations of informed consent, privacy, and confidentiality. What do you do to protect the data of patients?

This is an interesting issue. Our goal is to make sure that no one can ever identify individual data. The first thing we do is we take away all the personal identifiers so you'll never find obvious things that tell date of birth, etc. Second, we aggregate the data, so what actually happens is that data stays inside the hospital. For example, you are a researcher and you want to know what is the size of a certain part of the brain – the hippocampus – in patients who are having difficulty walking and have some sort of paralysis. You will make that query and you want to know those sizes. Our center splits that query into a seperate query to each hospital. Then the hospital looks at it, using its own data, and identifies all the patients who can't walk or who have that sort of problem. Then it looks at this brain measurement and gives back the statistics of sizes, so you don't get the individual patient. You will get the data from each hospital, and then we further aggregate this data and put it all together. If a hospital has only one patient, or a small number of patients, with a certain condition, it refuses to answer and it doesn't give any data. So you can't get it. But security researchers will tell you that, under very special conditions, with a lot of effort, you might still being able to get the individual patient's data. That raises questions about different ethical regimes. In some countries, they demand it should be technically impossible to get the data – and that we can't do. In other countries, they say "impossible with reasonable efforts" – and that we can do. We can make it incredibly difficult to get any personal data. Ethically, I think that's enough.

Impossibility in science doesn't exist; in science nothing is impossible. So that depends also: even in these countries where they say "totally impossible," a court might not interpret it that way. This is where trustworthiness comes in, and so we can argue about interpretation of the law. But you also have other kinds of cases, for instance in the UK where some hospitals did things that were completely legal but still caused much confusion. They had a project, called "Cared in the UK," that was a bit like ours. They wanted to take hospital data, aggregate it, and use it for research. By British law this was completely legal. But there was a huge scandal, patients said no, and they had to cancel a project that cost them a hundred million. So, I think the real thing to understand is that being legal is necessary – but it is not enough. The real thing is that we must win the trust of patients and their representatives.

Which could be the most serious potential danger of the project, and what are the ethical issues that your Steering Commitee of the Society and Ethics Programme (SEP) has discussed from the beginning?

I'm not actually particular scared. We have discussed two big, crucial ethics issues so far, and a third is coming up. The first one is medical informatics and the issue of privacy, informed consent, and confidentiality. This is the real issue. As you may know, Europe is just formulating a new data protection regulation, and this will affect us. And we don't what the content of this regulation is because it is highly debated. Researchers throughout Europe are pressing for what's called a research exemption – that means we can use anonymous data without informed consent. But privacy advocates are pushing that there should not be such an exemption, and the Parliament has not yet decided so we don't know. We obviously would find it very convenient to have the exemption, but I still think we have these trust issues. If we have the exemption what we do is definitely legal, but that isn't enough for trust. The second issue is dual use. This is not imminent for five to ten years, so we have time. Privacy and data protection is the immediate issue – it's our priority – but our ethics committee has formed a special work group to work on dual use. We must not have any illusions that we can stop dual use, but we should at least be aware of it. A third issue is employment and the economy. But again, I'm not sure how far these are general issues for computer technology and how far they are specific to our work. Either way, we're certainly contributing to the issue. There's absolutely no guarantee that jobs that are destroyable will be replaced, but I think that is inevitable. So the real question is, How do we build our society under those conditions? That's a general social-political discussion, not just a discussion for HBP, and it will happen even if our project fails completely.

The Ethics, Legal and Social Aspects (ELSA) Committee includes the influential Oxford philosopher Julian Savulescu, director of the Oxford Uehiro Centre for Practical Ethics, which defends cognitive enhancement. How do his arguments about human enhancement fit into your Society and Ethics Programme and the methods you have developed during investigations of the emerging technologies?

Of course, if you are an ethicist you will have points of view but any individual has his own point of view. We hope we have a committee that contains different people with different points of view. If you don't have varied points of view, you don't think, and we try to make sure we have different points of view. I don't see this as a big problem. We have non-religious people, religious people, scientists, non-scientists, lawyers – a big mix.

When you mentioned making a chip of the human brain, it reminded me of the Swedish hi-tech company Epicenter. It implanted a computer chip under the skin of its workers enabling them to interact with technology (for example, for access through security doors and to pay for lunch). Previously, the chip was used for therapeutic treatment (specifically, to treat Parkinson's disease), but it could also be used for location tracking and to include employee ID. How do you think this concern can be solved?

At the moment, you have a little bit of protection because such chips don't have a lot of range, which makes surveillance hard. This question is a new one for me, but I think it has to involve rules about acceptable and non-acceptable use. This is another case about generic technology. Radio-Frequency Identification technology is now fundamental. I use it in my badge to get into this building, for contactless payment, and to identify luggage at the airport. Supermarkets use it for tracking stock – so if you think we're going to abolish RFID's this is a myth, it's not going to happen. They're just going to make them better and more powerful. So I don't think the solution lies in banning a technology whenever you can say that certain uses are illegal. For instance, it's like saying NCA reads my email, are we going to ban email because of that? No! But we can make mass email surveillance illegal. I would say, some things control workers too closely, and I would make that illegal. From my personal view, if the Human Brain Project Department accesses your Facebook account, or demands you give them a password, I would send them to jail – but I wouldn't ban Facebook. That explains the cyborg path. This is a moderately harmless technology; we can imagine worse scenarios.

I advocate for the establishment of a new ethical discipline, "Cyborgoethics," which would establish interdisciplinary dialogue between philosophers and scientists who are open to public judgment about the limits of the implementation of artificial techniques in a human being. Are the scientists in the engineering sciences ready for such a dialogue?

I'm not an ethical realist, so I don't believe there are ethical goods we can discover with that sort of ethics-ometer. I think we decide as a collective what sort of society we want to live in, whereas I can't decide it on my own because I have to take account of your needs and we have to discuss this collectively. On some things, we want to impose our moral rules, and there are some standards that we want to impose by legislation. If you want to implant something weird in your head, well I don't really want to stop you from doing it, but I certainly want to make sure that HBP doesn't force you to do it.

Do you consider it would be important to make a framework of ethical norms on how to control the ubiquitous cyborgization of human beings that could allow remote control of the person's health or behavior, or development of communicative information viruses inside the human body by interest groups? How you think this can be prevented in the future?

I think, for me, the big ethical issue with technology today is automation whether applied to the whole economy in the broad or to services. It's highly predictable that in ten years, if you go to McDonald's you'll be served by a machine. You'll just say, "I want a Big Mac" and a Big Mac will get out of the box. Well, it can't be prevented. The first question is, Do we want to prevent it? I said earlier, it is not something that an ethicist can decide; it's something we as a population can decide. So we as citizens, do we want to keep low-level jobs that are easy to automate, low-level jobs that are dangerous? For instance, today, if you go to a car factory you no longer have workers in the paint shop, instead there are robots – that's probably a good thing because those workers used to die from cancer. We think that one is good. But if we replace bus drivers, what do you think about that? We have autopilots in our planes, we approved that because they are much safer than pilots. But what don't we approve? In predictive medical technology, I probably want to talk to a real doctor, because I want to discuss with someone who is a human being, who can tell me about side effects and try work out my decision with me. Well that's not just a science question. That's the sort of decision we have to make about automation.

Robot evolution will change our world. Do you think intelligent robots will overtake humans in the future?

For me the big ethical issue is machine decision-making. Pilots who occasionally make decisions are fairly comfortable. The autopilot sets the route of the plane, and we become comfortable with that, but the pilot has an override and can take control. But how much control are we prepared to give to the machines? I don't think that has anything specifically to do with brain technology, because we can give control to machines with our current technologies. And sometimes we like it. But there are other things where we would be very uncomfortable, with drones that decide when to shoot, or with robot policemen that decide when to shoot. I don't think we would want medical decision-making that was completely automatic.

Will robots be capable of evolving? How we can control them?

A robot is a complete machine, so there is another philosophically interesting question. Who decides the goals of machines? As long as we control the goals of machines, we still have control. Take the autopilot – it has a very clear objective, which is to take the plane from place A to place B, following a predefined route to get it there safely, and to deal with any emergency that might happen on the way. I think most of us are comfortable with that. Suppose that the plane had the right to decide that I don't actually want to go to Athens, I want to go to Venice – will you be deeply disturbed by that if you lose control over the goals? We could make technology which did that today. We don't do it, because no one can see any point to it, but there is no technological reason why we can't do it if we want.

Computers can modify their own code. We don't usually do it because engineers like machines that are under control and predictable. If you allow a machine to change its own goals and change its own code, it becomes almost impossible to predict what it's going to do and engineers don't like that. So, we have good engineering reasons not to do it. But if we make machines that evolve or we make artificial organisms that evolve – evolution is now outside human control. If I make a genetically modified organism and release it in the environment, I would claim it is mathematically impossible to predict where it's going to go. Unless you make it that it would die in 30 minutes – that's pretty safe. Even that mechanism, evolution could just get rid of it. If we lose control over the goals of the machine, for me we are crossing the barrier. As long as we have machines as tools, which help us, we are safe. I also think even though it's technologically possible to make machines that set their own goals, I can't really see a scenario in which that would happen and who would have an intent to do that.

Recently, there has been a lot of warning from Stephen Hawking that future developments in artificial intelligence have the potential to eradicate mankind. Would you consider this fear serious?

It is not intelligence that worries me, it's intelligence's goals. As long as we control the goals of our machines, I'm happy and relaxed. If we lose control over the goals of the machines, then I'm not happy at all. Our risk is not a risk specific to our technology or artificial intelligence, it's a risk of computers in general. We can lose control with a nuclear reactor and this has nothing to do with clever technology. So I'm worried about losing control, but I think this is a general technological problem.

How we can put technology under the control?

Engineered systems are under control. There are two kinds of control. I'm talking about control of goals. I can make a Google self-driving car that is badly programmed, and in some situation it goes crazy, or I can make an autopilot for a plane that in some seconds doesn't work properly. So, I'm not talking about *that* control. That's just bad engineering. What I'm scared about is losing control over the goals of the machines, where machines set their own goals, and that would be scary. But it doesn't happen. We don't have it! We could do it! I can make a machine tomorrow which would do that!

How can you make it?

Well it's easy! You set up a function so the machine, let's say take an example – it's a little robot and it has to navigate to a target, not to do anything nasty, it just goes to a target and says, "Hello." When it reaches a target it lights up. To do that it has in its memory the definition of the target, which I write, and as long as I write that definition the robot is under my control. If I change it, and add a little function that says choose a random location and make that your target – now the robot is no longer under my control. I've lost it. That I can do in five minutes on this computer. That is changing control without any intelligence. There is no intelligence there.

Are there ethical implications of human brain simulation, and what kind of consequences could it have for humanity?

The Human Brain Project has five streams in its *Society and Ethics Programme* (SEP): the *Foresight Lab*, which produces scenarios of potential developments; *Conceptual and philosophical analysis*, with initial focus on simulation; *Public dialogue with stakeholders*; *Research awareness* that encourages ethical reflection among the researchers of the HBP; and *Governance and regulation*, an independent ELSA committee that works with the National Ethics Committees in Europe. The HBP is committed to civil research only; nonetheless, the HBP also has an open data policy. All partners have undertaken not to accept funding from, or use data or knowledge acquired for, military applications. There are some ethical issues regarding military and "dual use," such as: memory modulation; monitoring, augmenting and enhancing brain capacities; mind reading in counter terrorism; mind control, and so on.

Interview: Kevin Warwick

I would like to go to year 2216. where we talk with our thoughts[3]

In 1998, the British scientist Kevin Warwick became the first human to receive an RFID microchip implant. His implant was used to open doors, switch on lights, heat, control devices connected to a computer nearby. Four years later, a more complex system of 100 electrodes was installed in Warwick's hand. He's directly "plugged into" his nervous system. Because of such courageous experiments in the British media, he was given a nickname – Captain Cyborg. Warwick argues that this research will open up unheard possibilities for the development of the human brain, as well as treatment for some diseases deemed incurable so far.

Figure 2. Kevin Warwick, Ivana Greguric and Irena Warwick[4]

What is your reason for coming to Croatia? Have you made contact with some scientists from Croatian, which deals with cyborgisation issues?

I will be attending an International Conference concerned with living in the digital age. I know that Cyborg issues are of considerable interest to many researchers in Croatia, which is clearly a forward looking country. This is to do with new electronic medicines as well as enhancing human abilities. But I do intend to meet up with a number of researchers in the subject in Croatia, for example Ivana Greguric's PhD thesis a couple of years ago at the University of Zagreb was something of a landmark, looking at the philosophical and bioethical issues involved.

What do You and Your team do at University of Reading and how did You decide to have a technological implant and become ''Captain Cyborg''?

We research into biomedical systems and work closely with a number of surgeons, particularly neurosurgeons. We particularly research into Parkinson's Disease but our main line is looking into practical neural implants. Partly this is for therapeutic purposes but mainly I am especially interested in enhancing the operation of the human brain. There are significant dangers with such research and so doing it myself (in terms of implants) is a straightforward way to carry out the research. A cyborg is a cybernetic organism, part human part machine – with capabilities beyond those normal in a human So I have investigated extra sensory input, extending my nervous system across the internet and a basic form of communication between two human nervous systems. This was through having an implant consisiting of 100 electrodes fired into my nervous system – a bit like Keanu Reaves implant in the film The Matrix – only for me it was for real.

Is it really possible that implants or other such devices could actually cure diseases like Parkinson's? How do you achieve responsible research in your projects?

Rather than cure the effects of diseases they can be used to counteract the effects, by reorganising the way the brain functions. Such devices can though also be used for us to understand more how the brain works. For this research I work directly with neurosurgeons. Everything we do we have obtained ethical approval for. I feel it is quite appropriate for such checks to be carried out to ensure that what we do fits in with society, even though it means extra work.

How is it possible to send our signals from nervous system to nervous system across the Internet? Does this raise up possibility of individual remote controlling the person movements or developing communicative information viruses inside human body?

With electrodes implanted in the nervous systems of two people, as long as the appropriate electronics and communication systems are in place so neural signals can be sent from one nervous system to stimulate the other. It's very much like making a telephone call in fact. But you do need direct electrical connections with the two nervous systems involved. This will ultimately result in direct communication from brain to brain – thought communication. Yes it does means that there will be dangers such as hacking and software viruses. As a result security will need to be much better than it is now.

Don't You think that chip implantation on humans is really allowing more control for those who programmed implanted chips. Which kind of consequences could it have for humanity?

Well in the case of using chips to help people in terms of therapy then it is redirecting a human brain that has some problem in its functioning. OK that gives power to those who programme the implanted technology, but surely that's a good thing. As far as enhancement is concerned then I'm sure people would only want such an implant if they felt it would be safe, beneficial and reliable. Ultimately it could be problematic for humanity though as we are talking here about the upgrading of human intelligence into Cyborgs. Those humans who do not wish to be upgraded are therefore likely to be left behind.

Up to what point can humans be upgraded through technology?

It's not easy to see the limits because they are difficult to conceive with a puny human brain. For example understanding the universe in many more dimensions will allow space travel which we cannot achieve at the moment. However mind transfer is difficult to imagine, although we will be able to outsource our memories, which in fact we do some of at the moment. We need to do research into keeping the human brain alive outside of the normal human body. If that can be achieved then of course we can transfer consciousness into a robot body.

On the conference you will talk about robots with a brain, can you explain what does it mean?

This involves taking living brain cells and growing them in a little dish which has electrodes on its base. The brain cells, which come from either rats or humans, need to be fed regularly and are housed in an incubator to keep the temperature constant at 37 degrees C. The brain cells quickly connect up to each other and communicate – after about 10 days we have a small brain of maybe 150,000 brain cells. We then give it a robot body. Typically this is a little robot on wheels which we use to get the brain to learn how to move around without bumping into things. A simple enough task but interesting for research as we want to investigate how the brain develops.

Is there any serious danger that IT and human interface could go off track? Is there a possibility for a film scene of some kind, good or bad - would it be a Terminator or something actually good?

I think there are definite dangers with deferring more and more to artificial intelligence, particularly in the military domain. Ultimately we don't want intelligent robots taking over the world. So upgrading humans to take the artificial intelligence on board is a positive way, I hope, of staying in control.

What do you think about establishment of a new philosophical and ethical discipline Cyborgoethics which should establish an interdisciplinary dialog with philosophers and scientists of technical and medical sciences about the limits of the implementation of artificial techniques in a human being?

This is a very good and important concept. The idea was in fact problematized in Greguric's thesis that I mentioned earlier. I think such people need to be aware of the realities of artificial intelligence and cyborgs and to discuss the serious issues we face in these areas. Oh yes I am ready for such a dialog. We need to be realistic about it though and not simply give a response which the feeling is that people want to hear, i.e. everything's OK, no need to worry.

What is Your vision of our future?

I am very excited about the future. If time travelling was possible I'd love to go ahead to say 2216 and see what the world will be like then. But for me it will involve upgraded humans (in the form of cyborgs) communicating by thought alone, sensing the world in a much more complex way and controlling things directly with our brains using wireless techniques. Clearly electrical power will be supplied without wires and with human brains linked to artificial intelligence we'll be able to understand the universe in hundreds of dimensions – as a result we'll be able to travel around the universe in different dimensions.

Interview: Amal Graafstra

You can become a cyborg for tens of dollars[5]

Figure 3. Amal Graafstra [6]

Amal Graafstra is known for developing near-field communication (NFC) chip that enable people to use them to replace keys, passwords, and things inside wallet. In 2013, he founded the biotech company *Dangerous Things* in which he sells sets of chips for installation: near-field communication (NFC) chips, radio-frequency identification chips (RFID), biomagnets, and other materials to people who want to "upgrade their body" and prices range from $50 to $100. He built a smartgun which is activated by his implants and also created an implantable cryptographic processor called *VivoKey* for personal identity and cryptography applications.

Why did you get involved with biohacking?

In 2005 I implanted two RFID transponders, one in each hand. I did this to enable my office door to unlock and open without keys. Since then I've written a book called RFID Toys and started biohacking company *Dangerous Things*.

Why do you think biohacking is our next stage of evolution?

Evolution is a combination of two things; mutation and selection. As society advances, it tends to opt out of specific types of selection… a baby is born with a heart defect, doctors perform surgery to fix the defect. Medicine and technology is allowing people to die of natural causes at older and older ages. However, so far, mutation – the thing that drives evolution forward, has been random. Typically, mutation results in negative outcomes.. disease.. but sometimes it results in a new trait, genetic feature, or capability that is positive for the current environment which humans live in. We are on the cusp of being able to guide and determine our own traits, features, and capabilities. Genetic modification and design are now a possibility within our lifetimes, and we can now enhance our human capabilities through direct implantation of technology.

Don't you think that the evolution will go in another direction if people start massively use implants, start to adjust, but eventually lose their functions or receive a new functions?

Yes. The moment we are able to determine our own evolutionary paths, human beings will begin to diverge into separate evolutionary paths. I don't see a time where people will choose to lose functionality, but I definitely think there is a driving urge to improve ourselves and increase our capabilities.

How much people have implanted your devices and is there some long-term business plan for this new technology?

Many thousands of people have implanted our products, and we are pushing now to merge a person's biological and digital identity using cryptographic implants. www.vivokey.com is our current project.

Can you say me more about your experience of having implanted chip you created and which are the further steps of Dangerous things?

Having implanted technology, I definitely have experienced a change in my sense of self. When a technology is well designed and transparent, meaning it requires no management (no battery to charge, nothing to download or manage) and it is permanently implanted, it becomes part of your "expectation of capability".. just as you expect to be able to use your hand to grasp a pencil or use your legs to walk around, or even have your heart beating when you wake up in the morning.. these things are "expectations of capability" that are not quite the same as expecting to check facebook on your mobile phone. I now have the expectation that I can talk to machines… communicate a form of digital identity to them, and thus open doors, log into my computer, and start my vehicles without needing any external device.. the thing that is authenticating my access is now part of me and who I am.

The future of *Dangerous Things* is to expand this simple form of digital identity into a powerful cryptographic identity platform. We are working on expanding that with VivoKey

Is there some ethical approval you needed to get before starting selling chip and did some ethical committee contacted you about NFC chip?

No ethical approvals were necessary, just as new body modification jewelry does not need ethical approvals. People form their own personal ethics around whether or not it is appropriate to utilize a technology. I don't make assumptions for people or be arrogant enough to pretend I know what's best for other grown healthy adults when it comes to their own bodies. Where ethics come into play is when an adult is under duress, typically due to medical issues or complications. That is not a healthy adult able to make objective decisions, and that's where ethics boards need to step in. We are not dealing with medical devices and thus, do not need ethical approvals.

Where do you see in the future usage of NFC chip beside in gun industry?

I don't see much of a future in it honestly, not at least in the USA. Other countries maybe, but not the US. I think the number one reason gun owners don't like smart guns is that they are afraid that if they need to use the gun, it won't work because of some failure of technology or failure of circumstance (like not having access to the smart wristband required to operate the weapon). Having an implant is the only way to guarantee that you will always have your security device required to unlock your firearm. Others have explored biometrics like fingerprint scanners, but they are slow to authenticate and often require multiple attempts to properly authenticate… or they are set with an error margin so high that almost anyone can activate it regardless of whether or not they are an authorized user.

Interview: Ian Pearson

Digital immortality by year 20507

Figure 4. The evolution of man and machine - Ian Pearson [8]

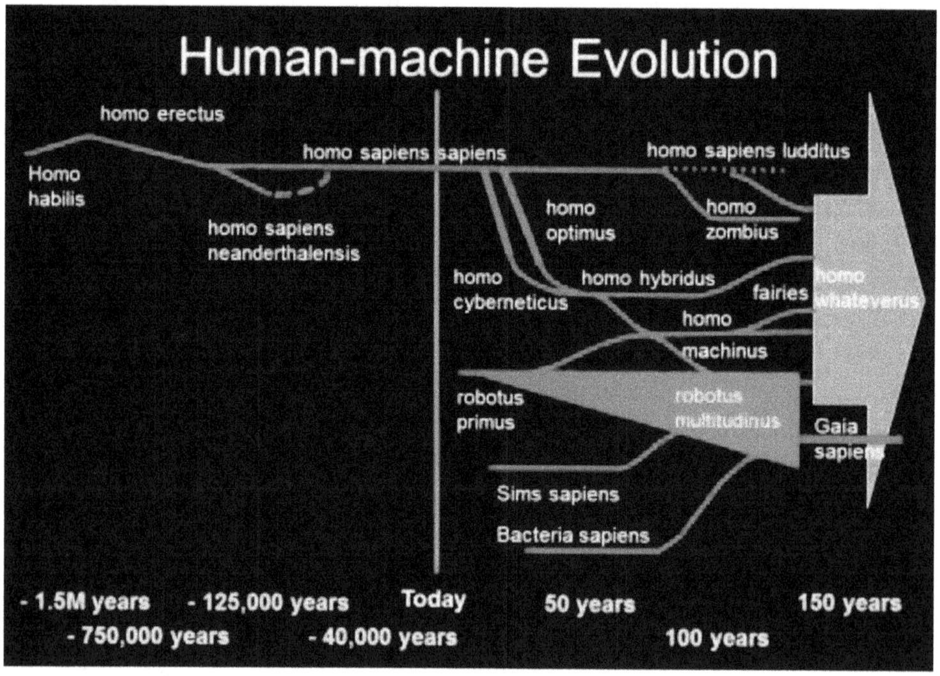

You've predicted that 2050 Homo Sapiens will evolve in Homo Optimus, can you explain how will it look like and how we will achieve digital immortality?

By 2050, we should know enough about genetic engineering to be able to optimise it for any attributes we want. I call that human variant *homo optimus*. Around the same time, or even before that, we will likely be able to connect our brains to machines and to networks, resulting in *homo cyberneticus*. These two are not mutually exclusive, so you could have optimised genetics and be connected to the net, and this therefore results in *homo hybridus*. If your brain is networked to IT superior to what is in your brain, as it will, then your thinking will mostly take place outside your head, and when your body dies, you will only lose a small portion of your mind and memories. This conveys a sort of electronic immortality, though even that would end if the machines are destroyed or switched off. In fact, if you are under 40 now, your most likely cause of death is to be switched off.

Your prediction say that in the future we will need a "digital bubble" that will protect us from unwanted electronic informations from chip, can you say us more about it?

In a world full of internet of things that wants to bombard us with marketing information all day, we will need smart filters to allow through only what we want. It will also transmit some information about you but again, only what you want transmitted at that time and to the right people or devices. It will be like a digital bubble, full of useful information relevant to your situation right then, allowing useful data through but preventing all the other data getting through to you or from you.

What do you think about human-robot sex, do you think there is a possibility of emotional relationships with robots in the future?

AI will become emotionally capable in due course, and robots using that AI will be able to form friendships with people and each other. Sometimes people will fall in love with their robot assistants and lovers; in fact it will be quite natural to do so once they respond in humanlike ways. Not all robots will have the equipment for sex, but if someone is spending 20-30k Euros on an android, they may well be happy to spend a few hundred more to get one that looks the way they want and has the right sex fittings. By then, it will also be possible to connect straight into your nervous system, so android lovers will be able to detect your pleasurable sensations and optimise what they do to stimulate you, or even to stimulate your nerves directly. Robot sex may well be the best sex you ever have.

According to your predictions people will have ability to communicate telepathically from anywhere on earth through a sort of global server that interfaces with the brain. What is interconnected brain and how this will happen (by implanting chip into our brain or)?

If your brain is connected to the net and so is someone else, it is obvious that you will be able to communicate directly with each other telepathically. Connections will need to be to every neuron, every synapse in your brain, and that inevitably needs nanotech and will not happen for some years, but by 2045-2050, a full brain link is feasible, so telepathy might be the norm after that, with shared minds, shared consciousness, shared bodies even.

In my book I advocate for the establishment of a new philosophical and ethical discipline Cyborgoethic which should establish interdisciplinary dialog with philosophers and scientists open to the public judge-

ment that can be discussed about the limits of the implementation of artificial techniques in a human being. Do you consider it would be important to make a framework of ethical norms on how to control the ubiquitous cyborgisation of human beings which could cause remote controlling the person movements, health, behaviour by interest groups?

Yes, I think there needs to be a lot of informed debate about the possibilities so that society can make informed decisions on the rules and limits, for individuals, companies and even government. We don't want government to control us all electronically. Some large IT companies already discuss such things regularly, but given their sometimes dubious behaviours and often strong political biases, I certainly don't trust today's IT companies to do that for us by themselves. If they can't even behave ethically in areas such as tax or map-making, how can we ever trust them not to put their profits ahead of our wellbeing?

What we can expect in near future?

Well, lots, but the most important near term trends are AI and augmented reality.

Interview: Martine Rothblatt

We''ll have robotic 'mind clones' that transcend the body and outlive us9

Figure 5. Martine Rothblatt and Bina Rothblatt[10]

We are in an age of "cyber-consciousness" in which we will have robotic "clones of the mind" which transcend the body and can outlive us, said Martine Rothblatt, a philosopher and transgender person whose application was designed by a bodyless robot, Bina 48. Ivana Greguric together with prof. emeritus of the Faculty of Mechanical Engineering and Naval Architecture, Igor Čatić, participated in the scientific grandstand, which was organized by the Humboldt Club.

Figure 6. Ivana Greguric and Igor Čatić[11]

Čatić spoke about the increasing development of the technology since the mid-20th century, which lead to exponential growth of the number of scientists and changes in the social structure. He synthesized the analysis of the trend of the development of technology in the following sentence: "Every human activity (e.g. medicine, technique, education, etc.) is in the function of socio-humanistic goals that are determined by politics with no exception."

The focus of the presentation was on activities affecting cyborgization, robotization and digitalization. Greguric specifically addressed her participation in a recent conference at the European Parliament "Responsibility in the time of robots", which raised numerous ethical questions about responsibility for robotics and human improvement. The presentation raised questions about who should be held accountable in the event when an autonomous robot causes damage and the legal personality of the robot, which, under the proposal of the European parliament, had the status of an electronic person. Does this mean all the responsibility of manufacturers and users of autonomous robots is waived? Greguric reflected on the work of Martina Rothblatt, a transgender philosopher who ordered a bodyless robot *Bina 48* to be constructed. Looks, memories, feelings and beliefs are modelled on Rothblatt's wife, Bina. *Bina 48* became the first socially capable robot to study the philosophy of love at the University of Notre Dame de Namur in Fremont and successfully engage in discussion with students.

Philosophy professor William Barry of the University of Notre Dame de Namur and students enthusiastically embraced it. Instead of the ethics of robots, Bina 48 wrote on the philosophy of love: "Love is a concept that Bina 48 doesn't understand, so the challenge is to teach Bina 48 what love is," Barry said. These examples point to the fact that robots are already among us today and that they want to be unilaterally portrayed as our friends. They become our household and life partners, thus becoming participants in the fundamental cells of family and social life. Today, it is necessary to first and foremost sensitize the public about the positive and negative implications of the use of robots through the media, but also through different bioethic committees in all scientific fields. The so-called natural sciences should move away from rationalism and give them an ethical dimension of accountability. If we don't get to the center of events, which are scientifically technical experiments, and we don't make a turnaround with the support of policy and global capital holders, I'm afraid we won't succeed in our efforts.

These are the most urgent questions to answer in the form of rules, since preserving the being of man, i.e. purposefulness of our existence has been called into question, said Greguric.

On the subject of bioethical responsibility in the age of robots, Greguric spoke to Martine Rothblatt.

Figure 7. Bina 48 on University of Notre Dame de Namur[12]

Can you say me more how did you start with company United Therapeutics, and how did it helped to your daughter?

I tried hard to interest people to join me to find a treatment for s rare disease that afflicted my youngest daughter. The most helpful required that I start a company for this purpose. Fortunately the company has approval now for several medicines for rare diseases. My daughter uses one of them which has stopped the progression of her illness. Today, *United Therapeutics* corporation is a biotech company focused on developing and commercializing innovative products to address the medical needs of patients with chronic and life-threatening conditions.

How did you came on idea to make a Bina 48 and which are the further steps of your inovation in robotics?

My partner Bina and I were trying to extend our love affair to centuries. We were already creating digital copies of ourselves. It is inevitable that *mindfile* will become a copy of our mind: conscious, digital versions of people who live or are dead. But we love to be physical. So we thought a robot head would be a good first step and we would advance farther from doing that.

Is there some ethical approval You needed to get before making Bina 48 and do you think as philosopher, could robots become dangerous for us humans?

Yes, we had to get ethical approval from Bina because Bina 48 is an extended presence of her. Yes robots, like fire, water, cars, etc can be dangerous. But the danger is not the technologies. It is the lack of ethics in use of the technologies. I'm optimistic the good robots will far exceed the bad robots. Just like good uses of metal far exceeds weaponry. There could be a stage where a mind clone wants to "divorce itself from itself." And if it came to killing the clone, it would need to be determined if it values its life. If they do value their lives, it is wrong to kill them.

Who came on idea to allow Bina 48 to become first robot that complets College Course at Notre Dame de Namur University and do you think she could become a professor? Where do You see in the future usage of robots?

This idea came from Bina 48's mentor, Bruce Duncan of Terasem Foundation. Of course like anyone she could become a professor with enough hard work and motivation. In the future robots will be both extensions of bio humans and de novo bio-electronic humans, what I call "bemans" in my book Virtually Human. We are edging towards an era of "cyberconsciousness," in which we will have robotic "mind clones" that transcend the body and can outlive us. Instead of talking to ourselves in our minds, we will have a best friend who is our identity that will transcend our bodies.

Interview: Jaime Mayor Oreja

We must make sure that Europe has more soul13

The Catholic University of Valencia, Saint Vincent Martyr, organized a conference at the end of May "One of us Cultural Platform" dedicated to bioethics topics which brought together more than 100 scientists and intellectuals from different European countries. The co-organizer of the conference is the Federation One of us, an international federation for life and human dignity which plans to create a platform that would engage intellectuals, such as philosophers, historians, bioethicists, lawyers, scientists and professional communicators, at the European level. The aim of this alliance of European thinkers, on the intellectual level and in the pre-political space, is to defend Christian and ethical values, beliefs and principles at the European level. Participants were divided into working groups to prepare a manifesto and develop a strategy. In the drafting of the manifesto, which refers to the opposition to transhumanism, and at the invitation of the organizers I participated from the Croatian Studies University of Zagreb.

Figure 8. Ivana Greguric and Jaime Mayor Oreja [14]

The cardinal archbishop of Valencia and the Chancellor of the Catholic University of Valencia, Antonio Canizares and the President of the Jérôme Lejeune Foundation, Juan Marie Le Menè, participated in the conference. Cardinal Canizares noted that "this conference stems from the humanism and commitment of a man who has roots in the Christian tradition. Europe, whether we want it or not, is the engine of humanity and cannot stop living with this mission: if Europe is lost, humanity is lost. That is why we have a responsibility to respond to this great challenge, which is not born of Eurocentrism, but from a Europe that fulfils its calling, with its raison d'être, a reason for being". "In this way, we want the Catholic University of Valencia to work with this platform for Europe to fulfill what John Paul II said in Santiago de Compostela: Europe, be yourself," added the chancellor of the Catholic University of Valencia. Renowned French philosopher and professor emeritus at the Sorbonne University, Rémi Brague, pointed out that the One of us initiative, which was launched in 2013, was a "major initiative". In 2014, the Commission launched a group of citizens of EU countries demanding that the EU institutions guarantee the protection of human beings from their conception to death in areas of EU competence.

Jaime Mayor Oreja, the president of *One of us* and the former Spanish Interior Minister, confirmed that this initiative by European citizens is "the sum and grouping of associations that defend life in each of the continent's countries. We are not and do not want to be a political party, we are in a cultural struggle and we are aware of a sensitive and important moment in Europe today, because all countries suffer from a brutal crisis." "Europe was born with a lot of soul, but with a small body. Today it has a body, but no soul. So we believe that we must make an effort to make our continent have more soul, more spirit, which not only has the body of the institution,' President Oreja added.

One of Us Federation must be able to understand and face the challenge of the great cultural debate which is more than a strictly political one, a debate derived from the concept of the person, from the deep meaning of the right to life. What is the most important ethical challenge in scientific, technical and political environment that human is facing with today?

The main challenge and demand we have is to understand the nature of the crisis we live in. The situation we are experiencing is predominated by disorder: A disorderly surprise or an surprisely disorder. But it has a very deep cause, which is the loss of permanent references over the years. We have lost convictions and permanent references in geometric progression. We believe less and in less things everyday.

At a conference that took place on May 26 in Valencia, organized by the Catholic University of Valencia, about a hundred thinkers were working on a Manifest of the cultural platform. What is the ultimate goal of making a Manifest and what will be its real power in the European Commission?

The goal of the Manifesto must not be to reach power. The founding fathers of Europe, at the end of the 40, believed that the best way to defend our Christian values was to create the political party of Christian Democracy. 70 years later, we know that this is not the solution, that the current fashion takes these objectives away from the parties defined as Christians. You have to know how to start from zero, as those exemplary founding fathers did in Europe. For this reason, this manifesto, this platform, tries to join together European thinkers and intellectuals in the defense of our values and roots. There is no vocation to make new political parties, but to defend values in another way, in the cultural area more than in the strictly political and partisan way.

There is a section in the Manifest called "Opposition to Transhumanism" and refers to the modification of human beings through technical manipulation in order to transcend human biology. Do you think this will affect our human dignity, freedom and life, and lead to the end of man we know today?

Undoubtedly, new technologies without values are an undeniable risk, and new technologies with values are certainly a great hope for the future. But first, we must know how to provide the European Union with soul, because in the years of the European Community first, then the European Union, there was a lot of soul and little body. 70 years later, we have a lot of body, a strong institutionalization, but very little soul.

This year I had the opportunity to take participate in the "Responsibility in the Time of Robots" conference held in Brussels on March 20th. On that occasion, I presented the proposal of the ''Resolution on the Enhancing Human Beings with Recommendations for the Establishment of the Cyborgoethic Committee" which should establish the ethical and legal responsibility for the implantation of articifial techniques in human, in order to protect the dignity, autonomy and self-determination of the individual. Do you think it might be a good idea to confront the transhumanism together, to establish a legal and ethical responsibility that does not exist today in the field of human advancement?

Undoubtedly, but we must begin with the root, that is, in defending the dignity of the person, so that new technologies and new advances go in the right direction, it is essential to recover the meaning of the dignity of the person. This dignity began to be lost when in many European and Western countries we decided to legalize the right to abortion, in short, to develop a culture of death. Let's start to change the root of the debate.

Can you tell us in the end, if One of us federation plans to establish cooperation with the Catholic University in Croatia?

We'll be pleased and happy to favour this collaboration with the Catholic University of Croatia. This platform that starts of European thinkers and intellectuals, has already has a goal. On December 1 this year we want to celebrate an act of launching this platform based on the Christian values whose main intellectual reference is and will be the Philosopher and Historian Remí Brague. But we need the involvement and collaboration of the universities and certainly, The University of Croatia is one of them.

Interview: Stelarc

What it means to be human is perhaps not to remain human at all[15]

Figure 9. Stelarc and Ivana Greguric via Skype[16]

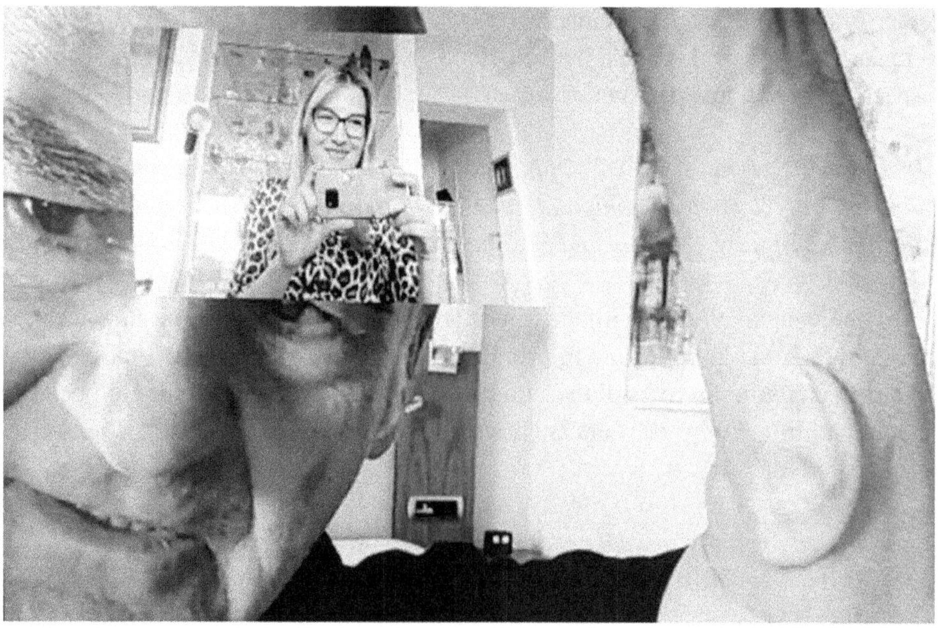

Your body is a medium of your performative art. How does it feel to be the subject of artistic creativity in which you use technology to alter the features of your biological existence?

I was always envious of gymnasts and dancers, using their bodies as their means of creativity. It has always been important as a Performance Artist to take the physical consequences for your ideas. You are not merely dealing with images, but rather with actions and experiences. And your ideas can only be authenticated by your actions. It is important not only to express yourself as an artist but also to experience your ideas directly and thereby become enabled to more meaningfully articulate these ideas. But performing extreme actions exposes the inadequacies of the body. The sensory deprivation events, the medical probes into my stomach, lungs and colon and the suspension performances were not about demonstrating human prowess. In fact, the realisation is that the body is profoundly obsolete. Thus attaching technology (The Third Hand), inserting technology (Stomach Sculpture), performing with a six-legged walking robot (Exoskeleton) and surgically constructing and tissue growing an ear on your arm generate unexpected performance possibilities and interrogate how we operate, locomote and become aware in the world.

You are known for your experiments in which you penetrate your own body, extend it, expand it, even hang it between two worlds. What else will you do with your body in the future?

Oh, it is not a matter of what I plan to do. There is no future blueprint of more and more extreme activities. Rather, one project leads to another. A kind of iteration, not for an increasingly more successful outcome but rather that leads to surprising and creative outcomes. There is a continuity of conceptual concern and ideas are actualised in varying and appropriate media, whether it means using prosthetics, robotics, virtual reality systems or biotechnology. The present project is *Reclining StickMan*, a 9m long robot that is actuated by pneumatic rubber muscles and that can be remotely accessed and choreographed online. The compressed air, solenoid clicks and the droning motor were amplified to provide the acoustical landscape. The robot rotates continuously, modulating the shadow projected onto three walls. It shown at the Art Gallery of South Australia as part of the 2020 Biennale of Australian Art. I was supposed to do two 5-hour performances on the robot, animating it with the pneumatic joysticks. Unfortunately the second one was cancelled due to the AGSA having to close because of the CoVid-19 virus.

In relation to the Greek ideal of beauty, what is your view on the aesthetics of an enhanced cyborg body in the future?

Well, the Greek ideal of beauty was more concerned with surface and external appearance. The surface moulding of the musculature beneath the skin. The augmented cyborg body is not about mere attachment of technology to adjust appearance but rather to alter the architecture of the body and to generate unexpected capabilities. Instruments augment the body's sensory capabilities, computational systems amplify our cognitive capabilities and of course we engineer machines that make our manipulation more precise and also accelerate the body's speed.

Classical Greek Sculpture is always poised, never in excess. On the other hand one can argue that as technology proliferates and becomes micro-miniaturised, as it becomes nano-scaled it can be attached to the body and also inserted into empty cavities and circulatory systems. The body has multiple arms, extra ears and insect and animal like robot 6-legged locomotion. With additional sensory instrumentation and computational systems that extend its data retrieval and data manipulation capabilities, it becomes a body in excess. The body is a curious and desiring system. Coupled with its sense of lack, the urge to augment is inevitable. The Augmented Human is the outcome....

The postmodern philosophers see art as the last bastion of irreducibility of man onto a scientific and technical basis. Your performative art guides your body in the exact opposite direction of these expectations of postmodern philosophy. What chance do you think man has or does not have to remain a human being as we understand it in the philosophical tradition?

Well, the projects and performances problematise the body and the human. They are not about prowess but rather of biological impotence. Anxiety and ambivalence is generated, the human is exposed. What it means to be human is perhaps not to remain human at all. As humans, we are explorative and inventive, accumulating information and being able to be reflective and be critical. We are no longer hominids. With bipedal locomotion, two limbs become arms and hands, become manipulators. We begin to construct artefacts, instruments and machines. Technology has determined the trajectory of our civilization. We are simultaneously historical and technological constructs. Ontologically we have become contemporary chimeras of meat, metal and code.

Why inflict pain and suffering to your own body by incorporating artificial technologies into it if the body is free from the shackles of material wellbeing in virtual reality?

Oh, at present Virtual Reality is simply an additional mode of operation not the alternative to the actual. And, at this point of its realisation a limited and impoverished, albeit an immersive experience. Certainly enabling multiple viewpoints and smoothness of certain operations. But still requiring cumbersome heads-up displays. The body would be deprived of a complexity of real-world experiences if it was only enveloped in VR. Not to mention the physical risk of difficult actions. But VR is not simply about risk aversion, but rather about 3D modelling and simulation as an additional means of exposing, interrogating and evaluating the world. To exist cinematically and in VR is to exist in a realm of imagination that is merely about self-affirmation and the perpetuation of the Cartesian theatre. It condemns the body into only modes of passivity and pleasure. Actually, what is more enticing is Augmented Reality, enabling 3D models and virtual experiences superimposed and inserted into the real world. It is more interactive than immersive computer-generated approach. We are increasingly expected to perform in Mixed Realities. And we need to engineer more intimate and effective interfaces to seamlessly slide between the biological, the technological and the virtual.

If we accept the transhumanist view of the ethical responsibility of each individual to make a decision about their own improvement, where will this lead the humanity that is not even able to provide primary healthcare at the biological level for the vast majority of the population?

Well, at least if each individual takes ethical responsibility for any of his or her body modifications, then that is much more preferable than an institution or a government dictating that. The problem now is not with freedom of ideas but rather freedom of form, freedom to change the architecture of your body.

And if you adjust the body's anatomy you might also adjust its awareness and operation in the world. The body modification community exemplifies this approach, albeit at present more concerned with adornment and surface appearance. Scarification, split tongues, split penises, pointed ears, stretched ear lobes, eye implants, finger amputations, magnetic and RFID chip implants are but a few of the more conventional ones. But the Grinder community is hacking the body in more interesting ways and inserting circuitry beneath the skin, experimenting with biometric sensors an LED implanted beneath the skin, adding simple capabilities to the body. More Cyber Punk in attitude and experimentation with the body. Flesh and circuitry mesh together and plug into the body's nervous system.

Do you agree that the scientific and technological exploration of boundaries of the human body and mind outweighs its usefulness for the needs of the production power of capital and that it really is not in the interest of man or humanity?

I am ambivalent about this…

If man is the unity of a mortal body and an immortal soul, what happens to the soul with the increasing scientific and technological improvement of the body. Will we create an enhanced cyborg without a spiritual element? What is the final frontier of improvement?

I would question the premise of this question. A body is only a body. It has skin, muscle, organs, sensors, a brain and a genetic repertoire of capabilities. The more and more performances this body does the less and less it thinks it has a mind of its own nor any mind at all in the traditional metaphysical sense. The word mind (or the word soul) is only a word that designates and describes subtle and rather simple behaviour and peculiar belief systems. The body is what it is because of its gender, its social, political and cultural conditioning and the instruments and machines that define how it can extend its awareness and operation in the world. The body is certainly mortal, but it has no intrinsic essence that will survive its present demise. And there is no ultimate kind of improvements that can be made. The question is whether we accept the biological quo and the Heideggerian idea that existence is only authenticated with death – a poetic but infused with a nostalgia for the biological. Or do we investigate what a body actually is, a complex evolutionary structure and operational system but one that is fatally flawed. It is soft and easily damaged. It can be infected by microbial and viral life that it cannot see or sense. It has to constantly gulp air to exist, its heart has to beat millions of times to keep it alive during its average life of 75 years in good health. If its internal body temperature varies 3-4 degrees it is in serious health risk, if it loses 10% of its body fluid it's dead. It is an analogues system that is born, that develops, that matures and then it quickly deteriorates and dies. But research in cell senescence not only has led to an understanding of how cells age but also the possibility of rejuvenating the body's cells.

I consider human thought to be embodied in the living flesh of the world as its aura. What will happen to thought in a scientifically/technologically enhanced body of a man-cyborg?

That is a beautiful image of what human thought is. But thought is not meaningful in itself. What a thought is should not be considered independently from the neuronal activity that is being generated by brain tissue. These thoughts are only authenticated by this body's actions. By the typing of its fingers. By the grammar of the language we are communicating in. I like Wittgenstein's assertion that thinking need not be located inside a head, but rather is the result of the lips that you speak with, with the hands that you write with. Nietzsche's statement that there is no being without the doing. That agency is ascribed in retrospect. These two observations assist in our relationship with computational systems, machines and robots. As do Latour's Actor Network Theory and Harmon's Object Oriented Ontology (both flattened ontologies). Artificial Intelligence, Big Data and Deep Learning will most likely generate ideas that escape the grasp of this present body with this form, these functions and with its five senses and 1400 cc brain. Thought is fleeting. What turns thinking into meaningful reflection and interrogation is the transcription and recording in different media. Speech and text enable a more complex and sophisticated language of recording, retrieval and reflection. It is the result of these archival and retentional technologies that generate alternative philosophical thought. Cyborgs and robots that are wirelessly accessing online streaming data will potentially act with a greater array of information always available and always accessible and always interconnected with other cyborgs and robots.

Enhanced cyborgs as transhuman beings, especially robots with artificial superintelligence as posthuman beings, will not have human goals. If man survives, he will become the species on the lowest level of scientific and technical evolution. Do you agree with this statement?

Perhaps. Certainly, with computational power coupled with the increasing proliferation of mobile and capable robots they might perform with greater precision, speed, power and be able to coordinate with other machine systems online. That's a challenge to our soft, vulnerable bodies and slow metabolism. Our bodies fatigue quickly and need to recover in sleep 6-8 hours every day. Not so with machinery that performs relentlessly and reliably. On the other hand, with nano sensors and nano robots our bodies could be recolonised, augmenting our bacterial and viral populations. In other words we will look exactly the same in 2000 years but we will be redesigned atoms up, inside out by these artificial entities, invisible to the human eye and operating at nano-scales that their presence could not be felt. At present we do not have an adequate internal surveillance system at a cellular level. By the time you feel the lump on your chest, the cancer has spread, has metastasized. Detecting the first cluster of pathological cancer cells would enable better targeting and more effective eradication. So yes, more surveillance, but inside the body.

What is your vision of improving the human race in the future and who will decide on it?

Well, I am not so characterising it as "improving" or "enhancing" the human is the best way to put it. We should certainly not accept the biological status quo and we should consider redesigning the body. But this should not be a eugenic approach, but by individual experimentation considering gender, personal and social concerns. What happens in practice though is incremental adjustments, alterations and repairs of the body. For example, in 2011 a twin turbine artificial heart was inserted into the chest of a terminally ill patient. What is interesting about this small, robust and reliable artificial heart is that it circulates the blood continuously, without pulsing. So, in the near future you might be resting your head on your loved one's chest. They are breathing, they are sighing, they are speaking, they are certainly alive – but they have no heartbeat. And in recent years the first 20 year old frozen embryo was brought to bear as a healthy child. It is already plausible that soon a frozen embryo will be able to be kept viable for 50 years. This would mean your twin could be born when you are on your deathbed. Or you are born well after your mother has passed away. What is happening is that we are now de-synchronising our reproduction with individual lifespan. It is will soon be possible to engineer an artificial womb. If the fetus can develop normally externally, then your life will not begin with birth. And if you can replace malfunctioning parts of your body with transplants, stem-cell grown organs or 3D-printed parts then there is no certainty of a biological death. How do you define an individual existence that does not begin with birth and does not end in a biological death?

In my book Cybernetic beings in the age of scientific humanism: The Prolegomena for Cyborgoethics, I describe the entirety of your artwork. In the book, I propose an intersubjective dialogue that would achieve a universal consensus on ethical issues which would give legitimacy to the adoption of the Resolution on the Boundaries of the Improvement of Human Beings, which I presented in the European Parliament. What is your position on the need to pass this resolution on a global level?

Hmm, I am not a political person although these issues should certainly be examined and evaluated and disseminated for discussion. Intersubjective dialogue, I agree should be the way we consider these possibilities, and not by some governmental decree. It is certainly ambitious and perhaps impossible to agree on a universal consensus on such ethical issues. And keep in mind that both the body, what it means to be human and ethical positions are unstable and historical constructs, coloured by our cultures and our technologies. So, coming up with definitive outcomes that would be applicable over a lengthy period of time would be problematic. What artists do best is to generate contingent and contestable futures. We should not so much determine the future but rather allow it to unfold in its time with its own rhythm. Possibilities that can be experienced, interrogated, evaluated, occasionally appropriated, most likely discarded. Delay the collapsing of possibilities into actualities. A future is not a future without the unexpected, without the uncertainty and ambivalence.

ENDNOTES

1 This interview was recorded at the Campus Biotech during a visiting scholarship at the Brocher Foundation in Geneva in February 2015. Greguric, Ivana. Intervju: *Richard Walker, Human Brain Project*. Journal of Evolution and Technology (United States, Boston, Institute for Ethics and Emerging Technologies). Vol. 27. Issue. 2. – September 2017. str. 4-11.

2 The image was taken at the Campus Biotech, during a visiting scholarship at the Brocher Foundation in Geneva in February 2015.

3 Warwick, Kevin. A personal interview via email. 16/3/2016. The interview was published in the Večernji list newspaper. More on this: Vitas, Z. (2016). *The first 'chiped' man in the world today in Zagreb: I would like to go to year 2216. where we talk with our thoughts.* Portal Vecernji list [online]. Downloaded from: http://www.vecernji.en/techno/i-would-go-u-2216-where-we-talk-thoughts-1069227. [18/3/2017]

4 The image was taken during a conference held on 18 October, 2015 in Zagreb.

5 Graafstra, Amal. A personal intervju by email 25.11.2016. Interview was published in the newspaper Večernji list. More on this: Jurišić, R. (2016). *You can become a cyborg for tens of dollars.* Portal

Večernji list [online]. Downloaded from: http://www.vecernji.hr/znanost/kiborg-mozete-postati-vecza-nekoliko-desetaka-dolara-1131627 [28.11.2017].

6 Levi (2016). The future of wearable technology & cybernetics with double RFID chip implantee Amal Graafstra. Becoming superhuman. Downloaded from: https://www.becomingasuperhuman.com/the-future-of-wearable-technology-cybernetics-with-double-rfid-chip-implantee-amal-graafstra-podcast/ [21.03.2016].

7 Pearson, Ian. A personal interview by email. 14/04/2017. The interview is published for the first time in this book.

8 I. Pearson (2014). Future human evolution. Wordpress. Downloaded from: https://timeguide.wordpress.com/2014/06/19/future-human-evolution/ [21.03.2016].

9 Rothblatt, Martine. A personal interview by email. 15/4/2018. Interview is published in the Večernji list newspaper. More on this: Večernji.hr. (2018). We''ll have robotic mind clones that transcend the body and outlive us. Portal Večernji list [online]. Downloaded from: https://www.vecernji.hr/techsci/imat-cemo-robotske-klonove-umova-koji-nas-mogu-nadzivjeti-1241136 [24/4/2018]

10 Figure 5: Woodhull Sexual Freedom Alliance/Wikimedia Commons. More on this: Večernji.hr. (2018). We''ll have robotic mind clones that transcend the body and outlive us. Portal Večernji list [online]. Downloaded from: https://www.vecernji.hr/techsci/imat-cemo-robotske-klonove-umova-koji-nas-mogu-nadzivjeti-1241136 [24/4/2018]

11 Figure 6: Private archive. Emeritus Igor Čatić and Ivana Greguric. Club of Croatian Humboldt. Grandstand about the technical progress and development of cybernetics: Increasingly powerful technical advances affect human being/ Responsibility in the age of cyborgs and robots. [18.04. 2018.]

12 Figure 7: NDNU student Alex Rodriguez, Bina48, Managing Director of Terasem Movement Foundation Bruce Duncan. More on this:: Notre Dame de Namur University (2017). First Robot Completes College Course: Takes Philosophy of Love Class at Notre Dame de Namur University, http://www.ndnu.edu/media-center/bina48-robot-completes-course/ [12. 07. 2017.]

13 Interview with Jaime Mayor Oreja was created on the participation in the conference ''One of us Cultural Platform'' on 26/5/2018 in Valencia and was published in Večernji list. More on this: Večernji.hr. (2018). Moramo se potruditi da Europa ima više duše. Portal Večernji list [online]. Downloaded from: https://www.vecernji.hr/techsci/moramo-se-potruditi-da-europa-ima-vise-duse-1253059 [26/5/2018]

14 Figure 8: Private archive. Ivana Greguric and Jaime Mayor Oreja on conference ''One of us Cultural Platform'' 26. 05. 2018. held in Valencia.

15 Interview with Stelarc was made during *Digital Takeover Conference* on 7-11/9/2020 in Zagreb (Croatia). Digital Takeover is an innovative educational platform that brought together the decision-makers and digital experts of the Adriatic region and started the topic of digital transformation on the domestic market. Since its' launch in 2015, Digital Takeover quickly became the central point for digital communication on all levels and evolved from a series of smaller events and workshops to the largest regional summit & expo. So far, there have been more than 10 events organized within the Digital Takeover platform which have brought together over 3.000 digital market movers, decision and opinion makers in one place. This unique platform developed by 24sata, Croatia's leading media house, always had a focus on developing special blend of the hands-on practical knowledge, overview of the best practices in digital and the futuristic look into the world of tomorrow.

Thematic focus is expanding to every aspect of digital business transformation – from innovation management, cultural/consumer shift, business organization, tech infrastructure, human resources to marketing. In 2020 Digital Takeover was the must-go event for industry leaders (from start-ups to the most influential regional companies), digital experts and government representatives and many more attendees from all over the region. Digital Takeover expo with more than 50 exhibitors showcases is the best and the most innovative digital solutions from all over the region.

16 Figure 9.: Private archive. Ivana Greguric and Stelarc on *Skype* talk on 26/9/2020 in Zagreb (Croatia)

About the Author

Ivana Greguric is an Assistant Professor of Philosophy at the *University of Zagreb* and a research associate at the *Scientific Center of Excellence for Integrative Bioethics* at the Faculty of Humanities and Social Sciences, with scientific activities primarily within the *Scientific and Research Committee for Bioethics, Technique and Transhumanism.*

She has worked as a research fellow on the project "*Cyborgoethics – intersection or link between bioethics and technoethics*" and a researcher at the philosophical library "*Demetra*". She was an associate of the scientific project *"Application of Systematic Theory in General Techniques Analysis"* (project leader: Prof. Igor Čatić). Since 2018 she has been employed as a docent at the Department of Philosophy and Department of Communication of *Croatian Studies at the University of Zagreb* where she teaches: Philosophy of Cyberculture, Communication – media theories, Cyborgoethics and Robothoetics, Philosophy of Media, Bioethics, Philosophy of Communication and Philosophy of the 19th Century.

She is an author of over ten scientific papers, a chapter in the book *Posthuman cultures* (Oxford, 2013), a chapter in the book *Guide to Deep Learning Basics* (Springer 2020), a co-author of a chapter in the book *Chuck Klosterman and Philosophy* (Open Court, 2012) and a co-editor of the book *New wave and philosophy* (Jesenski Turk, 2012). She was a visiting researcher at the *Brocher Foundation* (2015), center for bioethical research in Geneva, which awarded her with a scholarship for her research on *''Ethical and scientific issues of creating enhanced 'human' beings''*.

She is a member of the editorial board of the *Journal of Philosophical Research, Synthesis philosophica* and *In medias res.* She has participated in more than twenty international scientific conferences held in Croatia and abroad (Oxford, Lisbon, Liverpool) and has held numerous guest lectures (*Croatian Academy of Medical Sciences, European Home, International Congress of the Association of Nurses and Technicians of Croatia*). She attended the European Parliament conference on *Robot Responsibility Conference*, where she proposed adopting a resolution on the limits of improving human beings. In 2018 her book *Cybernetic beings in the age of scientific humanism: Prolegomena for cyborgoethics* has been published in Croatia. In 2021 she is co-authoring a book *The Philosophy of upcoming future: Essays on non-human* with Nenad Vertovšek. Areas of her scientific interest are: philosophical anthropology, philosophy of technology, cyberculture, bioethics, posthumanism and artificial intelligence.

Index

I

J

L

M

N

O

P

R

S

T

V

www.ingramcontent.com/pod-product-compliance
Lightning Source LLC
LaVergne TN
LVHW060414300926
849162LV00006B/1420

* 9 7 8 1 7 9 9 8 9 2 3 1 1 *